HARRISBURG INDUSTRIALIZES

ALSO BY GERALD G. EGGERT

Railroad Labor Disputes: The Beginnings of Federal Strike Policy
Richard Olney: Evolution of a Statesman
Steelmasters and Labor Reform, 1886–1923

HARRISBURG
INDUSTRIALIZES

THE COMING OF FACTORIES TO AN AMERICAN COMMUNITY

GERALD G. EGGERT

The Pennsylvania State University Press · University Park, Pennsylvania

338.09748
E29h

This publication has been supported by the National Endowment for the Humanities, a federal agency which supports the study of such fields as history, philosophy, literature, and languages.

Library of Congress Cataloging-in-Publication Data

Eggert, Gerald G.
    Harrisburg industrializes : the coming of factories to an American community / Gerald G. Eggert.
        p.   cm.
    Includes bibliographical references (p.   ) and index.
    ISBN 0-271-00855-5
    1. Harrisburg (Pa.)—Industries—History—19th century.
    2. Harrisburg (Pa.)—Economic conditions.   3. Harrisburg (Pa.)—
    Social conditions.   I. Title.
    HC108.H299E36   1993
    338.09748'18—dc20   91–47507
                                                                CIP

Published by The Pennsylvania State University Press,
Suite C, Barbara Building, University Park, PA 16802-1003

CONTENTS

LIST OF ILLUSTRATIONS

LIST OF TABLES

PREFACE

This undertaking is an indirect result of having taught both American business history and American labor history at The Pennsylvania State University for several years. That experience convinced me that to understand one the other must be mastered as well. More directly, the study grew out of a series of seminars aimed at introducing graduate students to the use of manuscript census schedules, city directories, and other such sources to measure social mobility in Harrisburg in the manner of Stephan Thernstrom's pioneering work on Newburyport, *Poverty and Progress: Social Mobility in a Nineteenth-Century City* (1964). With computer time provided by the Penn State Liberal Arts Research Fund, and with the invaluable guidance of Glenn Kreider and William McCane of the old "Liberal Arts Data Lab," I typed the entire federal manuscript population census schedules of Harrisburg for 1850, 1860, and 1870 into the University's mainframe computer. Glenn then generously gave of his time and talent in discussing what I hoped could be done with that body of information and working out the special programming needed to accomplish those objectives.

Eventually persuaded that the shortcomings of census occupational data prevented precise measurement of mobility, I struck out in a different direction. One particular graduate student, Robert G. Crist, who was beginning a second career in history, played a key role in determining that new direction. A descendant of several generations of Harrisburg-area residents, he not only knew the city well but seemed to be acquainted with everyone who knew anything about its past. Armed with information about the city and introductions to contacts in relevant banks and industries supplied by him, I expanded the study to its present scope. To the score or more students who labored in those seminars, but especially to Dr. Crist, this study owes much.

I am also indebted to many people for interviews about Harrisburg's early banks and industrial firms and for permission to use the corporate records in their custody: the late Wilson D. Lewis, onetime president and chief executive officer of the Dauphin Deposit Bank & Trust Company, who also gave me access to the papers of the McCormick Trust in the bank's custody; Robert J. Gentry (vice president and director of marketing), Alvan E. Sievers (public relations manager), and Walton J. Wolpert Jr. (marketing research officer), of the Commonwealth National Bank, for access to the records of the Harrisburg (later Harrisburg National)

Bank; G. F. Gilbert Jr., then senior vice president and secretary of the Harsco Corporation, for an interview and for a copy of a detailed history of his firm's predecessor, the Harrisburg Car Manufacturing Company; and to Peter H. Hickok of the Hickok Company. James Flower and Dr. Milton Flower allowed me to use an unpublished draft history of the Central Iron & Steel Company written by their mother, Lenore Embick Flower. Evan Miller kindly gave me access to his reminiscences and writings on early Harrisburg.

Several people made useful suggestions and saved me from blunders by reading and commenting on the manuscript or parts of it: Robert M. Blackson, Ira V. Brown, Robert G. Crist, Gary Cross, John B. Frantz, Philip S. Klein, Martha Anne Morrison, Peter Seibert, John Tuten, and Thomas R. Winpenny. I am also appreciative of my colleague and golf partner, Robert J. Maddox, who listened as I talked out ideas (no, not during his backswing), then asked questions that forced me to reexamine the more harebrained of them.

Funding assistance for the maps and illustrations was generously provided by the Department of History and the College of the Liberal Arts of The Pennsylvania State University. The excellent work of Tami Mistrick, a cartographer at the Deasy GeoGraphics Laboratory of Penn State, converted my rough sketches into the finished maps that appear in the book.

I wish also to acknowledge the contributions of a number of persons who assisted in converting the original manuscript into the finished book. Especially appreciated are two anonymous scholars who evaluated the manuscript for the Penn State Press. Not only did they read it (authors sometimes wonder about such things), but they understood it and gave pages of useful guidance and commentary that helped immensely in improving it. Finally, no author could ask for a more encouraging or thoughtful editor than Peter Potter, acquisitions editor at the Penn State Press, or for more careful copy editing than that of Peggy Hoover, whose impressive memory and sharp eye for detail added to the clarity of the text and the consistency of the footnotes. Such errors of fact or interpretation as remain are not the fault of any of these people; they are wholly mine.

INTRODUCTION

Time and again scholars of industrialization in the United States have turned to the textile mills of Lowell, the shoe factories of Lynn, the steel furnaces of greater Pittsburgh, and the assembly lines of the Detroit area to trace the rise and analyze the consequences of the revolution in production. This has been appropriate and was to be expected as a starting point. After all, those places and industries were where modern factories began or major new processes originated. They were also the sites of the greatest changes: where bells, time clocks, and factory whistles replaced weather, growing seasons, and necessity in setting the pace of work; where factory hands operating sophisticated machines took the place of skilled workers with simple tools; where profit-maximizing entrepreneurs pushed aside less efficient master craftsmen and merchant-manufacturers; where machine-tenders displaced artisan workers; and where armies of job-hungry immigrants and African Americans crowded against and antagonized older residents.[1]

Events at those places, however, were neither typical nor representative. As with the great-leader approach to political history, or trying to understand labor relations by concentrating on major strikes, conclusions based only on those experiences can be misleading. Because industrialization brought quick, dramatic, and pervasive changes in those communities, the implication has been that changes were sudden, revolutionary, and all-embracing wherever the process occurred. Similarly, the substitution of an alienated factory work force for a harmonious agrarian or artisan social order in those first or pioneering industrial communities suggests that such were the usual consequences of the new productive system everywhere.[2]

This was not the case. The industrial revolution occurred in at least two major stages. The first, or primary, stage occurred in places where a relatively few pioneering entrepreneurs introduced the seminal ideas, processes, or technologies that thereafter drastically altered the ways in which goods were manufactured. The little-recognized second stage occurred when those innovative processes spread out from the pioneering centers to hundreds of communities across the nation.[3] It was this secondary, derivative stage that marked the ultimate triumph of the industrial revolution. Second-stage industrialization not only carried the new methods to most corners of the nation, but also sometimes modified the impact and nature of that revolution. Factory production at those places was frequently less all-embracing, less harsh, less confrontational, and less alienating than in the pioneering centers.

Because the primary stage of industrialization has received most of the attention, it is often mistaken for the whole. Recent studies of places and industries less at the center of change, however, suggest the nature and significance of the secondary stage. They show that the patterns found at Lowell, Lynn, Pittsburgh, and Detroit were neither universal in application nor even the common experience of industrial communities.[4] Elsewhere, change proceeded at different rates, followed other lines of development, and produced dissimilar consequences. The studies indicate that as the factory system gradually spread to a large number of communities of varying sizes, the process was uneven. In some places, entrepreneurs almost immediately repeated the successful experiments at the pioneering centers. In other places they moved to the factory system cautiously or reluctantly. As might be expected, the latecomers usually were not innovators with new technological or managerial concepts to test; they were imitators concerned that they and their communities were falling behind and missing opportunities for gain. Their ventures tended to be adaptive rather than creative, prudent rather than daring, and aimed more often at local—or at best regional—rather than national markets. With perhaps more limited means, less-promising resources, and handicaps not faced by the great entrepreneurs elsewhere, their projects were modest. Consequently, the changes they brought to their communities tended not to be as profound, as upsetting, or as alienating.

Time and spatial sequences were also involved in the processes of city-building and industrialization. Before 1830, urbanization in the United States was confined to the five major seaports: Boston, New York, Philadelphia, Baltimore, and Charleston. The size of these communities was more a consequence of their roles as manufacturing centers than of their administrative and commercial functions. In those cities, American staples were assembled and processed for export to Europe, and European manufactures were distributed to the American hinterland. When industrialization began, it centered in those same cities and their immediate environs. Between 1830 and 1870, industry and urbanization were generally linked and together spread into the original hinterland, first into southern New England, upstate New York, and southeastern Pennsylvania, then westward into the Great Lakes region. From 1870 to 1910 the completion of the nationwide railroad network stimulated a much grander scale of production and tended to concentrate manufacturing in fewer metropolitan areas, such as the older prime cities of the Northeast, and in new centers—Pittsburgh, Buffalo, Cleveland, Cincinnati, Detroit, Louisville, Chicago, and St. Louis. Unable or unwilling to keep pace, by the close of this period many middle-size and smaller communities began to lose their hold on manufacturing and to shift to other means of livelihood.⁵

Harrisburg was one of the hinterland communities that moved toward urban status after 1830. Its initial growth was as a commercial and government center rather than as an industrial community. Though well located and generally prosperous, this medium-size town and state capital did not build its first industrial plant until 1849. In this it lagged behind many other comparable communities in the Northeast, though not behind other Central Pennsylvania communities. Until mid-century, industrialization within the state was limited to the areas around Philadelphia and Pittsburgh.⁶ Then, between 1850 and 1860, Harrisburg acquired a cotton textile mill, two anthracite blast furnaces and several iron rolling mills, a railroad car manufactory, and a firm that produced various kinds of machinery. Iron (and later steel) manufacture became by far the city's leading industry.

Except for the industrialist who developed machines for ruling bookkeeping forms, Harrisburg's entrepreneurs invented no unique industrial product or process, did not pioneer larger-scale operations, introduced no new managerial techniques, discovered no previously untapped sources of labor or raw materials, and introduced no new scheme for financing factory production. They also brought the community fewer wrenching adjustments than did their counterparts in the pioneering centers. In all this they were probably like many if not most small-to-medium-size American communities that industrialized in that era.

Uncertainty caused by the onset of civil war in 1861 briefly put a damper on manufacturing at Harrisburg. The city's location and transportation facilities, however, made it the natural staging area for the war's eastern front, and business leaders were quick to turn this to the city's and their own advantage. Whatever the war's impact on other communities, it proved profitable to Pennsylvania's capital city. Among other things, it enhanced the place as a major railroad center and enriched several of its business leaders. As the conflict slowly ground to its conclusion, Harrisburg's entrepreneurs fashioned plans and accumulated funds for a rapid expansion of their enterprises, especially those related to railroads: rail rolling mills, railroad-car plants, and shops for maintaining and repairing railroad rolling stock. Within fifteen years, Harrisburg's industrialization reached its climax.

As it turned out, manufacturing was only an interlude in the city's economic life, cutting across little more than two generations. After 1880 it settled into a long decline. As a few iron and steel firms elsewhere moved to operations on a national scale, Harrisburg plants either had to grow with them or risk obsolescence. For a variety of reasons, they changed little and remained relatively small in scale. As the city's iron and steel plants stagnated, no other industry or industries arose to take their place. At first slowly, then more rapidly, pursuits other than manufacturing provided livelihoods for more and more of the city's residents. Although deindustrialization was never complete (Harrisburg still has manufacturing plants), by most measures the city ceased being an industrial center after World War II.

Recent studies of industrializing communities fall into two groups. The great majority, employing the insights of labor and social history, have been concerned primarily with the impact factory production had on working people. In these, entrepreneurs receive little attention, and most of that is unflattering. A smaller number of studies have concentrated on the process from the perspective of industrialists. Workers in those studies, if discussed at all, usually have been lumped with various other factors against which entrepreneurs had to strive in their efforts to succeed. The two approaches, however, are dual aspects of a single process and ideally should be treated together. This study undertakes such a combining, employing the insights of both approaches without adopting the perspective of either. Its primary objective is to trace interactions among the groups most directly involved in the process. Entrepreneurs are seen neither as heroes nor as villains, and craft workers and factory hands neither as victims nor as suppressed revolutionaries.

The book is divided into two parts. The first deals with the process by which factory production came to Harrisburg; the second measures and analyzes the resulting changes in the lives of major segments of the population. Of necessity, aspects of business, economic, labor, and social history are utilized. The introduction to Part I provides a brief description of Harrisburg in the mid-nineteenth century on the eve of the construction of the town's first industrial plants. Then the first two chapters discuss the building of the essential infrastructure and the persons responsible for it. This development of transportation, banking, commerce, and other service components, and their sequencing and interrelationships, are traced and their roles in the rise of factory production are spelled out. The third chapter introduces Harrisburg's leading industrial entrepreneurs and the firms they created during the first decade of factory production. The impact of the Civil War, both in halting and then speeding up industrialization, is the theme of the fourth chapter. The post–Civil War completion of Harrisburg's industrial order (Chapter 5) and an analysis of the process (Chapter 6) conclude the first part of this book.

Part II begins with an introduction that describes the city after nearly a quarter-century of industrialization. This is followed by the first of three chapters that deal with "fathers and sons,"[7] Chapter 7, which examines the effects of industrialization on the entrepreneurial families and other business groups. Chapter 8 discusses the recruiting strategies the new industries used to secure workers. The study next turns to the city's traditional craft workers to determine how the coming of factory production altered their careers and the occupational prospects of their sons. This is followed by a chapter on Harrisburg's African Americans and other minorities, both fathers and sons. Two chapters address labor relations in Harrisburg between 1850 and 1900, while a third relates the labor policies of the industrialists to the political life of the community. Observations on the overall impact of industrialization on the community and its significance for our understanding of the process nationally make up the final chapter. The work concludes with a brief Epilogue, which sketches the fate of the entrepreneurial families and their firms and the subsequent shift of the city to a new economic base.

The Rise of
Factory Production

Fig. 1.   Harrisburg on the Eve of Industrialization, 1846.   Edwin Whitefield's "View of Harrisburg, Pa., from the West."

On the Eve of Industrialization

EDWIN WHITEFIELD, working from a hillside on the western bank of the Susquehanna River in 1846, prepared a drawing of Harrisburg, Pennsylvania, on the eve of its industrialization. The site the English-born landscape painter and lithographer selected was one frequently used for depictions of Pennsylvania's capital city: a view eastward from across the river. The scenic beauty of the waterway no doubt led artists to favor that perspective, but so did the fact that many of the town's principal buildings faced westward. Views of Harrisburg from the west seemed to look the town in the eye.

The lithograph portrayed a peaceful, semi-rural, idyllic residential community situated on a narrow flat along the river's edge. Unlike his watercolor of Johnstown, Pennsylvania, done that same year, White-field's panorama of Harrisburg showed no large mills, no factories, and no belching smokestacks. A large, parklike island and several islets divided the river in the foreground. Rolling hills to the east and a range of mountains to the north provided a backdrop. Dominating the community from its northern boundary was a large domed building. Stretching to the south was the rest of the town, composed chiefly of one- and two-story houses, a handful of larger structures, and, especially notable on the skyline, a tower and five cupolas (see Fig. 1).[1]

An 1850 map of Harrisburg provides a different perspective while confirming the basic features of Whitefield's sketch and identifying the structures he depicts. Harrisburg was situated on a triangular tract of land, its longest side stretching north and south along the river for twenty blocks. At the south the town was one or two blocks deep, at the north fifteen blocks. Two or three dozen business houses, churches, and public buildings were the town's outstanding structures. The rest were individual residences—nearly 1,400 according to the federal census of 1850.[2]

Whitehead's focus on the Susquehanna was appropriate for more than artistic reasons. Harrisburg began as, and long remained, a river town. From pre-Columbian times the Susquehanna had accommodated the canoes of Native Americans and later of the early European traders. Settlers, if headed west, often forded or crossed the river by ferry at Harrisburg. Those moving north or northwest followed trails along the Susquehanna until they reached the Juniata River, a branch leading westward. There they either continued north along the Susquehanna or followed the Juniata in a northwesterly direction. Once the frontier passed and agriculture developed in Central Pennsylvania, giant arks and keel-bottom boats brought annual harvests of grain by river from the hinterland to market at Harrisburg. Still later, charcoal furnaces and water-powered gristmills in the greater Juniata region shipped pig iron and flour over the same route. When the axes of lumbermen began to fell the virgin forests of northern and central Pennsylvania, great rafts of logs floated the river to Harrisburg or beyond each spring.[3] The Susquehanna provided the town with its fresh fish and, as the community grew, with much of its water supply as well. Springtime snowmelts and heavy rains brought great surges of water past the town and not infrequently into its streets.

The 1850 map shows a small, fenced burial plot at river's edge near the extreme right or southern limit of the town. There a large mulberry tree shaded the grave of John Harris, the first European to settle in the area. Harris received license to seat himself on the Susquehanna and to erect such buildings as he needed for his trade in 1705. When he first actually took up lands in the region is uncertain; his first known request was for 500 acres in May 1727. Inasmuch as settlers usually lived briefly on lands before making formal applications for them, it is likely he came to the area sometime earlier. The Proprietaries, in December 1733, granted him formal title to 300 acres and allowed him to acquire an additional 500 acres through one John Turner. Harris erected a stockade on his property, built storage sheds, dug a well, started to farm his lands, and opened trade with both Indians and settlers who followed him to the place.[4]

Meanwhile, in October 1733, two months before granting Harris his plantation, the colony's Proprietaries conferred on him "the sole keeping of the Ferry over Susquehannah River." Harris promptly sought permission to build a small house on the western shore of the river to accommodate travelers using the ferry and asked for 200 acres on which to erect the necessary buildings. He was permitted to build a house, but the

Proprietaries took the request for land under advisement. Harris, however, already held licenses to buy tracts of 200 and 300 acres on the western shore obtained from Samuel Blunston, agent of the Penns for lands west of the Susquehanna. When Harris finally received patents in January and March 1737, they were for tracts of 820 acres and 311 acres.[5] The cluster of buildings on the eastern shore was soon designated in official records as Harris's Ferry.

Although the structures Harris built had disappeared by 1850, near their site, and within rods of the settler's grave, stood a handsome stone mansion built in 1766 by his eldest son, also named John. The younger Harris had inherited much of his father's property in 1748, but only in his own later years did he found the town, laying out streets and platting, numbering, and selling lots after 1784. Two months before his death in 1791, the Pennsylvania Assembly conferred borough status on the community, which then consisted of perhaps 130 houses. By 1850, ownership of the mansion had passed from the Harrises to Thomas Elder, lawyer, banker, and one of the major architects of Harrisburg's economic infrastructure.[6]

The tower and five cupolas on the southern half of the skyline in the 1846 lithograph marked important structures in the town. The tower and northernmost cupola were atop, respectively, the Dauphin County jail (recently erected in the style of a Norman dungeon) and the graceful colonial style county courthouse, built in 1799 and used until 1860. A francophile Pennsylvania Assembly in 1785 had carved the new county from what previously had been the northwestern portion of Lancaster County and named it in honor of the heir-apparent to the French throne. The Assembly also designated the settlement as the county seat, but called it "Louisbourgh" after the king of France. Not liking the official name, the town's founder used the designation "Harrisburg" in all deeds he conferred. Eventually he won out when the Assembly incorporated the place as the borough of Harrisburg in 1791.[7]

The remaining cupolas from left to right belonged to three of the town's larger churches, the Lutheran, the German Reformed, and the Presbyterian, and to the somewhat smaller Union Bethel Church of God. Two churches without cupolas also stand out on the riverfront: St. Stephen's Episcopal Church, with a low but solid square tower, and, about two blocks to the right, the Baptist church of Harrisburg, a large structure with neither tower nor cupola. Not noticeable on the skyline because of its location between the river and higher ground was St. Patrick's Roman Catholic Church. At the time it was the only church in town with a cupola surmounted by a cross. Less-imposing structures were used for worship by Methodist, German Reformed, United Brethren, and African Methodist Episcopal congregations.[8]

Harrisburg's various denominations, especially the older churches, reflected the early ethnic makeup of the community. Although the first John Harris was an Englishman from Yorkshire, relatively few of his compatriots came to the Harrisburg area. Not until 1826 were there as many as six families of Anglican background to form an Episcopal congregation, though people of English stock did belong to the Presbyterian, Methodist, and Baptist churches. After 1727, large numbers of Scots-Irish pushed into the region. Among them were the Elders, the Forsters, the Graydons, the McAllisters, the Maclays, the McCormicks, the Murrays, the Rutherfords, and the Weirs, to name but a few. Zealously Calvinistic Presbyterian in religion, their place of worship at Second Street and Mulberry Alley had by the mid-nineteenth century become Harrisburg's most prominent and fashionable church. Coming soon after, and in about equal numbers, were the Germans, including the Boas, Brua, Bucher, Buehler, Eby, Egle, Fager, Fahnestock, Hummel, Kelker, Kunkel, and Wiestling families. Divided between the Evangelical Lutheran faith and the German Reformed faith, they shared a common place of worship until going their separate ways in 1814. The Lutherans erected a new church, Zion, at Fourth and Blackberry Alley. By 1842 that congregation, in turn, divided over the use of the German or English language in services. Those who preferred German established St. Michael's Church in the southern portion of the borough. The Reformed congregation remained at the original site at Third and Chestnut and, in 1821, erected Salem Church on that site. Five years later, John Winebrenner, a pastor of Salem Church, left to organize a new denomination, the Church of God, which built Union Bethel Church.[9]

The other major ethnic congregations were made up of African Americans and the Irish. The former had been brought to the community as slaves by the first settlers. Their principal church, Wesley Union, occupied a small building on Tanner's Alley behind the State Capitol. In 1827 St. Patrick's was built on State Street, between Second and Third, to serve a Catholic congregation that became increasingly Irish over the years. The remaining denominations, the Methodists, the Baptists, and the United Brethren, attracted middle- to lower-class groups with various ethnic backgrounds. At first less prosperous and less given to ostentation, they worshiped in plain buildings unmarked by steeples, towers, or cupolas.[10]

The domed building dominating the northern half of the skyline and towering over the community was the State Capitol. Government office buildings flanked it on either side, and a large wooden arsenal building stood on public grounds to the immediate south. To secure incorporation as a borough, the younger John Harris, in platting the town, set aside on the prominent rise at its northern limits more than four acres for use by the Commonwealth. The Assembly voted to make Harrisburg the seat of state government in 1810. Two years later, the legislature began holding its annual sessions in the Dauphin County Courthouse and in 1822 moved into a newly erected capitol building.[11] Being the political center both of Dauphin County and of the Commonwealth of Pennsylvania contributed to the town's growth. From 1,472 residents in 1800, Harrisburg grew to 2,990 by 1820 and to 5,980 by 1840. In addition to hosting annual meetings of the legislature, the town sprouted a number of related activities: boardinghouses and hotels to accommodate legislators and those who curried their favor, and printing establishments to print and publish the laws and reports of government and to issue newspapers projecting the views of the various political factions of the state. The town also attracted more lawyers than were usually found in communities of similar size.[12]

Notable both in Whitefield's lithograph and in the 1850 map of Harrisburg were the bridges, utilities, and means of transportation that came into being between 1812 and 1850. These constituted part of the infrastructure on which Harrisburg would build its industrial plant. The lithograph shows two bridges, a camelback covered bridge for horse-drawn vehicles and pedestrians, and, to the south, a recently completed railroad bridge with a steaming train headed away from the town. Both map and vista show the town's new reservoir—which resembled a low, flat-top pyramid—located just north of the Capitol, and a pumping station on the river that supplied the reservoir with water. Not visible in the vista were the banks in the center of the town or the long sheds at Market Square, where farmers regularly brought fresh vegetables and dairy products for sale to the townspeople. Also lost from view, because it ran along Paxton Creek at the borough's eastern boundary, was the town's primary transportation and commercial corridor.

The 1850 map and an 1855 lithograph (by an anonymous artist who made a bird's-eye view of Harrisburg as if facing south from above the Capitol) showed both the market sheds and the corridor. A canal, part of the State Works, which opened for business in 1830, ran north and south through the corridor and served as the town's principal artery of transportation. Later two railroads were constructed along its banks: the Harrisburg & Lancaster from the southeast, and the Cumberland Valley, which came over the bridge from the southwest.[13] Near the southern boundary of the borough they met and proceeded side by side to a wood-frame depot at Market Street, where both lines ended. In 1849 the Pennsylvania Railroad Company absorbed the Harrisburg & Lancaster into its system and extended the line northward through the borough. A few miles above Harrisburg the line crossed the Susquehanna by bridge, proceeded north to the valley of the Juniata, and thence westward to Pittsburgh and points west.

A traveler who in 1850 entered the corridor from the south and east by either canal or railroad would have passed through open fields, woodlots, and farmsteads as far as the borough limits a block below Paxton Street. There, on the right, stood a large sawmill. From that point on, the traveler would have seen, in the narrow stretch of land between the railroad to the west and the canal to the east, Harrisburg's half-dozen or so major business enterprises—first, an iron rolling mill with fifteen employees, a lumberyard and sawmill, a recently erected gas-works that brought gas lighting to the town, and the new Novelty Iron Works, which cast decorative iron grills for homes and business establishments. Next came a number of facilities for servicing the railroads: a machine shop, warehouses, and loading docks. On the east bank of the canal stood a brickyard and a firm that built and repaired canal boats. After a stop at the depot, the traveler would have looked out on merchant warehouses and a large wood-planing mill belonging to two Harrisburg contractors. Near the northern limits of the borough both canal and railroad passed between the community's first modern iron furnace, which went into blast that year, and a small iron foundry. Because it stood along the riverfront some ten blocks to the west, the traveler would have missed Harrisburg's other major industrial facility, a cotton mill, then under construction. Neither it nor the blast furnace was even in the planning stage when Whitefield sketched his panorama, four years earlier.

Data from the federal manuscript population census schedules of Harrisburg for 1850 provide yet another perspective on the community on the eve of industrialization: information about its residents. But census schedules, as most other representations, have flaws. In some ways they resemble a snapshot photograph, freezing the community's ever-changing population for a moment in time. Just as before and after a photograph is taken subjects move freely as dictated by necessity, chance, or whim, so the people of Harrisburg shifted from place to place before and after being visited by the census taker. Blurs on photographs, caused by movement as a film is exposed, are akin to errors of omission or duplication in the census caused by people who move while it is being recorded. An alphabetical listing of everyone in the Harrisburg population census schedules for 1850 reveals that three families and eight single persons apparently moved while the census takers were at work. For example, Henry Angee, age twenty-six and a tinner by trade, his wife Suzanna, age twenty-three, and their son Edwin, two, who had just moved into a home of their own, had only a few days before lived at Fishburn's Hotel. Someone at the hotel told the census taker that among its residents were Henry Ange, a twenty-six-year-old tinner, his wife, Susan, age twenty, and a son, William, age one.

Another twenty-seven persons appear to have been listed both at their homes and at their places of employment. Fifteen were teenagers, such as Louise and Rebecca Brodbeck, who probably were working at their first jobs. The girls were listed both in the home of their father, a German-born shoemaker, and as servants in the homes of a lumber merchant and an Episcopal clergyman respectively. Another resident, a twenty-two-year-old butcher named John Kosur, was listed as a roomer at the Buehler Hotel but also appeared in the household of Nicholas Reems-hart, the victualler who employed him. Several blacks were also counted twice. Ann and Samuel Bennet, Phoeba Belt, and Augustus Coates were all listed at Herr's Hotel, where they worked, but they all also lived in homes of their own with spouses, children, and other family members.[14]

Because the census in 1850 was mute on the point, relationships of persons living in the same households in 1850 can only be deduced from gender, age, and family name. Apparently three-quarters of the towns-people lived as families, most of which were composed of parents and children. It was not unusual, however, for the families to include elderly parents, siblings of the head of household, and grandchildren. Many households also included servants, employees, or tenants who were not so designated by the census taker. Some of these, however, may simply have been family members with different last names.

According to the census takers in 1850, Harrisburg had 7,834 persons living in 1,376 dwellings. Most resided in individual homes, some in hotels and boardinghouses. Eighteen were temporarily housed in the county prison. Females outnumbered males 505 to 495 per 1,000. Nearly 84 percent of the town's residents gave Pennsylvania as their place of birth. Of those, well over half were probably born in Harrisburg or Dauphin County.[15] Only about 6 percent came from other states, and a little over 11 percent were foreign-born (see Appendix H).

Of those born in other countries, the 421 born in Ireland were the most numerous. Although they constituted slightly more than half the foreign-born in 1850, more than a quarter (121) were very temporary residents. They lived in five boardinghouses and apparently worked on the construction gangs that were building the Northern Central Railroad opposite Harrisburg that summer. The majority probably left with their jobs as construction moved north; only one remained in town to be listed in the 1860 census. The other sizable foreign-born group (more than 40 percent) were born in Germany. African Americans were the community's only nonwhites. Constituting more than 11 percent of the whole, nearly 90 percent of them listed Pennsylvania as their birthplace.

By late twentieth-century standards, the town's population was very young. One-quarter of the residents were less than ten years old, two-thirds were under thirty, and fewer than one-tenth were age fifty or older.[16] Even so, they were older than the statewide population. In that era young adults were flocking to communities like Harrisburg in search of their fortunes. Consequently, the town had 500 more residents between twenty and forty years of age, and 470 fewer residents under twenty than it would have had if its proportions had corresponded to those of the state as a whole.

Census takers in 1850 were supposed to list the occupations of all males age sixteen years and older. In Harrisburg they also included that information for 131 adult females and for 17 youngsters. Of 2,439 males who were age sixteen and older, the census takers listed no occupations for nearly 300. Of that number, 33 were students and a few were elderly and probably no longer working. Whether the others were unemployed or the information was not supplied to the census taker is not known. Of persons with occupations listed, more than a quarter (601) performed white-collar functions (see Appendix C). The more important of these included 13 of the entrepreneurs who fashioned the town's economic infrastructure, 22 retired "gentlemen" of means, 136 merchants and merchant-manufacturers, 82 professionals (chiefly lawyers and physicians), and 15 farmers and proprietors. There were also 41 government officials and managers and 43 semi-professionals, such as teachers. The remaining 42 percent (259) were storekeepers, salespersons, or clerks.

The blue-collar workforce of 1,725 persons made up more than 70 percent of all Harrisburg residents with listed occupations (see Appendix E). Nearly half (48 percent) worked at skilled crafts (carpentry, cabinetmaking, blacksmithing, and the like). Many were probably self-employed. Semi-skilled workers, including 117 employed in the town's first industrial-scale iron furnace, made up less than 7 percent of the blue-collar force, while 690 unskilled or unspecified "laborers" made up the remaining 40 percent. Personal servants, if differentiated from other unskilled workers, numbered 76 and made up about 3 percent of all persons with listed occupations. Because many servants were female and not subject to listing, this group was probably considerably undercounted. Among the 131 female residents with recorded occupations, the most common occupations listed were seamstress (56), milliner (10), and schoolteacher (17). Eight women kept boardinghouses or ran hotels. Seven, mostly widows, were storekeepers or merchants. Only 16 were listed as cooks or domestic servants, though that number would have been much greater had all females with occupations been listed.

At mid-century, Harrisburg's economy centered around trade and commerce. Those who were not merchants or merchant-manufacturers were their clerks and salespersons, or the skilled artisans and laborers who made or processed the goods to be traded. Because the town was also the state capital, a small portion of its population was involved with political administration. Another small group, the town's lawyers, served as intermediaries between citizens and government officials, lobbying for legislation, helping to draft laws and resolve disputes, and, when necessary, arguing cases in court. The rest of those who worked, as in most communities, provided necessary services for one another.

Census information on land ownership in Harrisburg belies the once-common myth that preindustrial America was an essentially classless society in which most people were able to acquire real estate. In fact, only a decided minority were landowners. Of 4,245 residents twenty years of age or older, only 451, or little more than one-tenth, held any of the nearly $3,560,000 worth of real estate owned by Harrisburg residents (see Appendix F). They included 347 men and 104 women. Among the propertied few, concentration was intense. Five individuals (1 percent) held one-third of all the land owned by people of the community, and 45 (10 percent) owned two-thirds.

White, native-born American males by far owned most of the real estate. Less than 10 percent was held by women. Nine out of every ten women landowners were widows or single. Even so, many of the widows had adult sons living at home, including professionals and businessmen, who might have been expected under the laws of that period to take charge of their mothers' property. In a dozen instances, moreover, women property owners had husbands, eight of whom were landless and four of whom were landowners. A relatively small number of foreign-born and African Americans were among the property owners. About a tenth of Harrisburg's total population was born abroad, and a slightly larger percentage owned land. Of those, twenty-six were German-born and eighteen were natives of Ireland. African Americans did less well. Although they constituted 11 percent of the town's population, fewer than 7 percent of landowners were black. Wealthier landowners were not likely to come from these groups. Among the top 10 percent, only three were women, four were foreign-born, and none was African American. Of the blacks, only one ranked as high as the middle 20 percent of property holders, and twenty were in the lowest fifth. Of 104 women with real estate, eight were foreign-born and three were of African descent.

It will be against this composite view of Harrisburg at mid-century that the impact of industrialization on the community and its people will in part be measured. Before reconstructing the process by which factory production came to Harrisburg, however, it is necessary first to trace the establishment and development of transportation, banking, and utilities in the area, the essential infrastructure on which the new form of production would rest.

Economic Foundations
of the Community

Geography played a major role at each stage of Harrisburg's evolution from frontier outpost to modern industrial city. It provided advantages that encouraged certain promising lines of development, and it also imposed limitations that eventually thwarted fulfillment of many of those promises. The town was situated at a great natural crossroads, which made it a likely spot for development once Europeans came into the area. Located on the east bank of the Susquehanna, it lay some seventy miles north and west of the point where that river emptied into Chesapeake Bay. The river and its valley constituted one of the principal north-south river routes of the precolonial, colonial, and early national eras. As with most rivers at the time, transportation of goods went with the flow, in this instance southward. Northbound traffic usually went by foot or by packhorse over trails paralleling the river. At the same time, land travelers moving east or west through the great valley stretching westward from the lower Delaware River to the valleys of south-central Pennsylvania found easy fording or ferrying of the Susquehanna at that location.[1]

What eventually became Harrisburg began as a frontier outpost. John Harris's grant was situated at the intersection of the river and two major Indian trails. There he farmed, but he also operated a store where European traders exchanged cloth, metal goods, and whiskey with the Native Americans for furs. As the wave of Scots-Irish settlers began moving into the district, he applied to the Proprietaries of Pennsylvania for title to lands on both sides of the river and secured the rights to operate a ferry. Settlers heading west either forded the river at that point or crossed it on Harris's ferry, then pressed on to Carlisle, the westernmost British outpost. Those headed north or northwest took advantage of a last chance to buy supplies from Harris before following the Susquehanna and its branches to the north and northwest.

On and off for five decades between the 1730s and the end of the American Revolution, the Indians, the French, and the British struggled for control of central and western Pennsylvania. Periodically, one or another of the contestants would attack and burn out settlers who ventured into the hinterland above Harris's Ferry, sometimes driving them almost back to the outpost itself. As a result, the usual speedy evolution from a frontier subsistence economy to an economy based on agriculture was long delayed in the area north of Harrisburg and west of the Susquehanna.[2]

Accordingly, development of Harris's properties was slow. When Harris "the Settler" died in 1748, he left the portion of his property that became downtown Harrisburg to his son John Harris "the Founder," as the two are differentiated locally. The Proprietaries confirmed the younger Harris's right to operate the ferry in 1753. The new proprietor could do little to stimulate growth of the community, however, until peace returned. Officials considering a meeting with the Indians there in 1756 decided against the place because it consisted of a single house and had few conveniences.[3] Not until 1784 did the younger Harris turn to land promotion. That year he had his lands platted and numbered and began selling residential lots. Theophile Cazenove, a Frenchman who visited the town a decade later, described it as "one of America's little phenomena, in the matter of the rapidity of its rise." In 1785 it still consisted only of "the single house and farm of Mr. Harris," but now, along wide, unpaved streets, there were "about a thousand lots and already 300 houses neatly built in brick or 'logs and mortar,' 2 stories high, [with] English windows." He reported that there were three brick factories in the area. In the town were also thirty-two taverns and eighteen merchants "keeping in their stores European merchandise, and buying farmers' produce." By the time of Cazenove's visit, the town had become the seat of Dauphin County and acquired borough status. The county courthouse, then under construction, was large—the idea being that the state legislature would someday meet there.[4]

Harrisburg's economic focus shifted as the frontier families who passed through became settled farmers in its hinterland during the 1780s and 1790s. Gradually the village emerged as a trading center, and commerce became its primary economic function until the mid-nineteenth century. The people there made a living largely by receiving, passing on, or processing the products of farms and forests and supplying, in return, the simple needs of the growing farm population. The bulky traffic came south chiefly on the spring freshets, in large rectangular arks (which at journey's end were sold for lumber), or on smaller, sturdier keelboats that sometimes ascended as well as descended the river. According to Cazenove, about 200,000 bushels of wheat came down the river in 1793 to be ground into flour at the mills of "2 or 3 very rich millers" at Harrisburg and one at Middletown, nine miles farther south.[5]

Over the years, whiskey, clover seed, pork, a little pig iron and some coal, and great quantities of logs were added to the commerce. Part was unloaded and used locally or put on wagons to be hauled overland to Lancaster or Philadelphia. Dozens of boats at a time waited to unload at a red warehouse on the river at Paxton Street that served as the center for such transshipments. Goods not deposited at Harrisburg floated on downriver to other ports, where they were unloaded for shipment to Philadelphia or Baltimore. Most of the goods moving upstream continued to go by foot or packhorse over old Indian trails. It was not uncommon to see hundreds of packhorses (a single horse could carry only about 200 pounds) in files of twelve to fifteen, laden with salt and other merchandise, moving along the river at Harrisburg to a northward trail, or horse-drawn wagons seeking a fording place or the ferry that would convey them across the river to the trail leading to Carlisle and the west.[6]

Harrisburg might have remained just another sleepy, slow-growing interior river town and minor entrepôt but for its designation as capital of the Commonwealth on the eve of the second war with Great Britain. In keeping with its increased dignity, the town seemed to want and need everything at once: improved access for those who now must come to the seat of government, housing both for its growing permanent population and for temporary visitors, mills and shops to produce goods, and such service facilities as banks and public utilities, to name only the most obvious. The process of acquiring these things was neither simple nor linear. At one time entrepreneurs focused on transportation, at another time on banking, at a third time on public utilities, but more often on most or all at the same time. Undertakings were not discrete, but interrelated chronologically, financially, and in terms of personnel involved. Major panics in 1819 and 1837, and a sharp but brief panic in 1857, further complicated the process. Three eras of marked growth (1796–1819, 1825–37, and 1842–60) were separated by two periods of depression (1819–24 and 1837–41). Each economic boom coincided with the development of a promising new form of transportation. Each also made substantial contributions to the town's economic base and added new entrepreneurs to the community's small band of economic leaders.

The First Period of Growth, 1796–1819

Improved transportation probably ranked first among the various needs, not only of Harrisburg but also of the entire Susquehanna watershed. As production in the valley expanded, the river's inadequacies became increasingly troublesome. An early nineteenth-century traveler, with unconscious humor, summed it up. The Susquehanna, he declared, would be "one of the most useful rivers in the world" but for its "falls, shallows and rapids which impede the navigation."[7] Boats could descend the Susquehanna and its principal branches only as far as the Conewago Falls near Middletown, a few miles below Harrisburg. Except for small craft, however, even that navigation was confined largely to the spring months, when melting snows and heavy rains swelled the stream. Many places were hazardous—for example, at Rockville, a little above Harrisburg, there were shallows and numerous protruding rocks. From the Conewago Falls to Chesapeake Bay, the Susquehanna was not navigable. Rock shelves lay hidden beneath the surface, and a drop of 150 feet within those fifty miles made the stream swift-flowing and treacherous.

As a result, no significant metropolis or seaport developed on the Susquehanna. To find an outlet, goods proceeded by river as far south as possible, then moved overland either to Philadelphia, with its harbor on the Delaware River, or to Maryland's enterprising and fast-growing seaport, Baltimore.[8] Residents of the valley called for two types of improvements: clearing and deepening the channel of the Susquehanna, and constructing canals and turnpikes that would connect the valley with one or both of the great market cities.

In the last decade of the eighteenth century and the first fifteen years of the nineteenth, turnpike roads and bridges were the most common form of transportation improvement. Most often these were constructed by private groups under charters from the state legislature that authorized them to collect tolls. Interested parties would organize and petition the Assembly for authority to form a corporation to undertake a particular project. If approved, the act usually designated specific commissioners to enlist "subscribers" who pledged to purchase stock in the firm. The organizers and commissioners, who were frequently the same people, tended to be wealthy and prominent business leaders in the community. After subscribing to shares themselves, they solicited their friends, neighbors, and business acquaintances for additional investors. When a sufficient amount of stock was pledged and a specified portion of its value was paid in, the corporation received a charter, organized formally, elected officers, and began transacting business. In many instances the state of Pennsylvania itself contributed to projects, most often by purchasing stock in the firm.[9] Once organized, the corporation's officers, after inviting bids by contractors, awarded contracts for the actual construction. Because private corporations constructed and maintained the resulting facilities, they collected tolls, paid off construction costs, and pocketed any profits.

Baltimore merchants, anxious to tap the trade of the valley, promoted roads into the lower counties of Pennsylvania that bordered on the Susquehanna. At first Philadelphians assumed that fellow Pennsylvanians, out of loyalty, would direct their business to the Pennsylvania metropolis rather than to Baltimore, whatever the conditions of transportation. When it turned out that valley residents, including those at Harrisburg, only wanted markets and did not care whether at Philadelphia or Baltimore, the state began to promote the construction of turnpikes between the river and Philadelphia.[10]

The earliest of these new turnpikes favored Harrisburg's rival river ports to the south, Middletown and Columbia. In 1792 the Assembly had chartered the Philadelphia to Lancaster Turnpike Company to improve the primitive road between those two major points. By the time it was completed in 1794, another company had been authorized to extend the road ten miles west to Columbia on the Susquehanna. Columbia was only 80 miles from Philadelphia; Harrisburg was 105 miles. However, because the Conewago Falls still blocked river traffic below Middletown, the Assembly, in 1796, chartered a road from Lancaster to Middletown.

Early in the new century, the Assembly began authorizing turnpikes to Harrisburg. The road from Downingtown via Cornwall Furnace, though chartered in 1803, was not completed until 1819. The Berks & Dauphin County Turnpike (from the Schuylkill River in Reading to Hummelstown, east of Harrisburg) was authorized in 1805 but never completed. Once Harrisburg became the state capital, the Assembly authorized turnpikes linking it with its neighbors in all directions: two turnpikes in 1810, one southeast to Lancaster, the other south to Middletown; and three in 1814, all from the western end of the Harrisburg Bridge, the first south to York, the second southwest to East Berlin in Adams County, and the third west, then southwest, to Carlisle and Chambersburg. The most ambitious turnpike project, connecting Harrisburg and Pittsburgh, was first authorized in 1806. When that produced no road, a second company, chartered in 1811, considered alternate southern and northern routes. Eventually both were built, but each by several companies, not one. Construction of the southern route, following the general course of U.S. Highway 30, was completed by five companies between 1815 and 1820. The northern route, along what later became U.S. Highways 322 and 22, was completed by several companies the next year.

The new roads improved transit notably. People moved more quickly and more conveniently. Freight was hauled in Conestoga wagons drawn by six horses instead of on packhorses. Even so, the roads brought small profits to those who built them. At the same time, tolls were too high for hauling most bulky freight. That traffic continued to float the Susquehanna to Middletown before moving onto turnpikes for Philadelphia. Once the mile-long Conewago Canal was completed around the falls of the Susquehanna in 1787, river traffic continued south to the turnpike at Columbia or on to Baltimore. To encourage the flow of goods to Philadelphia instead of to Baltimore, the Assembly incorporated the Union Canal in 1811. Seventeen years later, in 1828, that seventy-eight-mile waterway ran from the Schuylkill River in Reading to the Susquehanna at Middletown.[11]

In many ways as important to Harrisburg as the turnpikes was the authorization in 1809 of a toll bridge across the Susquehanna. Three years later the legislature chartered the Harrisburg Bridge Company to build and operate the structure. The law set the capital of the firm at $400,000, to be raised by issuing 20,000 shares at $20 each. The company was to retain 3,000 shares, the dividends on which were to go into a sinking fund to make the bridge eventually toll-free. Profits in excess of 15 percent net were also to go into the fund.[12]

Many people, including those pledged to buy stock, doubted that any bridge could be built over so wide and seasonally unpredictable a river as the Susquehanna. When the company persisted, subscribers to the stock resisted paying. The state, which pledged to buy $90,000 worth of stock (4,500 shares), pushed construction by paying each of its four equal installments only as portions of the bridge were completed. When the company tried to borrow the needed funds, no one would take its notes, so Bridge Company officers borrowed large sums in their own names to pay for construction. They received repayment as stockholders honored their subscriptions and state funds were paid in. The initial cost of the bridge was $192,138. Of that, $90,000 (47 percent) came from the state, $57,700 came from the sale of stock to individuals, an additional $5,277 came from forfeitures by subscribers who failed to honor their pledges in full, and $39,161 was from loans.[13]

The original forty-foot-wide wooden "camelback" structure on stone piers began at the end of Market Street and stretched to Forster's Island (City Island) midway across the Susquehanna. A second span of nearly the same length completed the crossing, a total distance of two-thirds of a mile. Customers began paying tolls as early as October 1816, several months before all work was done. Once completed, the bridge brought that portion of Cumberland County immediately opposite Harrisburg firmly into the town's economic orbit.

Fig. 2.   A View of the Harrisburg Bridge from the Western Shore.   The "camel-back" toll bridge, designed by Theodore Burr, stood from 1816 until destroyed by flood in 1902. It was replaced that year by a steel bridge. Tolls were collected until 1957.

However difficult its initial financing, the Bridge Company in time began paying its way. For the first four years, all the tolls went to paying off debts. Dividends began in 1821 and during the subsequent twenty-four years averaged about 6.3 percent a year. Repairs amounted to only $2,000 a year. Eventually the Bridge Company became one of Harrisburg's more venerable institutions. Owning its stock or, preferably, being an officer or director, marked one as a member of the community's elite. The bridge, rebuilt several times over the years, was finally replaced by a modern bridge in 1902. It did not become free of tolls until it was purchased by the Commonwealth of Pennsylvania in 1957.[14]

About twenty persons were directly involved in getting the Harrisburg Bridge Company under way. Of those, Robert Harris, Thomas Elder, and Jacob M. Haldeman stand out. Harris was one of the commissioners who solicited stock subscriptions. Elder and Haldeman became early shareholders, members of the first board of directors in 1812, and successive presidents of the company—Elder from 1816 until 1846, Haldeman from 1846 until his death a decade later. Because of their work on behalf of the bridge and a wide variety of other economic activities, they became the nucleus of a growing group of energetic local business leaders who, over the years, were to build Harrisburg's infrastructure.

Robert Harris, a son of John Harris "the Founder," was born in the village prior to its incorporation as a borough. He engaged in farming until his father's death, then managed the business affairs of the Harris estate. Following the lead of his grandfather and father before him, he took great interest in promoting the town's development. One of his earliest public acts involved a large contribution to a fund to buy and tear down a dam on the Susquehanna (the stagnant millpond behind the dam was allegedly causing illness in the community). He also served on the commission that relocated the capital at Harrisburg. He was especially active in promoting improved transportation facilities in and around Harrisburg, being named in the acts of incorporation of the Conewago Canal, the Berks & Dauphin County Turnpike, and the Middletown to Harrisburg, and Lancaster to Harrisburg turnpikes. He also was an incorporator of the Harrisburg to Pittsburgh road and was one of the commissioners who surveyed the route. In 1814 he appeared as an incorporator of the turnpike to Carlisle and Cumberland. Twice in the 1820s Harris represented the area in the U.S. Congress. He spent his later years as a genial elder statesman whose home served as a meeting place for legislators, prominent citizens, and business leaders.[15]

Thomas Elder, the son of another prominent early settler, was born in Paxton Township just over Harrisburg's eastern boundary. His father, John, was the celebrated "fighting parson" of Paxton, a leader of the "Paxton Boys" who slaughtered Indians in the area. He carried a musket with him into the pulpit. Too old to fight the British during the Revolution, he helped raise troops and served on the local Committee of Safety. Son Thomas, upon completing his schooling, read law and was admitted to the bar in 1791. A colonel in the militia, Elder volunteered to help suppress the Whiskey Rebellion. Except for a term as attorney general of Pennsylvania (1820–23), he devoted himself to business interests: law, banking, and a variety of development ventures.[16]

The third of the early builders, Jacob M. Haldeman, lived in Cumberland County across the river from Harrisburg. His father was a prosperous Lancaster County landowner and sometime investor in the Philadelphia China trade and other business opportunities. Jacob worked in his father's various enterprises until he was twenty-five. Then, with $20,000 borrowed from his father, he purchased an iron furnace at the mouth of the Yellow Breeches Creek, opposite and a little below Harrisburg, in 1806. He added rolling and slitting mills to his facility and soon established himself as one of Pennsylvania's substantial ironmakers. During the War of 1812, Haldeman supplied iron for gunmaking to the federal arsenal at Harper's Ferry and in 1814 became one of the incorporators of the turnpike to the Conewago Canal and York. After repaying his father within six years, he expanded into land ownership and grain milling. The depression of 1819–24 enabled him to buy at bargain prices the farms and other real estate of people who were overextended. By the mid-1820s he was a major landholder in the area. Despite residing in Cumberland County until 1830, Haldeman played a major role in Harrisburg affairs. That year, he and his family crossed the river to live in the borough.[17]

Although not involved in the Harrisburg Bridge Company, William Calder Sr. emerged about the same time as a fourth major builder of Harrisburg's transportation infrastructure. Born and raised on a farm in Maryland, he decided by the age of twenty-one to follow another line of work. Moving to Lancaster, he drove stagecoaches for others until he could acquire a line of his own. Once in business for himself, he contracted to carry the U.S. mail as well as passengers. When the state capital moved to Harrisburg, Calder shifted his operations there, using a livery stable on Market Square as his headquarters. He steadily improved his services with larger coaches, more and better horses, and shorter relays. Eventually he owned all but one of the lines operating out of Harrisburg. At one point, fifteen of his coaches left town daily, one each to Columbia, York, Northumberland, Pottsville, Gettysburg, and Philadelphia, four to Pittsburgh (two each over the northern and southern routes), two to Reading, and three to Lancaster. Steadily expanding, he became "the very Napoleon" of stage proprietors and mail contractors in Pennsylvania. Before the coming of canals and railroads, he maintained 1,000 horses and innumerable stagecoaches at posts across the state.[18]

Even as its transportation problems were easing, Harrisburg acquired its first bank. Next to bridges, turnpikes, and stage lines, further development of the community required banking facilities. Banks frequently helped finance transportation improvements, but the town's growing merchant community also depended on banks for amassing capital, providing a medium of exchange, discounting notes, depositing surpluses, and making loans. These needs were first met in 1809 when the Pennsylvania Assembly authorized the Philadelphia Bank to establish branch offices in towns not already accommodated by branches of the Bank of Pennsylvania. The Harrisburg branch, officially designated the Office of Discount & Deposit, could lend up to a total of $300,000. Moses Musgrave became its first cashier, Robert Harris and other worthies served on the local board of directors, and Thomas Elder and Jacob M. Haldeman were among its early depositors and frequent borrowers. The flow of commerce into Harrisburg confirmed the need for a branch bank. Musgrave wrote at the time: "In lumber and flour alone this town does a business of $300,000 annually, the busy season being Spring and Fall, when the river is high."[19]

In 1811 the charter of the Bank of the United States expired, leaving state-chartered and private institutions to provide banking services in the United States. Pennsylvania had four state-chartered banks, all in Philadelphia. One, the Bank of Philadelphia, operated branch banks in Easton, Lancaster, Pittsburgh, and Reading. Another, the Philadelphia Bank, had branches in Columbia, Harrisburg, Washington, and Wilkes-Barre. There were, in addition, a number of unchartered banks, including at least four in Philadelphia and others in Pittsburgh, Chambersburg, Lancaster, and York. A demand for more financial facilities led the legislature in March 1814 to authorize forty-one new state banks. Of thirty-nine companies that applied for charters under the law, all but two opened for business. Largely because of the Panic of 1819, however, only twenty-six were still in business in 1822.[20]

Moving quickly to form a locally controlled bank, a group of leading citizens in Harrisburg secured a charter in May 1814. Popular response was enthusiastic both in the town and in nearby Cumberland County. In the latter, Jacob M. Haldeman, after buying 50 shares himself, persuaded neighbors and friends to take up another 235 shares. Subscriptions in that area were helped by the bridge then under construction, which promised easier access to the capital city and to the bank about to be started. Although the subscription books were open only six days, a total of 235 persons bought 6,427 shares of stock with a par value of $50, exceeding the authorized capital of $300,000 by more than $21,000.[21]

Ownership of the Harrisburg Bank was not highly concentrated. Two dozen people, including Robert Harris, Thomas Elder, and John Forster, each held 100 shares, with a par value of $5,000, and controlled a little over a third of the stock; 50 other people, with between twenty-five and ninety-five shares each, controlled another third; and the remaining third was held by 161 small shareholders. As might be expected, seven of the thirteen members of the first board of directors (including the president) and the cashier came from the major stockholding group, four directors came from the second group, and two from the third. When the bank organized in June, Harris and Haldeman were elected to the board of directors. The others included Henry Beader, a coppersmith who was recorder of Dauphin County; Christian Kunkel and John Peter Keller II, wealthy merchants who also regularly served on the borough council; and John Shoch, an innkeeper and treasurer of Dauphin County. Four represented nearby communities: Haldeman from Cumberland County, Thomas Brown from Paxton, John McCleery from Halifax, and Isaac

Hershey from Londonderry. The presidency fell initially to William Wallace, a wealthy ironmaster and chief burgess of Harrisburg; the cashiership went to John Downey.[22]

During the first half of the nineteenth century, "country banks" such as the one at Harrisburg were relatively simple affairs. Shareholders met annually to elect (and more often to reelect) the board of directors. The directors, in turn, selected the president, the cashier, the clerks, and other employees and fixed their salaries. Day-to-day operations were conducted by the president or the cashier, or both, depending on strength of personality, business sense, amount of personal investment in the bank, and other interests of the two. Periodically, perhaps once a week, the directors met to decide which notes offered by businessmen would be discounted and which would be rejected. They also decided which unpaid loans would be extended (and on what terms) and which would be taken to judgment. In practice, the longer a president or cashier was in office, the more complete his authority became. Under a strong president or cashier, the board's work dwindled to accepting recommendations and sometimes even ratifying actions already taken by the executive officer.

The physical facilities of such banks were simple at the start. The Harrisburg Bank, for example, opened for business in the parlor of its cashier's home. It soon contracted for a house to be built equipped with a vault and space for conducting business. Then, in 1817, it purchased and moved into the quarters previously occupied by the Office of Discount & Deposit. That two-story building, which remained the home of the Harrisburg Bank until 1854, consisted of a banking room and a small directors' room on the first floor, and living quarters for the cashier on the second floor.[23]

From the beginning the Harrisburg Bank helped finance the transportation improvements so vital to the growth of the community's trade and commerce. As early as September 1816 the cashier requested a number of banks in neighboring communities to send paper; the Harrisburg Bank's funds had "become diminished" because its "local situation" had required it to aid both the Harrisburg Bridge Company and the Harrisburg to Middletown Turnpike. Its ties to the bridge company were particularly close. The bank held considerable blocks of Harrisburg Bridge Company stock throughout the pre–Civil War period and frequently extended loans. Seven of the bank's early directors simultaneously were directors of the bridge company; Cashier Downey, with the permission of the bank's board of directors, served as treasurer of the bridge company without pay; and Thomas Elder, president of the Bridge Company and owner of 100 shares of bank stock, became president of the bank at

Wallace's death in 1816.[24] Among the bank's earliest transactions were frequent discountings of bridge company notes and loans to the Commonwealth of Pennsylvania to get the state to honor its own warrants to the bridge company that the bank had accepted as deposits.[25]

Securing money to lend was the Harrisburg Bank's first task. As soon as the board of directors organized, they dispatched Downey to Philadelphia to secure the printing of $500,000 worth of bank notes—$100,000 at once, the balance within three weeks. They also agreed to accept the notes of certain out-of-state banks and to make arrangements for "friendly intercourse" with other banks. Finally, they notified federal revenue collectors that the Harrisburg Bank stood ready to act as a depository for their funds.[26]

Another early concern of the Harrisburg Bank was that Philadelphia banks were trying to smother it and the other new country banks in their cradles. Within a month of opening, it received notice from the Philadelphia Bank's Harrisburg branch that only the Harrisburg Bank's own notes or notes issued by Philadelphia banks would be accepted. In response, Cashier Downey wrote to his counterpart in the new bank at Reading, urging cooperation and an exchange of paper to protect their institutions from the city banks. The Harrisburg Bank also hosted a series of conferences with other country banks to take "measures of self-defense against the intolerance of the city banks." These sessions, held in Harrisburg in December 1814, February 1816, and February 1817, protested the behavior of Philadelphia banks, discussed possible legislative remedies, and studied problems stemming from the general suspension of specie by Pennsylvania banks between September 1814 and 1817. On another front, the Harrisburg Bank persuaded the country banks between Philadelphia and the Allegheny Mountains to issue uniform notes (except for the name of the bank of issue), which it supplied to them.[27]

Early in 1815, Cashier Musgrave of the Philadelphia Bank's Harrisburg office intensified the pressure. He notified the Harrisburg Bank that his institution held a large quantity of its notes and wanted to know how soon and in what manner they could be redeemed. This was a legitimate device that Philadelphia banks employed to protect themselves against inadequately backed issues of notes by country banks. It could also be used to embarrass banks whose notes were perfectly sound by draining away their limited assets on short notice. The Harrisburg Bank replied that it had no Philadelphia paper at the moment but would give any other paper it had. By July, Musgrave had more than $30,000 worth of Harrisburg Bank notes he wanted to redeem. With considerable effort, the Harrisburg Bank scraped together enough Philadelphia, New Jersey, Delaware, and Southern state paper to redeem them. Ironically, a few weeks later, after sending

Cashier Downey to Philadelphia to persuade banks there to accept its notes, the directors of the Harrisburg Bank authorized him to reject notes of banks located west of the Susquehanna.[28]

New troubles afflicted the bank when its first cashier resigned in 1815 and its first president died in 1816. Thanks to the great need for banking facilities in the region, the bank prospered nonetheless. After only four months of operation it had nearly $95,450 in assets and issued a 33-cent dividend per share, equal to 8 percent a year on the capital stock paid in. By 1816 it was discounting notes for customers not only from Harrisburg and from Dauphin County but also from most adjacent counties and from as far away as Crawford County in the northwestern corner of the state. Dividends continued at 8 percent until the Panic of 1819.[29] Meanwhile, new officers, President Thomas Elder and Cashier John Forster, began long tenures. Elder would serve thirty-seven years, until his death in 1853; Forster served from 1815 until he quit in 1833 after quarreling with Elder.

Forster's appointment as cashier marked his entry into a leading role in building Harrisburg's economy. He was a native of Elder's birthplace, Paxton Township, and was married to Elder's niece. He cut short his schooling at the College of New Jersey at Princeton to serve as a volunteer in putting down the Whiskey Rebellion in western Pennsylvania. During the War of 1812, he fought in the defense of Baltimore and earned the honorific "General." The citizenry also chose him to represent them in the state senate between 1814 and 1818. Although he originally prepared for a career in law, Forster's main interests became mercantile pursuits, banking, and land acquisition.[30]

Elder and Forster began forcefully. In 1817 they purchased the business and offices of the branch of the Philadelphia Bank for just over $245,000. This made the Harrisburg Bank the town's only important banking facility.[31] That same year, the Treasury Department designated it and four other state banks in Pennsylvania depositories of federal revenues. The arrangement provided not only additional funds but also power over other banks. Revenue collectors were instructed to accept only the notes of state banks that were acceptable to the four depositories and to the second Bank of the United States, which had been chartered in 1816. These advantages were lost, at least temporarily, by Elder's dabbling in politics during the gubernatorial election of 1817. His efforts to defeat those favored at Washington led the secretary of the treasury to transfer the federal government's funds to the nearby bank at Swatara. Even so, the Harrisburg Bank enjoyed a strong credit line with the Bank of the United States. As of June 30, 1818, its obligations to that institution amounted to more than $62,000.[32]

The onset of a major business panic in 1819, however, considerably slowed its operations. Despite the strain of the general collapse of business, the Harrisburg Bank, unlike many other state banks, survived. So did the various turnpike and canal companies that were part of Harrisburg's growing infrastructure. All were in place to take advantage of the next upturn in the economy in the mid-1820s.

Second Period of Growth, 1824–1837

Although the depression began to abate as early as 1821, major new ventures in the Harrisburg area did not appear until construction began on the State Works. Anxiously watching the development of New York's Erie Canal, the Pennsylvania legislature in 1824 commissioned the exploration of a canal route between Philadelphia and Pittsburgh. Completion of the Erie Canal in 1825 led them in 1826 to authorize beginning the State Works even though the exact route and other details would not be hammered out until 1828. In the end, the Main Line took the form of a railroad from Philadelphia to Columbia on the Susquehanna twenty-nine miles below Harrisburg. There it connected with a northbound canal running parallel to the Susquehanna, passing through Harrisburg east of the capitol building, and proceeding twelve miles to the mouth of the Juniata River. Although a branch would continue northward along the Susquehanna, the main route turned westward, following the Juniata to Hollidaysburg. From there, using a series of inclined planes known as the Allegheny Portage Railroad, it crossed the mountains to Johnstown. A canal along the Conemaugh, Kiskiminetas, and Allegheny rivers completed the route to Pittsburgh.[33]

The coming of the State Works to Harrisburg occasioned the rise of two more of the town's economic leaders, Michael Burke and Simon Cameron. Burke had come to the New World from Ireland at the age of eighteen and worked briefly at an extensive fishery in Newfoundland owned by an uncle before striking out on his own. After moving to Lockport, New York, he clerked for a contractor until 1824, when he secured a contract for building part of the Erie Canal. That completed, he moved to Akron, Ohio, to work on another canal. Then, in 1829, with the Pennsylvania State Works under way, Burke obtained a contract for the portion of the canal between the towns of Mexico and Lewistown on the Juniata River. At that time he fixed his permanent residence in Harrisburg. Despite Burke's foreign origin and his Roman Catholicism, he won ready acceptance in his new home. In the 1830s, for example, the voters elected him to the borough council. When work on the canal was complete, Burke

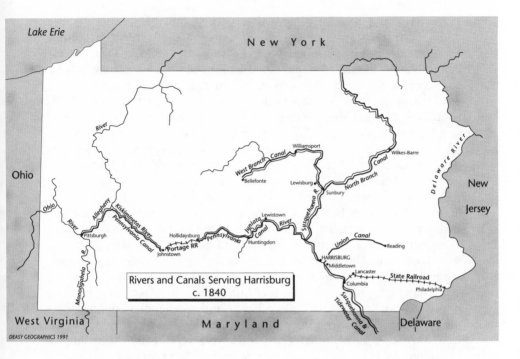

Fig. 3.   Rivers and Canals Serving Harrisburg, c. 1840.

remained in Harrisburg and made a fortune in construction projects and a variety of other business activities.[34]

The second rising leader, Simon Cameron, though only a sometime resident of the town before 1860, became one of Harrisburg's (and Pennsylvania's) more ambitious and controversial personages.[35] He began life in poverty. His father, a custom tailor by trade, could not fully support his family, and some of the children had to find other homes. A Jewish physician in Sunbury adopted Simon, and when his patron died in 1814 Cameron apprenticed himself to a printer. Soon he was working for a succession of newspapers in Pennsylvania and, briefly, for one in Washington, D.C. Between stints of editing, he organized successful political campaigns for the Jacksonian Democrats in Pennsylvania. Both activities involved him deeply in lobbying, intrigue, and political maneuvering. He told a legislative committee in 1842 that he had been "at Harrisburg every session [of the Assembly], more or less, since 1817. I have been upon the most intimate terms with the Legislature. Many years, I have known every member of the Legislature."[36]

In 1820, with help from an uncle, he bought a small newspaper in

Harrisburg and promptly merged it with another Harrisburg paper, Charles Mowry's *Pennsylvania Intelligencer*. Mowry, adept at politics, brought state printing contracts to the firm and, in 1826, won election to Pennsylvania's Board of Canal Commissioners, the body responsible for constructing the State Works. Profits from state printing provided Cameron with capital for his later undertakings; his ties to Mowry no doubt contributed to his successful bidding for contracts to construct three large sections of the State Works. Cameron quickly learned that government could be milked for profitable contracts and that having friends in high places was vital to the process. To him, business and politics were but the two sides of the same coin.

Although frequently in Harrisburg, Cameron did not live in the town between 1830 and the 1860s. In the earlier year, he recruited a crew of Irish laborers in Philadelphia and took them to Louisiana to construct a canal between the Mississippi River and Lake Pontchartrain. Ever alert to opportunities, he wrote that "everybody" in New Orleans was making money "at every business." Were his canal venture to fail, he would "be tempted" to go into some other line. "It would be a great place for a brewery—not one within 1500 miles. A tallow chandler could make a fortune soon. Tallow 3½ cts a pound—candles 20 cts." Raw materials were very cheap, manufactured goods were very costly. According to Cameron, nobody there liked to work, preferring to depend on the labor of blacks.[37]

By 1832 he was back in Pennsylvania. Settling in Middletown, he invested in real estate, held a partnership in an iron furnace, and engaged in a wide variety of other business and political ventures. His most important activity was to help found the Middletown Bank in 1832. Owning seventy-eight shares, he became cashier, a position he held until 1857. Cameron quickly moved to assume full control. Among other things, he courted the favor of the influential and wealthy Gratz brothers of Philadelphia, making loans to them without security and without prior approval of his board of directors. Although embarrassed by the brothers' repeated failures to repay on schedule, Cameron did not lose his position. Instead he used the Gratz brothers to advance his goal. "I have not yet taken possession of the Bank," he wrote to W. E. Gratz in December 1838, "but I have spoken to the people about your interests." In March 1841 he was seeking $10,000 to $12,000 to buy out directors who differed with him. The "views we have often interchanged" could be carried out, he promised, with "the capital of this little Bank, unimpaired, and in our hands."[38]

Meanwhile, Cameron lobbied vigorously in the legislature for Pennsylvania banking interests. Never had legislation he supported run into such trouble as in February 1838, he wrote Simon Gratz. Legislators promised one thing in private, then under party discipline spoke and voted quite differently on the floor. In accepting "the offer" of Gratz's institution, it was only fair, he noted cryptically, that Philadelphia banks incur some expense in the fight for measures for their safety. "You know a man cannot live at Harrisburg upon the wind." Despite not residing in Harrisburg, Cameron's widespread interests there soon made him one of its busiest entrepreneurs.[39]

Once the State Works connected Pittsburgh and Philadelphia in 1834, attention shifted to another canal project. In 1835 Jacob Haldeman and Simon Cameron were among the incorporators of a private company, the Susquehanna & Tidewater, which built a canal along the west bank of the Susquehanna River from Wrightsville (opposite Columbia) to Havre de Grace, Maryland, at last opening passage of the Susquehanna to Chesapeake Bay.[40] As matters turned out, that project marked the end of efforts to solve the area's transportation problems by water routes. River and canal transit could not meet the growing demands of commerce, much less the needs that modern industry would impose within a decade.

In spite of quickly becoming the major artery of traffic across Pennsylvania, the State Works had far less impact on Harrisburg's development than might have been expected. It no doubt contributed to the town's population growth from 4,312 to 7,834 between 1830 and 1850, an increase of a little over 80 percent, but even at their peak just before the coming of railroads, the canal and related enterprises such as boatbuilding created relatively few jobs in the community. In the 1850 federal census, only 55 of 2,310 persons with listed occupations were specifically identified with canal-related occupations. Forty-five were boatmen and seven were boatbuilders.[41] At the same time, canals offering speedier and more convenient travel had some impact. For example, they cut into William Calder's stagecoach business so much that he set up a packet-boat line over the canal for customers who preferred to travel that way.[42]

Potentially more important was the rise of railroad companies. The lines built in the mid-1830s, however, were short, connected nearby communities, and had limited capacity. Only later would they merge or be absorbed into greater railroad systems. In the meantime, Harrisburg's chief business leaders promoted a number of these small companies. In 1831 Jacob Haldeman and John Forster were among the incorporators of the Cumberland Valley Railroad to run from the western end of the Harrisburg Bridge to Carlisle and Chambersburg. Promotion, at least in this instance, did not mean financing. Initial efforts to raise capital for the line failed, requiring a legislative revival of the line in 1835. The

Fig. 4.   A Scene Along the Pennsylvania Canal.   The Eastern Division of the
Canal, which included the portion serving Harrisburg, operated from 1834 until
1901. Almost from the beginning it faced competition from the railroads.

second subscription drive in three weeks raised the $210,000 needed to
begin construction. Philadelphians provided 60 percent of the funds, and
residents of Chambersburg, Carlisle, and Mechanicsburg put up most of
the remainder; the people of Harrisburg coughed up but $450.[43]

In 1835 the legislature also authorized extension of the Portsmouth &
Lancaster Railroad to Harrisburg. Originally chartered in 1832, the line
was now renamed the Harrisburg, Portsmouth, Mt. Joy & Lancaster
(hereafter Harrisburg & Lancaster). By 1837 the legislature had char-
tered two more railroads, the Lebanon Valley and the Harrisburg &
Sunbury. William Ayres (who became a major builder in Harrisburg) and
Simon Cameron were among the incorporators of both lines; Jacob
Haldeman was an incorporator of the line to Sunbury. The inclusion of
Philadelphia incorporators by the two lines indicated that part of the
capital for building them came from that city.[44]

Although chartered later than the Cumberland Valley Railroad, the
Harrisburg & Lancaster was the first to reach Harrisburg. In August

1836 a small horse-drawn train rolled into the borough from Middletown. A month later a train pulled by a steam locomotive, the "John Bull," made the same trip. Railroad traffic between Harrisburg and Lancaster did not begin until a tunnel near Elizabethtown was completed in 1838. Meanwhile, users of the unfinished Cumberland Valley Railroad had to shift to stagecoaches at the Harrisburg Bridge to complete their trip into the capital. A railroad bridge over the Susquehanna, authorized in 1836, made a through trip into Harrisburg possible by January 1837. In August that same year the Cumberland Valley line reached Carlisle; by November it had reached Chambersburg.[45] By the end of the 1830s, Harrisburg had both water and rail facilities running in all directions from the town.

Until 1834 the Harrisburg Bank remained the town's only major financial institution. A relatively small clique ran the bank. William Wallace, its first president, died during his third year in office, but his successor, Thomas Elder, served for the next thirty-six years. Jacob M. Haldeman, after thirty-six years as a director, became the third president for four years. A number of the other directors also enjoyed long tenures in this period: Thomas Brown, thirty-four years; Joseph Wallace, thirty-two; John Shoch, twenty-eight; Peter Keller, twenty-seven; Abraham Oves, twenty-five; and John Geiger, twenty. Eleven others (Peter Brua, David Ferguson, John Fox, Robert Harris, Frederick Kelker, Isaac Hershey, Christian Kunkel, Michael Lebkicker, Luther Reily, John Roberts, and John Wood) enjoyed tenures of between eleven and sixteen years. Only about twelve men served fewer than ten years, and several of those died in office. The occupations of most of the long-term directors are known. Geiger, Kelker, Keller, Kunkel, Oves, and Wallace were merchants, Haldeman was an ironmaster, Elder and Roberts were attorneys, Fox was a drover and sheriff of Dauphin County, Shoch was an innkeeper, Harris was a farmer and a congressman, Reily was a physician, and Brua (Simon Cameron's father-in-law) was a carpenter. Most were also landowners.[46]

It is not surprising that this group recognized the link between improvements in transportation and community growth and development. In one way or another they gave the bank's support to almost every transportation company in the area. The bank purchased stock in the Peters Mountain Turnpike Company, which ran north from Harrisburg along the Susquehanna. It supported the Harrisburg Canal Company (whose president was the bank's cashier, John Forster). A substantial loan of $100,000 made to the commonwealth for construction of the State Works in 1826 was followed by numerous smaller loans over the years for the same purpose. The bank extended credit to two private canal companies, the Union Canal and the Susquehanna & Tidewater.

The Harrisburg & Lancaster, Cumberland Valley, and Lebanon Valley railroads also received help. The other principal customers were merchants who depended on the bank to provide credit to "carry on the Spring and Fall business." This meant loans for laying in stocks of goods ahead of busy periods, to be repaid at the end of the season, if profits allowed, or extended to a better year.[47]

The establishment of a second bank in the town in 1834 ended the Harrisburg Bank's near monopoly. In April of that year, the legislature authorized the Harrisburg Savings Institution. Eleven years later the bank changed its name to the Dauphin Deposit Bank, which remains its name today and which it will be called in this study. In some ways the early history of the Dauphin Deposit paralleled that of the Harrisburg Bank. Both had to struggle at first to amass the cash needed to begin operations and to carve out a place against stronger rivals. Although both also were founded just before a major business panic, they managed to survive.

As usual, leading citizens were the initial incorporators: contractor Michael Burke; newspaper editors James Peacock and Mordecai McKinney; physicians Luther Reily, Christian Seiler, and Edmund W. Roberts; and attorneys Herman Alricks, William Ayres, Charles C. Rawn, and James McCormick, who practiced in Harrisburg but resided in Cumberland County.[48] Between eighty-five and ninety shareholders initially provided the authorized capital of $50,000 by buying 2,000 shares of stock at $25 par. Charles Carson (identified variously as "gentleman" and "farmer," and one of Harrisburg's principal property holders in 1850) held sixty-nine shares, making him the single largest individual shareholder. Other major holders (with fifty shares each) included George Mish (ironmonger), William Dock (owner of a candle factory), and John Knepley (another "gentleman," "farmer," and "grocer," depending on date and source). Other prominent individual shareholders, with fewer than fifty shares, were attorneys William McClure and George Harris (grandson of the town founder, son of Robert who helped start the Harrisburg Bank, and solicitor of the Middletown Bank); a grocer, William Cattrell; and John Cameron (a brother of Simon). Far and away the largest block of stock, however, was that held by Simon Cameron's Middletown Bank. Its 850 shares (nearly 43 percent of the total) greatly exceeded the combined holdings of all the other major shareholders.[49]

Cameron's reason for wanting a financial base in Harrisburg can only be surmised. The community was developing faster than Middletown, and perhaps he saw it as affording greater opportunity for investment. Given his extensive lobbying activities, his loans to legislators, and his political involvement with the Democratic party, however, Cameron may simply have wanted easier access to funds at a bank with which he was not readily identified. So far as can be determined from the few officers and directors of Harrisburg banks in 1835 whose political affiliations are known, those at the Harrisburg Bank tended to be Whigs, and those at the Dauphin Deposit were apt to be Democrats.

The presidency of the new bank went to an attorney, William McClure. For the first several years, Simon Cameron's interests were represented on the seven-member board of directors by George Harris (1835–43), John Cameron (1835–38), and Simon Cameron himself in 1839. Opening for business in a booming economy, the new bank got off to a good start. On the first day of business (September 28, 1835) it received more than $10,000 in deposits, including $1,000 deposited by James McCormick, who would become the bank's second president in 1840. By October 9 its assets totaled $22,595; a month later the total was $35,331; and by November 1836 they stood at $93,339, the highest they would reach until after the worst of the Panic of 1837. To attract depositors, the Dauphin Deposit, in addition to its regular savings accounts, which paid the then standard 3 percent interest, created special accounts paying 4 percent. These deposits had to be made for a minimum of four months and could be withdrawn only on four weeks notice, and they received no interest for the period after notice was served. In June 1841, when the second (and only extant) special account book opened, 197 depositors were carried over from the first book.[50]

The Dauphin Deposit paid its first dividend seven months after opening. In 1837 its total dividends amounted to 10.5 percent, only 0.5 percent less than those of the Harrisburg Bank. The first comparable data on the relative strengths of the two showed that in November 1839 the total assets of the Dauphin Deposit ($108,937) amounted to about one-eighth those of the Harrisburg Bank ($968,829).[51]

From 1837 on, depressed business conditions again slowed economic growth and activity in Harrisburg. Once more, despite widespread business failures during the Panic years, Harrisburg's transportation and banking companies all survived into the next period of prosperity. In that period the building of Harrisburg's industrial infrastructure was completed.

The Maturing Infrastructure

A general recovery of the economy began in 1842. During this third period of economic expansion, Harrisburg's transportation, banking, and utilities infrastructure moved to maturity. Railroads emerged as the dominant means of transportation and began breaking down the economic isolation of inland communities. The town's two major banks continued to thrive and by 1860 had additional competitors. Also, the community at last built itself a facility that assured residents of a regular and dependable supply of water, and a private firm introduced gas lighting to homes, public buildings, and the streets.

Railroad-building in the area entered a new phase by the end of the 1840s. Although companies continued building short new lines that one by one tied every community in the region with every other, larger schemes were afoot. By purchase, merger, and lease of small lines, combined with new construction as needed, organizers were forging companies that each operated hundreds of miles of railroad under a single management. Meanwhile, freight began to surpass passengers in importance to the railroads as more powerful locomotives were developed to pull more and larger cars. It was as a promoter of railroads that another of Harrisburg's important builders, William Ayres, first began to move to prominence.

Placing Ayres among the chief builders of Harrisburg's infrastructure might seem questionable. Unlike the others, he never had more than modest means, much less a fortune. His membership in the group instead rests on his catalytic role in securing legislation, promoting, and in some instances arranging the finances for railroads, banks, and utilities important to Harrisburg's development. A farmer's son, Ayres took employment with a Harrisburg merchant at the age of twenty-eight and soon married into the prominent Bucher family. After two years he returned to his birthplace north of Harrisburg, where he farmed and kept a hotel. He also became a justice of the peace and a major in the militia. Ayres was thirty-six when he returned to Harrisburg to study law, meanwhile continuing to serve as justice of the peace for two townships and the borough of Harrisburg. He soon developed a wide acquaintance in the "upper end" of Dauphin County that, coupled with an ability to speak German, furthered his career. Among his clients were county officers and turnpike companies. Serving in the legislature from 1833 to 1835, he worked closely with Thaddeus Stevens in the battle to create a public school system for Pennsylvania. From 1838 until 1841 he sat as a member of the Harrisburg borough council. He was a Whig in politics and a supporter of William Henry Harrison for President in 1840.

As already noted, Ayres was among the incorporators of a number of early small railroad companies in the Harrisburg area: the Lebanon Valley in 1836, the Harrisburg & Sunbury in 1837, and the Harrisburg & Pine Grove in 1842. During the 1850s he promoted and served as president of the Huntingdon & Broadtop, which opened the coal fields of Juniata County, and later of the Harrisburg, Hamburg & Easton, which Ayres envisioned as part of a major New York to New Orleans route, one "laid down by Nature for a direct communication from the northeast to the South."[1] Far more significant, however, was his work for the Pennsylvania Railroad. A number of prominent Harrisburg residents took an interest in the proposed new line connecting Philadelphia and Pittsburgh that would pass through their town. A special convention held in the city in January 1846 included an array of leading residents, including William Ayres and Michael Burke, as well as lawyers, merchants, newspaper editors, and physicians. The body adopted a resolution, drafted by Ayres, favoring the "central route" for the new railroad and calling for incorporation of a company to build it. Meanwhile, without charge, Ayres worked "on the hill" to secure passage of the act that created the Pennsylvania Railroad Company that same year. In addition to being one of the original incorporators, Ayres sat on its board of directors in 1852 and 1853—the only nineteenth-century Harrisburger to be so honored.[2]

Work on the line between Harrisburg and Altoona began in the summer of 1847, and Harrisburg contractor Michael Burke, with two partners, began construction of a bridge to carry the line across the Susquehanna. The bridge was completed by the end of 1848; trains could pass to Lewistown by September 1849, to Huntingdon by the following June, and as far as Hollidaysburg by September. By December

1852 one could go all the way to Pittsburgh, but in the mountains the line of the Portage Railroad, which was part of the State Works, had to be used. With the completion of the celebrated Horseshoe Curve west of Altoona in 1854, the Pennsylvania Railroad reached Pittsburgh over its own lines. A year later, by leasing a number of existing lines, the Pennsylvania extended to Chicago and was on its way to becoming the nation's premier railroad company.[3]

Every advance of the Pennsylvania Railroad stirred interest in Harrisburg. When the directors of the company decided against an "opening fete" to celebrate completion of the line to Lewistown, nearly 200 young people arranged on their own a special "first" excursion to that neighboring town. In addition to the excitement of riding the train, they listened to speeches, banqueted, and enjoyed fiddle and bow dancing before returning.[4] Passenger travel over the line soon exceeded expectations of the railroad's managers. The *Harrisburg Telegraph* reported on April 6, 1854, that cars leaving the town frequently were full, resulting in disappointed customers farther up the line who were unable to get seats. Although a train the previous day had been the largest yet to leave the city, the reporter believed an extra train should have been introduced.

From the point of view of the town's economic development, however, the increase in freight traffic was probably more important. As early as 1853, freight revenues on the Pennsylvania exceeded passenger revenues. By 1860 the ratio was nearly three to one. On a single day in March 1854, some 325 freight cars passed through Harrisburg eastward, carrying an estimated 1,300 tons of freight. To haul the same tonnage between Philadelphia and Pittsburgh by horse and wagon, the *Harrisburg Morning Herald* reported, would have required a four-mile-long procession of 650 wagons and 2,600 horses and would have taken between twelve and fifteen days. Railroad trains regularly required only thirty to forty hours for the same trip.[5] Because canal boats on average carried thirty-five tons when crossing the Portage Railroad, it would have taken thirty-seven boats to haul the same 1,300 tons of freight between Pittsburgh and Philadelphia over the State Works. Although the trip could be completed in eight days, the average time was closer to twelve days. Moreover, the canal system would have had to step up its pace substantially to handle the increased traffic. On average, only about seventeen boats a day passed any given spot on the canal in either direction, and several of those were passenger boats.[6]

At first nearly every issue of local newspapers reported new speed and tonnage records over the line, new equipment passing through Harrisburg for use on the Pennsylvania and other railroads, promotions of new railroad ventures, and occasional accidents. Although such reports grew less frequent once their newness wore off, the impact of the Pennsylvania Railroad involved more than novelty. Harrisburg began to enjoy the benefits of being on one of the nation's principal trunk lines between the East Coast and the Midwest. The State Works between 1837 and 1846 had carried, on average, 490,000 tons of goods a year. As early as 1855 the Pennsylvania Railroad was hauling three-quarters as much, and four years later it was transporting 754,354 tons. In 1852 the gross revenues of the line east of Pittsburgh amounted to nearly $2 million; by 1859 they were more than $5.3 million. At Harrisburg alone the number of passengers arriving and departing swelled from 103,000 in 1853 to nearly 145,000 two years later, and then declined sharply during the Panic of 1857. Freight tonnage to and from the town was more impressive, increasing from 6,400 tons in 1853 to 32,000 tons in 1857, but it declined to 29,000 tons the next year.[7]

Philadelphia interests had promoted the State Works, and later the Pennsylvania Railroad, to preserve that city's commercial control of trade with the west. To serve similar ends for Baltimore, business leaders there continued their efforts to lure the trade of Central Pennsylvania and the west away from Philadelphia and down the Susquehanna to Chesapeake Bay. During the 1820s they experimented with steamboat navigation on the river. When that failed, the Susquehanna & Tidewater Canal was built to accommodate other vessels. Railroad penetration of the region began about the same time. Both Maryland and Pennsylvania chartered small lines paralleling portions of the river: the Baltimore & Susquehanna, the York & Maryland, the York & Cumberland, and the Susquehanna Railroad (from Nanticoke to the New York boundary).[8] Joint acts of the two state legislatures created the Northern Central Railroad in 1854 and authorized it to absorb the small companies and to lay any additional track needed to forge a single railroad from Baltimore to Canandaigua, New York, via Harrisburg, Sunbury, Williamsport, and Elmira. Well located strategically, the line might, with connections, tap into the Great Lakes trade at Buffalo, offer a direct route for anthracite from northeastern Pennsylvania without passing through the City of Brotherly Love, and divert a portion of the Pennsylvania Railroad's eastbound traffic from Philadelphia southward to Chesapeake Bay.[9]

From its beginning, the Northern Central was embroiled in the wars among the trunk lines. President J. Edgar Thomson, even as he pushed the Pennsylvania Railroad westward, searched for potential connections to the east and south. He early established a link with New York City over the Camden & Amboy, but his limited control of that line was not enough. The Northern Central, which crossed the Pennsylvania at Harrisburg, could be used to develop connections with both New York and Baltimore. It might even drain off some of the traffic of the Baltimore &

Ohio Railroad (B&O). John W. Garrett of the B&O moved first, however, joining forces with the Philadelphia & Reading (another rival of the Pennsylvania) to acquire working control of the Northern Central. Garrett wanted to open a connection to New York for his line over the Northern Central.[10]

Simon Cameron, an original incorporator of the Northern Central, devised a scheme to shift control of that line to Pennsylvanians. In addition to helping incorporate numerous railroads in the area, he served as president of the Harrisburg & Lancaster Railroad in the late 1830s and of the Lebanon Valley and Susquehanna railroads in the 1850s.[11] While serving in the U.S. Senate (1845–49 and 1857–61), Cameron soon grasped the potential strategic value of the Northern Central to the nation's capital. From the mid-1850s until the Civil War, all traffic to Washington from the Northeast (New England, New York, New Jersey, and eastern Pennsylvania) passed over the Camden & Amboy and the Philadelphia, Wilmington & Baltimore to the city of Baltimore. Similarly, most traffic from the west and northwest passed over the main line of the B&O to Baltimore. From there a single-track branch line of the B&O provided the only connection with the nation's capital. If Cameron could wrest the Northern Central away from the B&O and other Maryland interests, he could break those monopolies. Accordingly, in the fall and winter of 1858–59, he and his brother-in-law A. B. Warford bought enough stock to secure Cameron's election to the board. Elected with him were his brother William Cameron, William Calder Jr. (son of "the very Napoleon" of Harrisburg stagecoach line operators), and Jacob S. Haldeman (son of the recently deceased Jacob M. Haldeman). Even so, the officers of the company and the majority of directors remained Marylanders. In June 1860 Simon Cameron proposed that Thomson and Thomas A. Scott of the Pennsylvania join him in the effort to complete the takeover.

To counter Cameron, the B&O and its allies promptly bought up large blocks of Northern Central stock, but the purchasers lost their nerve when Lincoln's election in November produced a period of financial stringency. As the Marylanders dumped their shares on the market, Cameron and his forces stood ready to buy. At the March 1861 meeting of Northern Central stockholders, Baltimore & Ohio interests controlled only 10,000 shares; the Pennsylvanians controlled 21,000. Cameron and his allies took over, electing Warford president of the company; Simon Cameron's brother, John D., vice president; and Simon's son, J. Donald, to the board.[12]

J. Donald Cameron, by then only thirty years old, became president of the line in 1863. He was prepared for the task. He had planned to make his career in law, but he changed his mind while reading for exams in his father's banking office in Middletown. There he overheard numerous conversations and soon found himself immensely interested in and informed about his father's many business affairs. Simon Cameron took Donald into business with him, entrusting him at a tender age with considerable responsibility. Donald, for example, was nineteen when he took his father's place as cashier of the Middletown Bank. His marriage to James McCormick's only daughter, Mary, in 1856 began a family alliance with the McCormicks that would dominate Harrisburg's industrial scene in the Gilded Age.[13] Meanwhile, as we shall see, control of the Northern Central by Pennsylvanians proved vital to the defense of Washington during the Civil War.

By 1860 Harrisburg was well served by railroad lines in all directions. The Pennsylvania Railroad connected it with Pittsburgh, Chicago, and the Great Lakes region to the west and with Philadelphia to the east. The Cumberland Valley Railroad operated south and west to Carlisle, Chambersburg, and the Valley of Virginia beyond. The Northern Central gave Harrisburg ties to Baltimore and Washington to the south, and the upper

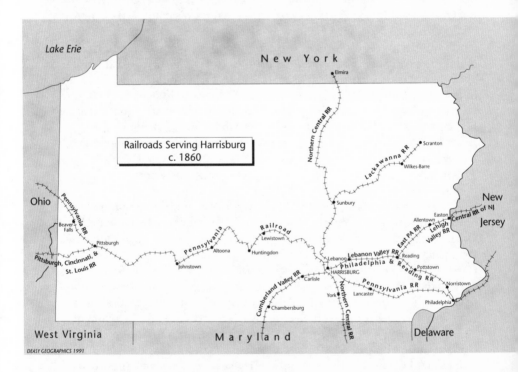

Fig. 5.   Railroads Serving Harrisburg, c. 1860.

Susquehanna Valley and central New York to the north. The Lebanon Valley Railroad ran eastward from Harrisburg to Lebanon and Reading, while the Dauphin & Susquehanna, running northeast to Pine Grove and Pottsville, provided access to the anthracite coal fields.[14]

In contrast to the relatively small impact of canals on its economy, railroads brought significant growth to Harrisburg. Both the total population and the number of persons directly employed by railroads increased markedly. Total residents of Harrisburg swelled from 7,834 in 1850 to 23,104 by 1870, an increase of nearly 195 percent. Because much industrialization was taking place in the same years, only part of the gain can be directly attributed to the railroads. Nonetheless, railroading itself employed substantial numbers of Harrisburg residents. The town in 1857 became headquarters for the Middle Division of the Pennsylvania, and extensive yards, repair shops, and roundhouse facilities would soon be built in the city east of the Capitol. By the next census, 153 (3.6 percent) of all persons listing occupations in the federal census indicated that they were railroad employees. By 1870 the figure was 620 persons, more than 8 percent of the total. Not included were many persons listed by skill or simply as "laborer" who in fact worked for the railroads. A further impact, as will be seen, was the rise of railroad-related industries such as the Harrisburg Car Company, which in 1860 employed 140 persons.[15]

During the 1850s the Commonwealth of Pennsylvania decided to get out of the business of owning bridges, turnpikes, and canals. The Pennsylvania Railroad and its subsidiaries acquired most of the State Works in 1857. Although the main route of the canal between Columbia and Hollidaysburg at the front of the Alleghenies continued in use for the rest of the nineteenth century, the State Works west of Hollidaysburg closed down by the end of the Civil War.[16] When the state offered its Harrisburg Bridge Company stock (with a par value of $90,000) for sale early in the 1850s, Harrisburg entrepreneurs James McCormick and Jacob M. Haldeman purchased it for $9,000.[17]

In the 1830s, James McCormick, the founding patriarch of one of Harrisburg's more important families in the late nineteenth and early twentieth centuries, began to move to the fore of community builders. For several years, in the manner of Jacob Haldeman before him, McCormick lived across the Susquehanna while engaging in business in Harrisburg. Although his father had died when he was four, James's mother had tutored him for the College of New Jersey at Princeton. After graduating with honors, he returned to his native county to read law. Admitted to the bars of both Cumberland and Dauphin counties, McCormick established his office in Harrisburg. At first, his substantial earnings went into farmland along the Susquehanna in Cumberland County and into Harrisburg real estate. In the late 1830s he moved into banking, being elected to the Dauphin Deposit's board in September 1838 and to its presidency two years later. He filled the latter position until his death in 1870.[18]

Although McCormicks would dominate the Dauphin Deposit for more than a century (from 1840 until 1945) and operate it as a private bank from 1874 until 1905, it was not yet quite their private family preserve. As already noted, McCormick had not founded it, and at least until the mid-1840s Simon Cameron had considerable influence in its affairs. Apparently Cameron was not completely satisfied with the arrangement, however. Near the end of the legislative session of 1842 he had a proviso inserted into a measure regulating auctions and auctioneers "and for other purposes," that the directors of the Bank of Middletown, if they deemed it "expedient," might lawfully "remove" their bank to Harrisburg. Although the bill was signed into law, Cameron either changed his mind or was unable to persuade the directors to make the move.[19] The marriage of J. Donald to McCormick's daughter in 1856, however, renewed the tie to the Dauphin Deposit and intertwined the business affairs of the two families well into the twentieth century.

Even as the influence of the Middletown Bank and of Simon Cameron over the Dauphin Deposit Bank waned in the 1840s, it was replaced by that of the Haldemans and the Harrisburg Bank. In 1839 Robert J. Ross, son-in-law of Jacob M. Haldeman (a founder, director, and later president of the Harrisburg Bank), became cashier of the Dauphin Deposit. In January 1845 the Dauphin Deposit increased a standing loan of $25,000 from the Harrisburg Bank to $50,000 (a sum equal to the Dauphin's authorized capital) and a few years later raised it to $100,000. In May 1845 it hired Haldeman's son John as a clerk. In October of that same year, the board voted to purchase 300 shares of its own stock from the Bank of Middletown (reducing Simon Cameron's influence to 550 shares) and to sell it to Ross at par. Finally, in 1849 John Haldeman became a director of the Dauphin Deposit, a position he held for sixteen years before 1864. Sometime before 1856 (probably in the 1840s) Jacob M. Haldeman himself acquired 330 shares of Dauphin Deposit stock, a purchase that helps explain his family's growing role in that bank.[20]

The affairs of the Dauphin Deposit Bank were notably more closed than those of the Harrisburg Bank in this period. Its board was smaller (seven members rather than thirteen) and played a lesser role than did the board of the older bank. The minutes of the Dauphin Deposit's board meetings became increasingly terse and uninformative during McCormick's presidency. In the 1840s, many meetings were attended by

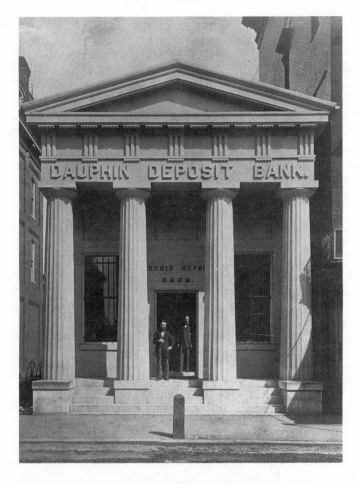

Fig. 6.   The Dauphin Deposit Bank Building in the Late
Nineteenth Century.   This Grecian-style building was
completed in 1839, a year before the first James McCormick
became the bank's president. Still in use, it has outlasted the
McCormicks by nearly a half-century.

only a bare majority. The entry for September 23, 1854, read "No
quorum for last year"; the same was true for 1855 and 1856. Clearly
McCormick and Cashier Ross were running the bank on their own.
Despite (or perhaps because of) lack of guidance from the board, the
bank did well. With an authorized capital one-sixth that of the Harrisburg Bank, in the 1850s the Dauphin had assets averaging about half

those of its rival's (see Table 2.1). Its annual dividends were higher and (according to the inventory of the estate of Jacob M. Haldeman) its shares were more valuable. The par value of shares of the Harrisburg Bank was $50 and that of the Dauphin Deposit shares was $25. On January 1, 1857 (the date of the inventory) Harrisburg Bank shares were worth $33 each and Dauphin Deposit shares were worth $60.[21] The banks were alike in that several directors of each served long terms in office between 1835 and 1859. One-fourth of the forty-four persons serving on the board of the Harrisburg Bank in those years, and 28 percent of the Dauphin Deposit's twenty-five directors, enjoyed terms of at least ten years.

The scanty reports in its minutes book and a lack of other extant records make it impossible to reconstruct the role of the Dauphin Deposit Bank in Harrisburg's economy during these years. Suffice it to say that its earnings permitted dividends to grow steadily from a low of 6 percent a year in 1843 to 10 percent by 1847, a figure maintained through 1852, when the minutes book stopped recording dividends.[22]

Meanwhile, the Harrisburg Bank in the 1840s and 1850s underwent considerable change. Nearly all the older directors who had served from the beginning were gone by April 1853, when President Elder died. Jacob M. Haldeman, one of the survivors, took over as president but lived only until December 1856. He was succeeded by another short-term president, William M. Kerr (a grand-nephew of Elder), who died in 1864. By good fortune the bank had a strong cashier, James W. Weir, appointed in 1844, who managed its affairs during what might otherwise have been a

Table 2.1.   Comparative Assets of Harrisburg and Dauphin Deposit Banks

	Harrisburg Bank (Founded 1814, capital $300,000)	Dauphin Deposit (Founded 1834, capital $50,000)	Dauphin's Assets as % of Harrisburg Bank's 17%
Nov. 1839	$ 986,829	$114,556	12
Nov. 1844	684,256	244,174	36
Nov. 1849	816,641	225,312	28
Nov. 1854	1,113,890	685,880	62
Nov. 1859	982,507	477,396	49
Nov. 1864	1,560,650	822,368	53

SOURCE: Pennsylvania Auditor General, *Annual Reports on Banks.*

chaotic period. More or less by default, the board came to rely on Weir for formulating policy and making decisions.[23]

The economy was on an upswing through most of the 1840s and 1850s. In addition to financing merchants, infrastructure projects, and, as will be seen, major industrial firms, the Harrisburg Bank in the early 1850s sought other outlets for investments. "As the means of the bank to discount are abundant and the offerings light," the board instructed Weir to negotiate in Philadelphia for loans on acceptable collateral.[24] Loans were also made to other area banks.[25] When the brief but sharp Panic of 1857 hit, the Harrisburg Bank imposed discipline on the local business community. To make all means at its disposal active, all accommodation borrowers were obliged to reduce their debts by installments as rapidly as possible. Loans that had "assumed a permanent character" were to be collected or taken to judgment. In 1858 the bank unloaded various canal bonds that it held, suffering a loss of over $10,000. In their place it bought U.S. Treasury notes.[26]

Harrisburg Bank dividends, which were 6 percent a year from 1844 through 1851, ranged from 8 to 10 percent through 1864, 10 percent being paid for seven of the thirteen years. In its first forty-four years of operation, the bank wrote off $133,563.21 in bad loans, a little more than $3,000 a year. Following the list of names and amounts in the back of a minutes book, Weir observed, " 'Lot of very bad fish,' the above."[27]

Throughout the period, the Harrisburg Bank actively assisted the maturing transportation facilities of the community. It made loans to the Harrisburg & Lancaster Railroad and to the state for surveying the route of the Pennsylvania Railroad. It made short-term loans to the Pennsylvania and two of its principal officers, Thomson and Scott. In November 1853 it began handling the railroad's payroll. Merchants too continued to be accommodated: "such amounts as they may need for the spring business, not exceeding $10,000"; "the usual facilities for their spring business"; "$5,000 for their purchases of lumber this spring." Loans also began to appear for insurance companies and for such civic enterprises as public schools, volunteer fire companies, and county and borough governments.[28]

In 1858 Simon Cameron at last became directly involved in Harrisburg banking. He and seven other partners formed a partnership known as Cameron, Calder, Eby & Company. Two years later, in January 1860, they officially named their firm the State Capitol Bank and opened for business at the corner of Second and Walnut streets. Cameron's two major partners, already active in building Harrisburg's infrastructure, now emerged as leaders. William Calder Jr., son of the town's enterprising stagecoach operator, had started working in his father's business at the age of twelve. By age sixteen he was already in charge of the elder Calder's "Pioneer Packet Line," which provided passenger service over the state canal system from Columbia to Pittsburgh. To forestall his son from leaving the area to go into business for himself, Calder sold out to him and retired. Young Calder added a livery stable to the stagecoach business and bought three or four farms in the Harrisburg area, where he raised horses. In 1857 Simon Cameron, at the time president of the Lebanon Valley Railroad, gave young Calder a contract for completing construction of the rail line. Somehow, probably because of the brief depression then under way, Calder persuaded a crew of 600 to accept work on the promise of eventual payment. "Without one cent of cash he completed the road" and in due course paid the workers. The next year he entered into banking with Cameron and became a director, with Cameron, of the Northern Central Railroad.[29]

Cameron's other partner was of Mennonite stock. Jacob R. Eby only "occasionally attended school" to the age of fifteen. After briefly clerking in a local store, he apprenticed himself to a carpenter for three years. Then, in 1836, he traveled down the Ohio and Mississippi rivers as far as Vicksburg, where he worked at his trade for eight months and saved $225 in the process. Returning to Columbia, he became a clerk in the firm of Cameron, Lauman & Clark, which was building a dam for the Susquehanna & Tidewater Canal. Before long a Middletown lumber merchant offered Eby a position and an interest in his firm. Six years later, in 1845, Eby sold off his interest "at great advantage" and joined with his brother to buy a thriving Harrisburg grocery and shipping business for $4,500. About the time the firm became Harrisburg's largest wholesale and retail grocery and pork-packing business, Eby joined forces with Cameron and Calder. The other known partners included George H. Small, cashier of the State Capitol Bank, and Cameron's son J. Donald, who remained cashier of the Middletown Bank. Three partners did not reside in Harrisburg and preferred not to be identified.[30] The known partners were soon to enjoy a profitable association that included a number of ventures other than banking.

The final major elements of Harrisburg's infrastructure to be constructed before the onset of industrialization were a waterworks and a gasworks. According to Harrisburg historian William Egle, that they appeared when they did was largely due to the persistence and hard work of William Ayres. Efforts to supply the community with adequate water for drinking, commercial uses, and fire protection began early in the nineteenth century. The first proposals called for a reservoir that would be supplied by diverting water from nearby creeks and streams or by confining the water of springs in the area. In the 1820s several of the town's leading citizens, including Jacob M. Haldeman, John Forster, and Robert Harris, subscribed to stock in a water company authorized by the legislature. Once 10 percent had been paid on the shares, the company organized with Forster as president. The project called for a canal to carry water to Harrisburg after being drawn from the Susquehanna at the village of Dauphin, nine miles above the capital. As the project got under way, the Pennsylvania Canal commissioners selected exactly the same route for their canal and condemned the property without compensating the water company. When the state Supreme Court upheld that action, the firm dissolved and its investors lost their initial payments on the stock. A second private company, chartered in 1833, again with Forster, Harris, and Haldeman among its chief promoters, failed to win popular support and dissolved.

William Ayres, elected to the borough council in March 1838, persuaded the council to seek legislative permission to construct a municipally owned and operated waterworks. After successfully lobbying the measure through the Assembly in March 1839, Ayres headed the committee that planned the project. Securing the assistance of competent engineers, the committee devised a plan that the borough adopted the next year. The council then designated Ayres and another council member to negotiate for land on which to construct a reservoir. Apparently a number of the town leaders, possibly including those who had previously tried to establish a private waterworks, were cool to Ayres's scheme. Disputes, arbitrations, and lawsuits followed, but the necessary land was secured for under $4,500.

Meanwhile, to fund the project, Ayres appeared in person before the directors of the Bank of the United States in Philadelphia (by this time operating under a state charter). Although the bank loaned the borough only $25,000, the council was sufficiently encouraged to issue its own certificates of indebtedness, called "borough notes," to pay for labor on the project. The notes were accepted in payment of town taxes and other borough obligations and for nearly three years constituted an important part of the town's currency. By 1843 Harrisburg had a waterworks equal to its needs, constructed at a cost of $120,000. Rapid growth of the city would force expansion of the system by 1868.[31]

At the end of the decade, Ayres took the lead in bringing about a privately owned gasworks. He and a number of other citizens secured a charter from the legislature in March 1848 authorizing a company with a capital of $75,000. It took more than a year to raise the $5,000 required to organize the firm, but in July 1849 Ayres became the first president of the Harrisburg Gas Company. The plant was built for $60,000, and some 25,000 feet of gas main were laid beneath the streets of the borough. By the end of September 1849 the plant was in operation. By February 1851 gas illuminated the streets and homes of the borough as well as the State Capitol.[32]

Only Ayres's efforts on behalf of a free bridge over the Susquehanna failed. He was no match for the powers who owned the existing toll bridge. The other major builders of the town's infrastructure do not appear to have been fond of Ayres, but his talents were appreciated elsewhere. Nicholas Biddle apparently saw the Harrisburg attorney as potentially useful to him. Although Biddle had retired as president of the Bank of the United States in 1839, he was instrumental in getting Ayres named to the board of directors in 1841. Biddle invited Ayres to serve, presented him with stock, and secured his election as "a country gentleman to complete the board of directors."[33] By June 1841, creditors had instigated suits against Biddle for alleged mismanagement during his leadership of the bank. Correspondence between Ayres and Biddle in 1841 and 1842 reveals that the Philadelphian had chosen well. He flattered his "country" director, urging him to take leadership of the board, and spoke of looking forward to long conversations with him whenever he was in Philadelphia. In return, Ayres stood loyally by Biddle at board meetings, supplied him with inside information, and acted on Biddle's suggestions both on the board and in the lobbies of the State Capitol.[34] Historian Egle described Ayres as "ever ready to labor and sacrifice for the public good." In spite of many opportunities to become wealthy, he "proved unflinchingly honest," "could never be tempted or bribed," and so "died poor."[35] It is not clear whether Egle was excusing Ayres's lack of fortune or commenting on the ethics of the town's other entrepreneurs. In any event, Ayres made a number of important contributions that paved the way for industrialization.

Although by no means completed in 1850, the infrastructure was by then able to support Harrisburg's first modern industrial plants. Many of those who were most responsible for bringing improved transportation, banking facilities, and public utilities to Harrisburg were its first industrialists as well. It was not that they were inventors with new ideas to test, or even that they were seeking businesses to manage. Their interest was in investment. Putting their money in manufacturing would have been no different to them than the earlier ventures into transportation and banking. It would call for the same business insights and willingness to take risks and would with luck produce similar or even greater chances for increasing their fortunes.

Harrisburg Industrializes

Five years ago Harrisburg was one of the dullest, drowsiest towns in the State, and we had reason to fear that our citizens would never awake from their Rip-Van-Winkle sleep. All of a sudden, however, capitalists began to invest their funds in various enterprises, and realized handsome profits, and from that time to the present, our growth has been rapid and healthy. On every side we hear the hum of busy industry, and the multiplication of factories, and furnaces, rolling mills and forges, flour and lumber mills, railroad car and machine shops, &c all show a degree of substantial prosperity, that five years ago we never expected to see realized in Harrisburg.
—Harrisburg Morning Herald, *June 23, 1854*

Allowing for a bit of editorial license, Stephen Miller of the *Harrisburg Morning Herald* was right. After years of quiet preparation, industrialization came to the town with a rush in the half-decade after 1849.[1] Those years witnessed not only the arrival of the Pennsylvania Railroad but also the building of the town's first blast furnaces and new, large-scale rolling mills, a cotton factory, a railroad car manufactory, a firm that produced various ingenious machines and instruments, and a large number of smaller shops and factories.

Both nationally and locally, conditions were ripe for enterprise in 1849. The times were relatively prosperous. The recent war with Mexico had ended victoriously, with the United States acquiring a vast new empire in the Southwest. Those new lands, coupled with the discovery of gold in California, soon provided the restless with a fresh impetus to "westering." The troublesome question as to whether slavery would be allowed in the new territories would be resolved by compromise, at least temporarily, midway through 1850. For those living in the Northeast, the next few years were a time for bold new ventures and economic development. The railroad-building craze was under way, accelerating the transportation of goods and people and changing the way Americans earned their livings and conducted their businesses.

In Harrisburg an infrastructure of transportation, banking, and major utilities was in place. The town was a fully developed commercial center serving a substantial agricultural hinterland. Its leading entrepreneurs and scores of prosperous merchants had money to invest, and local newspapers stood ready to boost almost any type of enterprise. What exactly triggered the town's sudden burst of industrial activity remains unclear. Probably the best explanation is that several overlapping, inter-acting, and reinforcing elements came together at the same time. One factor, certainly, was the entrepreneurs who had already helped create the infrastructure and now promoted, directed, and financed the first industries. However, the fact that the town's first manufacturing firm organized the very year the Pennsylvania Railroad reached the commu-nity was probably more than coincidental.

The process simultaneously followed several paths. For example, to-pography and geography would have brought enterprises such as the Pennsylvania Railroad to the community with or without the support of Harrisburg residents. The cotton mill appeared when local entrepreneurs who had money to invest, and a persuasive New Englander with capital and know-how, joined forces to introduce a new and unfamiliar indus-try. The rapid growth of ironworks took advantage of the industry's change from charcoal to anthracite fuel for furnaces and the need for larger integrated facilities. The decision to build railroad cars in Harris-burg stemmed directly from the ready availability of iron and lumber and the fact that the town was becoming an important railroad center. Still another development involved one man harnessing his mechanical genius and turning it into a machine-producing business. The surprising features were that these developments all began within a span of five years and were to a significant degree the work of a handful of entrepre-neurs and their allies and assistants.

In the winter of 1848–49, on the eve of the erection of Harrisburg's first factory, the Pennsylvania Assembly debated a general incorporation law for manufacturing businesses. The measure aimed at facilitating the cre-ation of corporations by granting limited liability to stockholders. Harris-burg newspapers, mouthpieces for the state's various political factions, followed the debate closely and voiced the views of their sponsors. The *Telegraph,* representing Whig views, favored the measure and blamed Democrats for breaking down "all efforts at collecting capital to build up manufactures" on the spurious grounds of encouraging "individual enterprise." As a result, the streets of Harrisburg and all the towns of Pennsylvania were "filled with poor suffering children, begging for bread" because of the lack of employment.[2] The *Harrisburg Keystone,* presenting the arguments of traditional Jacksonian Democrats, declared that the legislature "had no right to confer corporate privileges on any set of individuals" except when required by a sufficient public interest. Corporations for manufacturing purposes "create irresponsible aristocracy to control our elections, destroy the democratic party, the equality of our rights, [and] . . . check individual enterprise and responsibility."[3]

The editors soon turned to the example of Massachusetts and its celebrated textile mills. According to the *Telegraph,* Lowell working girls, far from having their wages constantly driven down as Democrat "locofocos" charged, were earning enough to buy pianos. Mill girls had purchased eight in the past six months! By contrast, Pennsylvania girls, engaged as "slavish kitchen maids or beggars," became acquainted with music only in the form of "low ribaldry and vulgarity of street hawkers and brothel visitors."[4] In response, the Democratic *Keystone* pointed to the rapid growth of "unjust and oppressive" inequality in Massachusetts. There the "free republican spirit and sturdy independence of thought and action" that characterized the laboring majority in Pennsylvania had been "broken and crushed" by the "iron will of soulless corporations."[5] The legislature passed the measure, and on April 7 Governor William F. Johnston signed it into law. Within weeks the first use of the law in Harrisburg was to establish a cotton mill.

The Pennsylvania Assembly was not the only institution that debated the merits of corporations and limited liability. In the summer of 1851 the directors of the Harrisburg Bank faced a fateful decision regarding their future support of corporate enterprise in the community. By then they had already made extensive loans to the new cotton mill, and the air was filled with talk of new manufacturing enterprises. Many of the town's business leaders, however, resented and distrusted corporations and questioned the safety of loans to institutions with limited liability. Harrisburg Bank director Valentine Hummel, a respected and wealthy lumber merchant (who would preside over the bank briefly after the Civil War), introduced a resolution that would have put the damper on any future loans to corporations. Whenever a corporation sought a loan in its corporate capacity, the bank's decision would be made by a recorded vote of the directors. Whoever voted for a particular loan would personally and individually be held liable for any losses to the bank. Under such a rule, while the usual loans to individual proprietors and partnerships would have continued, few if any would have been made to corporations. The board voted five to three against the motion.[6]

The Harrisburg Cotton Manufacturing Company

Early in 1849 a small group of Harrisburg business and professional leaders made plans to start a cotton textile mill. If newspaper commentary is to be believed, they had several motives: filled with local pride, they resented nearby Lancaster and Reading stealing the march on Harrisburg; they wanted to spur business activity in the town; and they sought to benefit the community by providing employment and income to "a class of the community—usually of no use in the production of wealth—i.e., women and children." Profit-seeking was at best a secondary consideration. Had they sought only profit, they would have gone into iron manufacturing, which would have employed but a quarter as many heads of family. This last explanation, offered after the mill was in operation and *not* making money, listed "the misrepresentations of parties anxious to secure the contract for building the mill" as yet another factor.[7]

Clearly, the original idea grew out of developments in nearby Lancaster but had more to do with the search for a profitable investment than local pride. In the mid-1840s a number of enterprising Lancaster merchants hit upon manufacturing as the means of revitalizing their community and enriching themselves. At precisely the right moment, Charles Tillinghast James, one of America's most successful industrial promoters, appeared on the scene to expound the virtues of textile manufacturing. James had extensive training and experience as an engineer and designer of textile-making machinery. During the 1820s he had worked in the Slater mills in Providence, Rhode Island, where he rose to the post of superintendent. He came to believe that steam power, for a variety of reasons, was far superior to water-driven machinery in producing textiles. Soon he was writing and lecturing on the subject, traveling through New England, the Middle Atlantic states, and parts of the Midwest and the South. In essence, his message was that steam-driven cotton mills offered small, economically stagnant communities an opportunity to expand, prosper, increase property values, and provide employment and steady incomes to the poor. Although the firms he envisioned were to be privately owned corporations, James touted them as community ventures involving the entire citizenry.

James visited Lancaster in the spring of 1845 and spoke to a group of leading merchants. His talk touched off a flurry of enthusiasm in the business community and the local press. By July seventy-five stockholders, including wealthy merchants, lawyers, and bankers, had organized the Conestoga Steam Mills. James himself held $1,000 worth of the stock. By 1847 the first mill was built and running, and James was planning a second mill. In both 1848 and 1849 the company paid dividends of 10 percent. By then the second mill was in production, and in 1850 the company built a third mill.[8]

The first indication of interest in Harrisburg appeared in the newspapers. In January 1849 the *Democratic Union* reported on the Lancaster mill and the dividends it paid. "When will the Capitalists of Harrisburg waken up to the importance of establishing manufacturing enterprises in our borough?" it asked. In March the newspaper noted that one of Harrisburg's leading contractors had visited the mills in Lancaster with a view to starting such an enterprise in Harrisburg. The paper went on to recite Harrisburg's advantages over Lowell and the other New England textile-mill towns: fuel (coal) for driving machines was half as expensive; cotton could be bought at the same price, less "the price of freight between here and Lowell"; foodstuffs ranged from 15 to 20 percent lower; and houses could be rented cheaper than in Lowell.[9] By the end of May, meetings were held in Harrisburg to discuss starting a cotton mill. The newspapers joined in the campaign to raise the needed capital for "Our Cotton Mill." By June 20 only half the $200,000 needed had been pledged. Given the experience of Lancaster, it seemed surprising to the *Union* that the committee soliciting stock subscriptions was doing so poorly. If capitalists would subscribe but 1 percent, or even one-half of 1 percent, of their wealth, the money would soon be in hand. That same day the *Telegraph* declared that if those with the means did "what they should and could" the proposed mill would be under construction before the end of the summer.[10]

When the subscribers held their organizational meeting on July 7 the prime movers became evident. James McCormick, lawyer and president of the Dauphin Deposit Bank, explained the purpose of the meeting and proposed resolutions to govern the organization. Isaac G. McKinley, co-publisher and editor of the *Democratic Union*, served as secretary. The five others elected to the governing committee were William Calder Sr., the stagecoach line operator; Dr. Luther Reily, the town's wealthiest physician; Philip Dougherty, a contractor who made his fortune building

canals and railroads; John H. Briggs, who had studied law with Mc-
Cormick and was active in Harrisburg borough government; and Daniel
D. Boas, lumber merchant. All except Boas were among the top 10
percent of Harrisburg real estate holders in 1850; four were among the
top 5 percent. The banking community was especially well represented.
In addition to McCormick, active participants included James W. Weir
and Robert J. Ross, cashiers respectively of the Harrisburg and Dauphin
Deposit banks.[11]

The committee to launch the enterprise met on July 11 with James,
who had recently become U.S. Senator from Rhode Island. Things thereafter moved quickly. One week later the committee bought a six-to-
seven-acre site for the mill along the river (John Forster's peach orchard
and ferry-house lot, now the corner of North and Front streets). Ground
was broken for the building on July 31. By December nearly a million
bricks had been used to raise the mill's walls, the slate roof was on, and
the windows were all in place. Workers would finish the interior during
the winter and have the plant ready for machinery by the opening of
navigation in the spring.[12] As committee member McKinley described it
in the *Democratic Union,* the location on the banks of the Susquehanna
at the northern limits of the borough was a place of "rare beauty," large
enough to accommodate from three to five mills if the business proved as
profitable "as we have every reason to believe it will."[13] At that point,
profit seems to have been on everybody's mind.

Neither the minutes book nor any newspaper article discussed the new
mill much in 1850. The cotton trade had suddenly fallen on bad times.
Early in 1851, however, James again came to Harrisburg, and the stockholders met to discuss chartering the company and starting the mills.
Charles C. Rawn, a lawyer and minor investor in the company, declared
that without concern for profit or loss but on "the ground chiefly of
benefitting the town and country and people employed" he favored "at
once going to work and keeping at it." The majority apparently agreed.
The state granted the firm a twenty-year charter, beginning May 1,
1851, and the shareholders reelected as directors the seven men who had
served as interim directors since 1849. James McCormick became president of the firm; John Briggs became secretary-treasurer.

In many ways the new facility was a transplant from New England.
James not only supervised construction of the mill (which closely resembled its northern counterparts) but also supplied it with machinery
of his design made in that region. Nine Rhode Island–born machinists,
housed at a Harrisburg inn, installed the machinery. On June 17, 1851,
the *Whig State Journal* reported that "about fifty female operatives
from some of the New England Factories" were expected to arrive

within the week to work in the mill. Their role probably was to train local workers and then return home. In any event, no cotton-mill employee in the census of 1860 gave New England as place of birth. To manage the mill, the directors selected another New Englander with experience in textiles, S. P. Mason. When Mason resigned in April 1852, Josiah Jones of Connecticut took his place.[14]

Except for James, who briefly held 200 shares in the company, local people financed and owned the cotton works. Indeed, with few exceptions, the 177 other initial shareholders came from Dauphin County. Fourteen, including James, held 835 of the firm's 2,000 shares. Four of the town's leading entrepreneurs were among the fourteen: James Mc-Cormick (100 shares), William Calder Sr. (90), Michael Burke (50), and Jacob M. Haldeman (50). The Harrisburg Bank, of which Haldeman was a director, held 50 shares, and William Calder Jr. held ten shares. The balance included many of the wealthier business and professional residents of Harrisburg and a few persons of modest means.

Fig. 7. The Harrisburg Cotton Manufacturing Company. Harrisburg's first factory resembled New England textile mills more closely than it did Harrisburg's other industrial mills. Notice the cupola and the abundance of windows.

Three subsequent lists of stockholders, only the last of which was dated (May 1865), show relatively few changes. Senator James disappeared from the lists, and McCormick's holdings expanded to 500 shares. All the leading entrepreneurs on the initial list, or their heirs, remained shareholders through 1865. The shares held by the top thirteen to fifteen shareholders on each list ranged from 58 percent to 61 percent of the total; the portion held by the town's leading entrepreneurs (and their banks and heirs) ranged from 39 percent to 41 percent.[15]

Although outwardly "everything" was being "conducted in the most admirable order," Briggs, the firm's treasurer, resigned after three months, claiming that his duties interfered too much with his private business. McCormick's brother-in-law William Buehler, an experienced businessman, major landowner, and operator of the Buehler House (one of Harrisburg's finer hotels), became the new treasurer. The next April the superintendency changed hands, from Mason to Jones, and in December the directors borrowed $40,000 and dispatched the supposedly busy Briggs south to buy raw cotton.[16] By May 1852 the company owed $9,000 each to the Dauphin Deposit Bank and the Harrisburg Bank. Meanwhile, James consigned another $20,000, still owed him for constructing the plant, to Robert Ross, cashier of the Dauphin Deposit, and Ross in turn consigned the debt to the bank. The company gave a mortgage for $38,000 to its treasurer, Buehler, and issued seventy-six 7 percent bonds at $400 each and assigned them to the two banks to cover outstanding loans. The debt was repaid—part in 1855, the balance in 1860.[17]

In its early years, debt plagued the Harrisburg Cotton Company. It neither consistently made money for the investors nor provided steady employment for its workers. Frequently the cost of raw cotton was high, the price for cotton cloth low. Twice more, in March 1853 and December 1855, superintendents resigned and were replaced. In 1854 the directors decided to shift from producing brown shirting material to osnaburg, which was used for grain bags. Orders for 830,000 bags for guano kept the plant running in 1859. Frequently the mill shut down. A breakdown or general repair of machinery led to a two-week layoff in August 1852, "a few days" in the spring and apparently longer in early summer of 1853, in August 1856, and much of the autumn of 1857. The operatives went on strike between October 18 and 26, 1853, a fire in the plant suspended operations from February 20 until March 14, 1855, and a lack of orders resulted in closing down altogether between mid-September 1857 and December 27, 1858. In July 1860 the board voted to pay Superintendent Davis $312.50 for still another period when the mill was down.[18]

Fluctuations in the value of the company's stock tell the story. In

March 1853 the company appraised its stock at 40 (par was 100). The Harrisburg Bank, in its annual November reports to the state auditor general, valued its shares at 60 in 1853, 100 in 1854, and 50 in 1855; the inventory of the estate of Jacob M. Haldeman listed it as worth 66 on January 1, 1857. Dividends too were disappointing: 6 percent in 1856 and 5 percent in 1860—the only dividends before the Civil War.[19]

Iron and Steel Industry

It was in iron manufacture that Harrisburg more substantially entered the new industrial age. All the essential ingredients for iron production (forests for charcoal, limestone quarries for fluxing material, and iron deposits) were plentiful in the general vicinity of Harrisburg. Before 1850, however, the only ironworks within the borough proper, or just over its borders, were forges, foundries, and nail works. When, after 1850, these expanded their scale of operation and mechanized more thoroughly, the changes were essentially evolutionary rather than dramatic or sudden departures from the past.

The first ironworking shops to serve the village and nearby farms appeared even before Harrisburg became a borough. Iron tools and hardware were among the few products that even the most self-sufficient could not produce on their own. As early as 1785, for example, Henry Fulton established a nailery on Front Street. His enterprise, "only a little remote from a smithy," closed when he moved west sometime before 1814. Other cold iron "factories," not much larger than blacksmith shops, appeared and disappeared over the next few years.[20]

The first significant ironworks in the area appeared in 1806 across the Susquehanna near the ore banks of Cumberland County. As already noted, Jacob M. Haldeman (whose wife, Eliza E. Jacobs, was the granddaughter of a Lancaster County ironmaster) that year bought and upgraded a charcoal furnace on the Yellow Breeches Creek and went into ironmaking on a substantial scale. In 1828 he sold the facility to Jared Pratt of Baltimore. Pratt used the facility until 1844, when he purchased a plant in West Fairview, a few miles to the north, and shifted his operations there.[21]

Pratt's West Fairview plant had been established in 1833 by two Harrisburg residents, Gabriel Hiester and Norman Callender. The partners had purchased land and water rights at the mouth of the Conodoguinet Creek, directly across the Susquehanna from the capital. Taking advantage of an eight-foot drop in the creek at that point, they constructed a 300-foot dam to ensure a steady supply of waterpower. There they built a rolling mill and began producing boiler plate and bar iron. Hiester's son, Augustus O., took over the mill in 1836, operated it through the difficult Panic years, and finally sold to Pratt and his son. The Pratts converted the facility into a nail factory. By 1850, capitalized at nearly $43,000, the factory employed 125 workers and annually produced more than 29,000 kegs of nails valued at $117,583.[22]

Completion of the Pennsylvania Canal in 1834 increased the scale of iron manufacture in Harrisburg proper by lowering the cost of hauling essential raw materials. In 1840 Hunt & Son erected the first major rolling mill in town at the intersection of Second and Paxton streets near the canal. They produced nails, spike rods, and sheet iron. Fire destroyed their facility, but in 1845 Pratt & Son built a new rolling mill on the property for turning out boiler plate. Depressed business conditions kept the works idle for several years. Needing funds in 1849, the Pratts turned to the Harrisburg Bank for a series of short-term loans. Jacob M. Haldeman endorsed their notes. In the manufacturing census for 1850, the Pratts reported a capitalization of $25,000, a labor force of fifteen, and an output of 500 tons of boiler plate valued at $45,000.[23]

The introduction of large anthracite furnaces for producing pig iron, the first of which appeared within the town in 1850, represented a notable new development for the community. Prior to that date, ironworks had obtained pig iron that was their raw material from charcoal furnaces distant from the town, such as Jacob Haldeman's. So long as furnaces operated with waterpower and used charcoal for fuel, they were of necessity located on streams close to hardwood forests.[24] This began to change when exhaustion of the forests drove up the price of charcoal. Experiments proved that anthracite could be substituted but required specialized equipment. Hard coal needed a hot rather than cold blast to ignite it and maintain combustion. Even so, anthracite furnaces reduced fuel costs by half and, because they were larger, produced more iron. A further change was the shift from waterwheels to steam engines for power. Together these new technologies made it possible to locate new furnaces in market centers (such as Harrisburg) on main transportation routes.[25]

It was fortunate that large-scale iron manufacturing came to Harrisburg after 1850. The 1840s had been marked by depressions in the industry at both ends of the decade. In the 1850s, ironmasters could take advantage of new technology in the field while enjoying a rising market. Labor was plentiful, and local businessmen and banks had money to invest. By then, canals and railroads connected the town with nearby sources of limestone and high-quality iron ore and with the major urban markets of the Northeast. Most important, the nation's best anthracite fields were only fifty miles or so to the northeast. Canals for transporting this new fuel to Harrisburg cheaply already existed, and railroads to the hard-coal district were under construction. Demand for nails could scarcely be met in the Harrisburg area, and rapid expansion of the railroads was calling for a wide variety of iron products.[26]

Outsiders did most of the pioneering in the early 1850s, with the town's established entrepreneurs watching from the wings. The Harrisburg Bank provided loans, some of which were endorsed by local businessmen, and a few of the town's builders assisted in other ways.[27] Only in the wake of the Panic of 1857 did one of the local entrepreneurs, James McCormick, enter the field. Within three years he controlled the greater part of the industry in Harrisburg. However, by then a pair of brothers from Chester County, Charles L. Bailey and Dr. George Bailey, were also making their mark as ironmasters in the community.

David R. Porter, an outsider who had been governor of Pennsylvania, teamed up with Harrisburg contractor Michael Burke to erect the town's first anthracite furnace. Burke built the furnace, Porter provided the know-how. Before entering politics, Porter had operated a large charcoal furnace in Huntingdon County and, more recently, had been connected with the Sarah Ann Furnace at Columbia, which in 1841 had converted from charcoal to anthracite. The Harrisburg Bank supplied much of the money.[28] The furnace went into blast in 1850. Coal recorded at the weigh station at Harrisburg indicated the close relationship between the new furnace and the canal. Between 1844 and 1848 the annual tonnage of coal slipped from 3,838 tons to 421 tons. As Burke & Porter accumulated stock and then began operations, the tonnage shot to 13,452 in 1849, to 30,452 in 1850, and to 60,158 in 1851.[29]

Iron ore for the new furnace came from a mine in Cumberland County owned by a director of the Harrisburg Bank. Capitalized at $123,000, the Porter Furnace (Burke soon left the business to devote himself to full-time contracting) employed 112 workers and turned out 3,500 tons of pig iron a year worth $80,500. During the Panic of 1857, Porter was forced to sell. By March 1858, Harrisburg agents of R. G. Dun & Company (forerunners of Dun & Bradstreet) reported that a number of judgments had been entered against the former governor. By the end of the year he had failed. "No prospects of recovery," "perfectly irresponsible," and "hopelessly insolvent" read subsequent reports. To protect its interests, the Harrisburg Bank bought the property in April 1860 and sold it to Absalom Price and William Hancock four months later. Porter moved to Texas but soon returned to Pennsylvania, where he died in 1867. In 1870 the furnace, then operated by Price & Sharp, turned out 5,700 tons of pig iron valued at $172,400.[30]

Meanwhile, in 1853, Lancaster businessmen George S. Bryan and David Longenecker erected the town's second anthracite blast furnace, the Keystone, just over the borough limits in South Harrisburg. That same year the Bailey brothers from Chester County constructed a rolling mill on Herr Street at the Pennsylvania Canal. Bryan and Longenecker, heavily in debt to the Harrisburg Bank, ran into difficulties during the Panic of 1857. Obliged to meet a prompt schedule of payments on their loans, they had no choice but to sell their property.[31] By contrast, the Baileys, who had Philadelphia backing, weathered the panic and were soon expanding.

During the next decade the Baileys emerged as one of Harrisburg's two major iron and steel dynasties—a distant second to the Mc-Cormicks. Sometimes as uneasy partners, usually as rivals, the Bailey and McCormick iron interests grew side by side until the end of the century, when they merged. The Baileys had the advantage of being in the iron business in Harrisburg before the McCormicks and brought with them years of experience. Their mother was a Lukens, a leading ironmaking family of Chester County. Their father, originally a farmer, became an ironmaster in his own right following his marriage. Charles and George, age thirty-two and twenty-five respectively in 1853, had been schooled at the Westtown Academy in Chester County. Charles subsequently clerked in a Philadelphia drugstore but left for the counting-house of the family furnace at Coatesville. George studied medicine at the University of Pennsylvania, earned an M.D., and then gave up practice in favor of ironmaking in Berks County.[32]

When the Baileys founded the Central Iron Works in Harrisburg, Morris Patterson of Philadelphia supplied $20,000 of the capital and received two-fifths interest in the firm. The Baileys put up $10,000 of their own money and managed the business. Charles held a two-fifths interest, and George a one-fifth interest. When the partnership was renewed in 1859, Patterson provided another $5,000 in capital. Under the agreement the Bailey brothers were salaried: Charles received $1,000 a year under both the original and the first renewal of the partnership; George received $600 under the original and $1,000 under the renewal. Profits or losses of the firm were shared in proportion to ownership.

Although they were newcomers to Harrisburg, by 1857 the Baileys were rated by a local agent for R. G. Dun & Company as "industrious men of first rate habits and very attentive to business." They were "considered safe men." Their new facility, a rolling mill long known as "Hot

Pot," stood at Herr Street and the Pennsylvania Canal. Its principal equipment consisted of four puddling furnaces and a train of rollers driven by steam. In 1860 the plant turned out 2,000 tons of boiler plate worth $152,000, consuming 720 tons of pig iron, 1,500 tons of blooms, and 2,500 tons of coal in the process.[33]

Meanwhile, James McCormick committed himself in the iron business. The Panic of 1857 broke several of Harrisburg's pioneering ironmasters. With easy access to information and money as president of the Dauphin Deposit Bank, McCormick picked up some bargains. He and Robert Ross (cashier of the Dauphin Deposit and son-in-law of Jacob M. Haldeman) tried to buy the Keystone Furnace from Bryan and Longenecker by assuming the owners' indebtedness to the Harrisburg Bank. Failing in that, they paid $90,000 for the works that had originally been capitalized at $123,000 and renamed it Paxton Furnace. Paxton No. 1 (there would later be a No. 2) was a blast furnace fueled by anthracite. In 1860 it used 15,000 tons each of anthracite and iron ore, and 3,500 tons of limestone, to produce 5,480 tons of pig iron valued at just over $100,000. Capitalized at $200,000, the furnace provided employment for forty workers.[34]

Dun & Company agents rated McCormick and Ross as "very prudent men" and estimated their worth at between $250 thousand and $500 thousand "in real estate and other good securities." When Ross died in 1861, McCormick bought full control from Ross's widow. That year, McCormick alone was reported as worth $500 thousand. By 1863, Dun's agents referred to him as "good as wheat," "rich doing well in business," and "A-1 beyond a doubt."[35]

McCormick, meanwhile, proceeded to buy up the remaining iron properties of the Pratts, now absentee owners living in Plymouth, Massachusetts. In 1859 he acquired their nailery in West Fairview for only $12,000 and immediately began upgrading it. By 1860, with 130 employees (5 more than Pratt had employed in 1850), the plant produced 60,000 kegs of nails worth $195,000—double Pratt's production a decade earlier.[36] In 1862 McCormick purchased the rolling mill in Harrisburg that the Pratts had built to replace Hunt & Son's old factory. This plant turned out nails, spike rods, and sheet iron. What McCormick paid is not known, but his object probably was to find a steady outlet for the iron he was producing at the Paxton Furnace.

McCormick needed someone with expertise to run his new properties. Unlike the Baileys, he himself had no practical working knowledge of ironmaking. He had sons, however—Henry and James Jr.—who were preparing to help run and some day take over their father's growing empire. Henry, the older of the two, began reading law but then decided instead to learn the iron business from the ground up. Following an apprenticeship at the Reading charcoal furnace in Robesonia, he became manager in 1856 of the Henry Clay anthracite furnace at Columbia. When James McCormick purchased Paxton Furnace in 1857, he promptly put Henry, by now thoroughly schooled in running furnaces, in charge. James, a year and a half younger than Henry, successfully completed his study of law and followed their father into law and banking.[37]

This left the problem of running the nailery and rolling mills. McCormick's solution was to turn to the Baileys, jointly operating the West Fairview Nail Works with Charles L. Bailey and leasing the Harrisburg rolling mill to the Bailey brothers. The manufacturing census for 1860 showed the industry's changed structure. McCormick, with Paxton Furnace and the West Fairview nailery, controlled 73 percent of the capital invested in iron; the Baileys, with the Central Iron rolling mill, controlled 15 percent, and four foundries held by small operators held 12 percent. Because of a depressed market for iron, two of the latter were not even in production at the time of the census.

The Pennsylvania Railroad Company

The Pennsylvania Railroad began as part of Harrisburg's transportation infrastructure but soon became part of the industrial sector as well. The company's first facility in town was the small station built in 1837 by the Harrisburg & Lancaster Railroad. As the company grew in size and complexity, headquarters cities had to be designated or built at regular intervals along the line. At these points, engineers and the train crews began and ended their runs. There also were located large freight yards, where trains were made up, and facilities for repairing rolling stock. In the early 1850s the company had expanded the village of Altoona into a major center because of its location at the base of the Allegheny front. But another rail center was needed between Altoona and Philadelphia, and Harrisburg became headquarters of the Middle Division. In keeping with its new dignity, a grand new Italianate station, complete with tower, was built in 1857. To finance freight yards, a roundhouse, and repair shops, Harrisburgers were asked to provide the land free as an "inducement" to locate there. In fact, no other site was likely. Harrisburg was the westernmost city midway between Altoona and Philadelphia; there the Pennsylvania line intersected with both the canal and a major north-south railroad, the Northern Central; and iron mills, the railroad-car works, and a supply of skilled workers were at hand. How much money, if any, the townspeople raised is not known, but work began on a new engine house capable of simultaneously servicing forty-four locomotives, on March 1, 1861. Machine and car shops opened not long afterward in the northern part of the city along the canal (by then owned by the Pennsylvania Railroad Company), providing work for at least 100 people.[38]

Fig. 8.   The Second Pennsylvania Railroad Station, Harrisburg.   The view of the station as passengers entering from the south saw it. A tall Italianate tower adorned the building's front, which overlooked Market Street.

The Harrisburg Car Manufacturing Company

Closely related to railroading but not directly tied to railroad companies was another major new industry, the Harrisburg Car Manufacturing Company. The founders of this firm, unlike those of the cotton mill, had no interest in creating a community enterprise designed to provide work for the borough's unemployed women and children. Neither did their undertaking evolve from more primitive businesses of a similar nature, as had some of the town's major new ironworks. It was a new enterprise, aimed at meeting the growing need of railroads for freight and passenger cars.

The idea for the car manufactory originally took shape in the mind of Harrisburg attorney David Fleming. As a young man, Fleming had financed his legal training by teaching school. For a while he taught in Baltimore, and later he worked for a contractor on the Baltimore & Port Deposit Railroad. In 1838 he returned to Harrisburg, and during the next decade he covered the state legislature as a reporter for four Philadelphia newspapers. In those same years he edited a local paper and read law, being admitted to the bar in 1841. In 1847 he became chief clerk of the Pennsylvania House of Representatives.[39] With the legislature chartering railroad companies at every session, and one line after another coming into Harrisburg, it was evident that the town would soon be an important rail center. Recognizing that these railroads would need a large supply of rolling stock, Fleming began promoting the idea of a company to manufacture railroad cars.[40]

Between 1820 and 1880 the Market Square stagecoach office of the William Calders, father and son, served as the meeting place of the "live men of Harrisburg." There "legislators, jurists, governors, soldiers, bankers, merchants and manufacturers . . . met in the reciprocities of social amenity" to consult with one another, pursue practical information, and arrange business enterprises.[41] Apparently Fleming's idea was discussed in the Market Square office because William Calder Jr. was the person who brought it to life. He, Fleming, Jacob M. Haldeman, Isaac McKinley (editor of the *Democratic Union*), A. O. Hiester (former manager of the West Fairview Rolling Mill and now associate judge of Dauphin County), William F. Murray (a Harrisburg lumber merchant), Thomas Wilson (a machinist born in New York), and Elias Kinzer (background unknown) formed a joint stock co-partnership with a capital of $25,000 to build railroad cars. McKinley became the company's first president.

Calder also secured the land on which the car works was built. It was absolutely essential that the firm be adjacent to a railroad. Rumors of the venture leaked out, and the town's "general committee of busy-bodies" advised holding out until the highest possible price could be extracted. Calder fought back. He may have been responsible for an article in the November 9, 1853, *Telegraph*. Harrisburg was rapidly improving and beginning to lose its "rustic looks," the article pointed out, but growth was now being limited by "the mistaken views of some capitalists . . . who own the vacant ground adjoining our borough." They put "a very high estimate on their property, and will not abate one dime in their demands." On a dozen or more occasions, this tactic had injured both their interests and that of the town by driving away outside investors. When the tightfisted property holders held firm, Calder quietly enlisted the assistance of his cousin, Captain John Murphy. Years before, Calder's father had brought Murphy to Harrisburg to work on his packet line. Murphy's father-in-law owned a nineteen-acre farm facing the Pennsylvania Railroad just north of the borough line. With money supplied by Calder, Murphy purchased the farm and then sold two and half acres to the partners for their car works.[42]

About this time the partners concluded they could go no further without someone with technological know-how to manage the business. Fleming undertook the search, going to the foremost car manufacturer of the day, Bradleys, in Worcester, Massachusetts. There he learned that one of their best men, William T. Hildrup, had recently moved to Elmira, New York, to set up in business for himself. Hildrup, just turned thirty, was a native of Connecticut. After completing a carpenter's apprenticeship, he had moved to Worcester, where he immediately found work at Bradleys. Over the next ten years he learned all phases of the business and was recognized as a master of the trade. Knowing that Bradleys bought the iron and coal they used from Pennsylvania and sold their cars to the Reading and Pennsylvania railroads in that same state, he decided to establish a firm in Elmira, somewhat nearer the Pennsylvania market. However, when Fleming offered to take him into the Harrisburg firm as a partner and general manager, Hildrup closed down and moved to the Keystone State.

Hildrup took charge, supervising the building of a wheel foundry, a castings foundry, a machine shop, a blacksmith shop, a planing and framing mill, and construction sheds. When completed, the facility could turn out nine eight-wheeled cars a week. Hildrup soon found himself not only running the business but also training young men to work as machinists. Because Harrisburg residents had little or no previous experience with industrial machinery, he established a free school during evenings in the winter where he taught line and freehand drawing and the elementary principles of theoretical mechanics.[43]

The firm had been in operation little more than three years when the sharp but brief Panic of 1857 seriously endangered its existence. Orders for cars dried up, the firm's debts amounted to $60,000, and the firm was holding another $60,000 in notes of now dubious value in payment for cars. Had anyone been willing to buy, the partners would have sold. Not having that option, they made the most of the resources that were at hand. A stay law, suspending judgments for twelve months, passed the legisla-

Fig. 9.   The Harrisburg Car Manufacturing Company.   This facility was built in
June 1872 to replace the original plant, which burned down two months before.
Running full, the new works provided employment for more than 700 persons.

ture and generally eased the situation for beleaguered businessmen. Meanwhile, the partners put their employees to work wherever jobs could be
found. For example, Calder, who held a contract for completing construction of the Lebanon Valley Railroad, put car company employees to work
building depots and bridges for that firm. Some constructed the new
Harrisburg depot for the Philadelphia & Reading Railroad, and others
worked on alterations of the State Works or engaged in making agricultural implements.[44] Better times returned in 1858, but not for long. The
onset of war in 1861 would again slow the economy temporarily.

The Hickok Eagle Works

William O. Hickok founded the Eagle Works, the last of Harrisburg's new
industries of the 1850s. The plant produced a variety of devices but special-

ized in ruling machines for printing business forms. As a boy of five, Hickok had suffered severe head injuries when he fell from a horse and was trampled. Surgery saved his life, but he never regained vigorous health. When he was eight his parents moved to Lewistown, Pennsylvania, where his father taught in an academy. The frail child remained with his grandparents in New York until the age of fifteen and received his education at an academy there. His inventiveness appeared early; as a boy he devised mechanical toys for himself and his brothers. In 1830 he joined his parents at Lewistown, briefly attended his father's academy, and for a while clerked in stores. When his father gave up teaching in 1836 to become a bookbinder and publisher in Chambersburg, William accompanied him and began an apprenticeship in that trade. Two years later the Hickoks shifted their business to Harrisburg. In 1841 the elder Hickok died, and William continued the business with a succession of partners. "Last night it blew," young Hickok wrote in his journal in 1844, the morning after his bindery burned to the ground. According to friendly accounts, Hickok, without capital and entirely on his own, by hard work and self-denial scraped together enough in two years to open a one-room shop on Market Street, where he made bookbinding specialties.[45]

Running in Hickok's mind was the idea for a machine that would efficiently rule paper for business forms. A Bavarian, Sigmund Adam, had first devised a way to line paper mechanically in 1800. An American named Hathaway improved on Adam's invention and by 1835 was selling a crude machine for ruling business forms. Familiar with the limitations of Hathaway's product, Hickok set out to build a better one. Although his "Improved Ruling Machine" was not patented until 1852, it received a diploma of honor from the Franklin Institute in Philadelphia in 1850 and a prize premium at the first Pennsylvania Agricultural Fair in 1851.[46]

A visitor to his small shop in July 1851 was surprised at both the sophisticated machinery (scroll saws, planing machines, lathes for turning wood and cutting screws, a gear-cutting machine, and a mortising machine) and the array of items being produced (ruling machines, standing presses, copying presses, seal presses, turning lathes for iron and wood, ice cream freezers), "and a great variety of other work." Although he had been making the ruling machines for only two years, they were immensely successful and already being used in "almost every State in the Union, and in Canada and the West Indies." In December of that year Hickok entered a partnership with Gilliard Dock and established the Keystone Machine Works. Together they produced "large scale tools and machinery of every kind from a screw to a steam engine." The partnership lasted only until September 1852.[47]

The glowing newspaper accounts notwithstanding, Hickok was having difficulty staying afloat. Dun & Company agents reported that he and a partner, Cantine, had failed in 1846 and that his property had been sold at sheriff's auction in 1848. Two years later he and a new partner, Barrett, faced pending judgments amounting to $3,500 that could "sweep his [Hickok's] personal property at any time." However, both partners were married, their business was good, and because they were "very industrious and economic" they had good credit. A year later Hickok's debts remained unpaid. Even so, he was doing well and his reputation as a bookbinder and machinist stood "deservedly high." His personal property was valuable, according to the reporter, "but I do not consider it worth as much as the judgments against him."[48]

In 1853 Hickok purchased land east of the Capitol at the juncture of North Street and the Pennsylvania Railroad and there built his Eagle Works. Although the plant specialized in ruling machines, it turned out a variety of products, many invented by the owner. Most were for the use of bookbinderies: standing presses, finishing presses, embossing presses, gilding presses, brass bound pressing boards, ten-drawer type cases, grinding machines, stabbing machines, gauge table shears, backing machines, hand stamps, office desks, cutting machines, and ink. In May 1855 a reporter from the *Morning Herald* described the Eagle plant in detail. In addition to the products previously reported, he noticed such products as Corinthian columns for a new building in Harrisburg, school desks to be used in Huntingdon County, and Hickok's own "celebrated Patent ruling Pens," which were used in "every quarter of the globe." One of the more popular moneymakers was a "portable" cider press designed by Hickok that sold for $35. It was little more than a binding press enclosed in a wooden fence that allowed the cider to run out between the slats. The press, which weighed 420 pounds, stood four feet tall and occupied an area two and a half by three feet. It could be worked by hand, by horse, or by steam power; it had a capacity of six to twelve barrels of cider a day; and, if the apples were ground in advance, it could be worked by a boy.[49]

Detailed payroll books and company catalogs make up the bulk of Hickok's extant records from this early period. The earliest payroll listed 16 employees in February 1851. Between then and the opening of the Eagle Works in 1853, Hickok's work force ranged from 15 to 22 persons. The new plant had 61 employees in June 1853, and 77, 68, 98, and 74 in successive Junes. By November 1857, depression had reduced the number to between 40 and 50, where it remained until the Civil War.[50]

Hickok apparently faced the same problem at his plant that Hildrup did at the Harrisburg Car Works: a shortage of skilled machinists and cabinetmakers. Accordingly, he hired a few middle-age artisans skilled in these crafts and a number of youths between the ages of seventeen and twenty-one who worked side by side with them to learn the trade. Many of the young employees later went to work as machinists for the railroad or other businesses. As a result, Hickok's firm helped upgrade the skills of Harrisburg's labor force. Payroll lists and census information revealed Hickok as remarkably unbiased in his quest for workers. He hired blacks (one was a blacksmith), Irish immigrants, a Hungarian, a Mexican, and a Chinese (there were indeed few of the latter three nationalities in Harrisburg in that era).[51]

The manufacturing census of 1860 showed the Eagle Works with a capital of $20,000 (ranking tenth among Harrisburg firms), turning out $32,353 worth of goods (ranking eighth), and with forty-two employees (making it the third largest employer in town). Dun & Company reports painted a more depressing picture, but one that demonstrated Hickok's tenacity and the faith of some townspeople in his ultimate worth. On February 1, 1855, Hickok was "much embarrassed" by debts amounting to $30,000. He claimed assets of $40,000, but much of that was "probably" in notes and accounts that would bring 50 cents on the dollar. Attempts to buy time did not work, and he failed on February 12. The next January, Hickok, "still insolvent," was "doing a large business." Claims could not be collected, but he kept going by "operating under another person's name" and paying cash. In June 1856 James McCormick bought up his property at sheriff's sale, and Hickok continued in business, using publisher Isaac G. McKinney & Company's name for credit. Until the Civil War, Dun agents gave scathing assessments of Hickok: "personally insolvent, very tricky and not at all reliable," "insist on cash payment or good security," "cannot recommend credit," "not responsible."[52]

In only five years Harrisburg's industrial foundation was in place. Where, in 1849, there were only a large shoemaking establishment with 21 employees and a rolling mill with 15 workers, by 1860 the Big Five factories had work forces ranging from 40 to 300. Hickok's Eagle Works, with the smallest annual output, produced $32,300 worth of goods; the cotton mill, with the largest, $200,000 worth. Although they would be remodeled, rebuilt, and expanded, the cotton mill, the car works, the McCormick and Bailey iron properties, and the repair shops of the Pennsylvania Railroad would remain Harrisburg's principal industries through the remainder of the century. But while still in their infancy they were to be tested by the crisis of the Civil War.

The Civil War Interlude

Four years of civil war first dampened then stimulated industrial development at Harrisburg. Initial shock at the outbreak of hostilities in April 1861, coupled with uncertainty over the future, slowed the economy much as the Panic of 1857 had. That stage passed quickly, however. The community soon became one of the nation's busier centers for amassing troops and supplies and forwarding them to the eastern front. Feeding, clothing, housing, and providing hospitality for thousands of soldiers and loading and unloading, storing, processing, and transporting unprecedented quantities of war materiel created innumerable jobs and brought added income to the city. Most of the new industries were able to turn the war to advantage. As profits increased and the horizons of the leading entrepreneurs broadened, plans for a rapid postwar expansion took shape.[1]

The city's significant contributions to the conflict stemmed from three factors: its strategic geographic location, its function as a major railroad center, and the fact that one of its leading entrepreneurs, Simon Cameron, was U.S. secretary of war during the opening months of the war. Although only indirectly related to manufacturing, these factors considerably influenced the course of the city's postwar industrial development. Of central concern here, however, were the ways the war affected industry, whether it hastened or retarded the process, and what new directions, if any, were the result of that experience.

Geographically, Harrisburg lay in a plain commanding the northern approaches to the lower Susquehanna River valley, which led south to Baltimore (and to Washington and Richmond beyond), and to the Cumberland Valley, which ran southwestward into Virginia's Shenandoah Valley. The city served as the gateway for Union forces, both those defending Washington and those launching offensive actions into Virginia. Except for troops and supplies moved by sea or directly along the coast, virtually everything used on the eastern front passed through Harrisburg into one or both of these channels. Similarly, aggressive Confederate armies attempting to encircle the nation's capital poured into these same valleys from the opposite direction, threatening Harrisburg on two occasions.[2]

Enhancing the city's location were its important rail connections. The nation's foremost east-west railroad, the Pennsylvania, came nearest to Washington at Harrisburg. There it crossed the Northern Central, which connected central New York and Pennsylvania with Baltimore. This vital junction alone made Harrisburg important to the war effort. However, the city also had rail connections to Reading and across New Jersey that provided an alternate route from New York and New England to the Baltimore-Washington area without passing through Philadelphia. Once hostilities broke out, Harrisburg quickly became and remained the key to the defense of Washington and the primary staging area east of the Appalachians for troops and supplies.

On April 15 (following the firing on Fort Sumter) Abraham Lincoln called for 75,000 volunteers to suppress the rebellion. Two days later, Governor Andrew G. Curtin called up 25,000 Pennsylvanians. As enlistees poured into Harrisburg, the governor took possession of a tract just north of the city for a military base. There Camp Curtin was built. In the beginning it was the only encampment in the state where the Commonwealth directly provided subsistence. By June 1862, with 20,000 soldiers, it ranked third in size in the union, exceeded only by encampments of 50,000 at Cairo, Illinois, and 40,000 at St. Louis.[3] During the first nine months of Camp Curtin's existence, an estimated 68,000 troops passed through en route to the front. At later stages it served as a hospital for the sick and wounded, as a prison for captured Confederates, and at war's end as the mustering-out point for veterans. In all it housed and processed some 300,000 soldiers during the war. Three smaller military facilities were also located nearby: Camp Cameron, a mile or two east of Camp Curtin; Camp Haly (named after Sara Haly, daughter of Jacob M. Haldeman) near the cotton factory; and Camp Couch, near Penbrook. Three others were just across the river.[4]

Of shorter-lived importance was Simon Cameron's service as secretary of war. During his ten months in office, either directly or through agents acting under his authority, he brought considerable business to Harrisburg and to a number of its entrepreneurs, including himself and his son. These contributions came in two areas: the steps he took to make the Northern Central Railroad the major artery for transporting

Fig. 10. Union Army Encampment Protecting the Harrisburg Bridge, 1865.
Fanny F. Palmer's "Harrisburg and the Susquehanna . . . ," Currier & Ives,
Lithographers.

troops and supplies to Washington and northern Virginia, and the contracts the army awarded to Cameron's Harrisburg business associates
for supplying it with horses, mules, and provisions.

Cameron became secretary of war in 1861 because he was a major
force in party politics in Pennsylvania. For two decades before the war,
he had adroitly used his wealth and business connections to acquire
political influence, laying the foundations of the notorious post–Civil
War "Cameron machine." Twice the Pennsylvania legislature had sent
him to the U.S. Senate: in 1845 (to fill the vacancy caused by James
Buchanan's appointment as secretary of state) and again in 1857. Originally a Jacksonian Democrat, he employed the tariff issue in 1845 to
outmaneuver his own party's favorite. Once he showed that he had the
support of Democrats wanting protection for the state's iron industry,

the Whig minority threw their vote to him. Cameron's opportunism led party regulars to shun him in subsequent Senate elections, so he flirted with the Know-Nothing movement in 1854–55, when the Know-Nothings briefly controlled the legislature. He might have won reelection to the Senate then had not more votes been cast than there were legislators present. In the uproar that followed, the legislature could agree on no candidate. Finally, a variation of Cameron's 1845 tactic won him reelection in 1857; he combined the votes of enough breakaway protectionist Democrats with those of united Republican party and American party legislators again to defeat the Democrat regulars.[5]

A Republican by 1859, Senator Cameron declared himself a contender for his new party's 1860 presidential nomination. Because of his reputation for "spoilmanship, opportunism, and political chicanery," few took him seriously. Even Pennsylvania's delegates were divided. Nonetheless, he led an important delegation. When Abraham Lincoln's managers held out the promise of a cabinet seat, Cameron threw his support to the Illinois contender. Following victory that fall, Lincoln, disturbed by reports regarding Cameron's character, refused to appoint him secretary of the treasury, the post Cameron wanted, and only reluctantly named him head of the War Department. Lincoln believed that Pennsylvania required and deserved a place in the cabinet and that no other Pennsylvanian would be acceptable to Cameron's powerful backers.[6]

From the beginning, critics charged that Cameron used his office to enrich himself and his friends. He had never separated his political and business interests, which he regarded as complementary and interchangeable, and certainly he saw nothing wrong in public service coinciding with his personal fortunes. Neither did he concern himself much with propriety or the appearances of propriety. By January 1862 Lincoln had to displace him. When the post of minister to Russia fell vacant, it went to Cameron. The abrupt removal, however, did not prevent a congressional investigation into a variety of complaints. Cameron's reputation and those of a few of his Harrisburg friends received an unfavorable drubbing, and in the end the House of Representatives censured him for his lax handling of army contracts.[7]

Admittedly Cameron went to the War Department under very adverse circumstances. Without time to settle into office, the new and untried administration was plunged into civil war. The War Department was in disarray and unprepared for the emergency. Shortly after Cameron took over, both the adjutant general and the quartermaster general left to serve the Confederacy. Meanwhile, Maryland, whose loyalty was uncertain, threatened to secede and separate Washington from the remainder of the Union. When volunteers, summoned from the loyal states by

Lincoln, marched through Baltimore en route to the capital, rioters stoned them and tore up railroad tracks.

Under the circumstances, Cameron improvised. It is not surprising that he turned to people he could trust to cope with the emergency. Those people, many with obvious abilities, too often were his personal friends and business associates. Among those with Harrisburg connections was his sometime ally, Vice President Thomas Scott of the Pennsylvania Railroad. Cameron called him to Washington to run the telegraphs and railways and keep open communications and transportation lines between the capital and the rest of the country.[8] To secure large numbers of horses and mules quickly, the army, perhaps at Cameron's suggestion, turned to William Calder Jr., the secretary's banking partner from Harrisburg. Calder, in turn, enlisted the help of his associate and partner in the Harrisburg Car Company, William T. Hildrup. Cameron's other banking partner, Harrisburg merchant Jacob Eby, obtained lucrative contracts to supply rations for the military. Congress eventually questioned all these arrangements.

The exigencies of war soon provided Cameron with a good reason to strengthen the Northern Central at the expense of its chief rivals. He later vehemently denied any corruption, explaining that because he knew how to make money he did not have to steal it. As he explained:

> When the war broke out I knew that this railroad from Baltimore to Harrisburg, the Northern Central of Pennsylvania, was bound to be good property; the soldiers and people devoted to the preservation of the Union traveling to Washington would necessarily be transported over it. The stock was then worth but a few cents on the dollar. I knew that from the very necessity of the case it would advance in value to par or nearly so. I bought large blocks of this stock, and told Mr. Lincoln if he would give me ten thousand dollars I would make him all the money he wanted. . . . [Lincoln's refusal] was his mistake; . . . the investment would have been perfectly legitimate and . . . he might as well have made a large sum of money as not.[9]

Correct as far as it went, the explanation completely omitted the measures Cameron adopted to ensure that profitable outcome. He had first taken interest in the line three years *before* the war, scheming with Thomson and Scott of the Pennsylvania Railroad, and other Pennsylvanians, to wrest control of the Northern Central from the Baltimore & Ohio. At the outbreak of hostilities, his brother-in-law was president of the line, his brother was vice president, and he, his son Donald, and his banking partners Calder and Eby were all on the board. Two years later his son became the line's president. When a change in Pennsylvania law in 1861 allowed the Pennsylvania Railroad to buy the stock of other companies, it bought up Thomson's and Scott's Northern Central shares, making that line one of its subsidiary branches.[10] To realize the line's potential, however, control of all trade between the northern states and Washington by the B&O and the Philadelphia, Wilmington & Baltimore (PW&B) had to be broken. A portion of the traffic from New England and New York passing over the Camden & Amboy to Philadelphia, and from there to Baltimore over the PW&B, had to be diverted to the Northern Central. Similarly, some if not all of the great flow of goods from the west and the northwest over the B&O had to be shifted to the Pennsylvania as far as Harrisburg, and then over the Northern Central to Baltimore. Finally, something had to be done about the B&O's possessing the only track between Baltimore and Washington.[11]

The secession of Virginia (which then included West Virginia) in mid-April halted traffic from the west over the main line of the B&O, forcing a shift to the Pennsylvania–Northern Central route. Cameron promptly ordered the army to take over the B&O's branch line between Baltimore and Washington and to guard the Northern Central from Harrisburg to Baltimore. Even before the latter order could be implemented, the Baltimore riots began. Torn-up track and destroyed bridges prevented rail traffic from reaching the nation's capital.[12] Cameron wanted to send troops over the Northern Central with orders to fight their way through Baltimore if necessary. Others argued against further provocation of antiunion sentiment in Maryland and persuaded him to send them over the PW&B to Perryville at the mouth of the Susquehanna. From there they went by ferry to Annapolis, where they resumed the trip to Washington by rail.

Cameron and Scott, believing supplies for the troops at Washington should be transported over their lines, moved to reopen the Northern Central promptly. Scott urged officials in Pittsburgh and Philadelphia to speed ammunition and supplies as far as possible over the route. The army, meanwhile, deployed troops along the Northern Central preparatory to marching into Baltimore. Once quiet returned to that city without military action, reopening railroads became the first order of business. Northern Central facilities were rebuilt with military aid; the PW&B and the B&O had to rebuild on their own. In spite of his hostility toward the B&O's management, Scott soon realized that the Pennsylvania alone could not move the immense traffic from the west. He would have welcomed a reopening of the B&O, but that did not occur until the end of March 1862.[13]

Meanwhile, in May 1861, Cameron moved against the PW&B. On his order, fifty carloads of troops leaving New York City each day were to be evenly divided between the Camden & Amboy (which connected with the PW&B) and the Central of New Jersey, which, through the agency of various other lines, joined the Northern Central at Harrisburg. The alleged reason for the order was the PW&B's inability to handle so many cars. Samuel M. Felton, president of the line, pointed out publicly that his road had carried 90 cars a day and could handle 200 if necessary. Other excuses followed: though seventy miles longer, the Northern Central route was unbroken and did not involve the use of ferries, as did the PW&B at Philadelphia; the PW&B route cost $6 per soldier, the Northern Central but $4; and the importance of developing two military routes by dividing the business. In the end Cameron and Felton compromised the issue, allowing the troop commanders to choose their own route. Although the PW&B subsequently carried 80 percent of the total, the Northern Central got 20 percent where previously it had none.[14]

The general rates established by Scott for hauling soldiers and supplies by rail came under fire by Congress for being too high. The investigators especially objected to charges over the Pennsylvania–Northern Central route. Instead of going continuously from Pittsburgh to Washington, which would have entitled the government to through rates, freight was unloaded from the Pennsylvania at Harrisburg and reloaded onto Northern Central trains, bringing the latter part of the trip under local rates. Moreover, the Harrisburg stop added more than $2 a car to the cost of shipping livestock. Animals were taken, at a charge of 80 cents a carload, to the Pennsylvania's stockyards a mile and a half from the city. There, an additional 25 cents was charged for unloading, and a $1 toll had to be paid for crossing the Cumberland Valley Railroad bridge. Supposedly the cattle could be kept at Harrisburg, safe from the enemy, until needed. In practice, most were promptly forwarded. Passengers ordinarily could be carried from Pittsburgh to Baltimore at a cost to the railroads of $1\frac{1}{3}$ cents a mile. Allowed 2 cents a mile for hauling soldiers, the railroads earned a profit of two-thirds of a cent per mile for each person transported. In fact, they realized considerably more because they usually transported troops in large numbers, in freight cars, at low rates of speed, at a cost of only 9 mills each.[15]

At least three Harrisburg entrepreneurs—Cameron's banking partners Calder and Eby, and Hildrup, general manager of the Harrisburg Car Company—were involved in supplying the army. One important need was horses. To secure them, the congressional committee of inquiry later complained, the army contracted not with "original owners and breeders of horses" but with "bankers, publishers, hotel keepers, distillers, dry-goods men, and men of all sorts of business more or less foreign to the subject of such contracts."[16] Calder was one of the "bankers" the committee referred to. The appellation, though accurate, was incomplete; he also bred horses, owned the livery business in Harrisburg founded by his father, and for many years operated a statewide stagecoach business. Having purchased thousands of horses in the course of his career, he knew horses, where they could be obtained, and what they cost.

During the war, Calder, with Hildrup's help, furnished the army with 42,000 horses, 67,000 mules, 5,000 oxen, "thousands of tons" of hay, and corn. He later claimed credit for fixing the prices paid by the government at $125 a horse and $117.50 a mule, "contenting himself with a moderate commission which was chargeable to the owner of the stock."[17] Calder's testimony at the 1862 congressional inquiry revealed a slightly different account. At the outset of the war, army officers met with him in Harrisburg. Who may have recommended him, or how the officers knew about him, he did not explain. In any event, he offered to lend them the 400 horses they needed immediately. They preferred to buy but were unwilling to pay the $135 he asked. Finally they agreed on $125. Calder proceeded to procure the horses—seventy-five from his own stock and the balance purchased for $120 each. Other contracts, for the most part at the same terms, followed regularly. By the time of the investigation, Calder had already furnished 1,500 horses on noncompetitive contracts and 6,500 on competitive contracts. Elsewhere in the report the committee concluded that the average price paid by the government for horses had been $125, the price received by original owners between $90 and $100, and the average profit to contractors never less than $25.[18]

Calder pointed out to the committee that he used his own money to pay "every man his money as soon as I received his horse."[19] He soon found it necessary to borrow. By April 1862 he had already borrowed nearly $800,000 at 6 percent from J. Donald Cameron (whether from Cameron personally or from the Middletown Bank, of which he was cashier, Calder never asked), giving drafts against his own property. Cameron deposited the loans in Pennsylvania bank notes in an account at the State Capitol Bank (in which both men were partners), and Calder wrote checks against the account. Patriot though he was, Calder demanded payment from the government in gold, which he turned over to Cameron. Cameron, in turn, profited from the exchange because the bank notes were valued at less than par. Allowing him the modest $5 profit per animal he claimed, Calder during the war cleared at least $570,000 from the sale of horses alone. If the committee's averages were accurate, he may have made up to five times that amount. He also profited from the sale of other livestock and hay and corn to the army, and, as he admitted, "benefitted the extent of his partnership share" from the bank's profits on the loans extended.[20]

Supplying provisions for soldiers stationed in the Harrisburg area offered other business opportunities. Apparently the commissary department at first purchased rations by contract with local merchants rather than by competitive bidding. Captain Beekman DuBarry testified to the congressional committee in November 1861 that he had purchased rations valued at $180,500 between May and November. Of that sum, $150,000 went to Jacob Eby. In addition to being a partner in Cameron's bank, Eby was Harrisburg's wealthiest merchant in 1860.[21] DuBarry testified that he had probably been in every store in Harrisburg and chose to buy from Eby because his store was "the only place in which I could find articles of daily consumption in very large quantities."[22] According to Eby, the contracts were not for any specified time or number of rations and could be canceled at any time. On one occasion he laid in rations in anticipation of need only to have the unit he was supplying move, causing him to lose $2,000.[23]

By September 1861 the army had introduced competitive bidding for rations contracts. That month Captain DuBarry let proposals for supplying 980 men stationed along the Northern Central Railroad. William T. Hildrup, who was *not* in the grocery business, won with a bid of 22 cents per ration. Two persons made lower bids, both for between 18 and 19 cents, but when DuBarry tried to call on them he could not find them. In October a Captain Donaldson submitted proposals for supplying 3,000 troops at Camp Cameron. At least three Harrisburg merchants bid: George W. Hummel (18 cents per ration), David McCormick (18½ cents), and Jacob Eby (19 cents). The contract went to Eby. McCormick, who claimed to have "very intimate" business relations with Donaldson, said the captain told him he had lost out because Hummel's bid was lower. When McCormick asked why Hummel did not get the bid, Donaldson replied that "he had positive orders from Washington to give the contract to Mr. Eby, without regard to price."[24] Eby, when questioned, acknowledged that proposals had been issued and bids submitted, but "they were never opened." The commissary officer had gone to Washington, where the order for competitive bidding apparently was countermanded. When he returned he told Eby to "go on with the old contract" at 19 cents.[25]

The impact of Secretary Cameron's activities on Harrisburg industrial developments was twofold. They considerably accelerated the city's inevitable development as the major wartime railroad center in the East, making it the logical location for large new railroad yards and repair shops. They also produced a flow of money into the community, particularly into the hands of entrepreneurs who would invest in new industrial projects at war's end.

The impact of the Civil War on Harrisburg banks, railroads, and industries was not uniform, but varied from one sector of the economy to another and, in some fields, from firm to firm. However, what Hildrup wrote of the war's effects on the Harrisburg Car Works seems to have applied to most industries. "At its outbreak, the shock embarrassed and prostrated manufacturing operations. The depression, however, was of brief duration, and little by little a reaction took place, until owing to a large demand for war supplies, a marked business activity prevailed."[26] On the negative side were new taxes, growing inflation, and the heavy demands of the army on personnel as well as disruptions in established patterns of trade and the priority given to military traffic over commercial traffic on the railroads. Hildrup noted the burden of wartime taxes after 1863. The final year of the war, 1865, bore especially heavily on his firm. On a total manufacture worth $259,000, the company paid $35,000 in taxes, leaving only $4,000 in net profits.[27]

Maintaining a work force was among the more difficult problems industrialists encountered. Again, the car works' experiences were probably not unique. The army's repeated calls for recruits "created a scarcity of able-bodied workmen" in the factories. The most vigorous laborers responded by enlisting, leaving behind those who were too young, too old, or otherwise unsuited for military service. Employers would slowly rebuild a satisfactory body of laborers, "a new call for recruits would occur, which in fulfillment of patriotic duty, was liberally responded to," and once again the factories were left with "a physically impaired work force."[28]

When Confederate forces neared Harrisburg in 1863, most enterprises temporarily ground to a halt as managers and workers alike turned to repelling the invaders. Hildrup, for example, selected the sites and planned the fortifications built on the western shore to protect both the bridges and the city.[29] At Hickok's Eagle Works, seventy-four employees appeared on the paybooks for the last two weeks of June, and sixty-eight appeared for the last two weeks of July. Only forty-three people received pay during the critical first two weeks of July.

A federal conscription law went into effect shortly after the battle of Gettysburg, further threatening the labor supply. Hickok was president of the city council when resolutions were offered on July 14 to appropriate $20,000 to pay the federal government "to exempt any and all citizens of Harrisburg who may be drafted and unable to pay the amount of such exemption." The rationale was that "this class comprises the workingmen who cannot now be spared from the City." Although defeated in mid-July, the measure was passed on August 1, and a similar measure on January 26, 1864, when the army required another 247 men.[30]

Despite these and other problems, the war years were good to most of Harrisburg's banks, railroads, and manufactories. Even the Harrisburg Bridge Company enjoyed one wartime bonanza. Following the battle of Gettysburg, it billed the War Department $3,028.63 in tolls for the soldiers who crossed it in defense of the city and its bridges. The bill was paid, and the stockholders that year received double their usual dividend on their stock.[31]

Both the Harrisburg Bank and the Dauphin Deposit Bank expanded their activities significantly during the war. The assets of the Harrisburg Bank increased nearly 59 percent between November 1859 and November 1864; those of the Dauphin Deposit increased more than 72 percent in the same period.[32] The 10 percent dividends paid by the Harrisburg Bank from 1862 through 1864 were the highest since the prosperous mid-1850s. During the war the minutes of the Harrisburg Bank became increasingly perfunctory, revealing little. The Harrisburg Bank did loan money to the Commonwealth of Pennsylvania to cover war expenses and to local governments to enable them to pay bounties to those volunteering for military duty. When Confederate raiders burned Chambersburg in a raid in the summer of 1864, the Harrisburg Bank contributed $800 to a rebuilding fund for the town. Only General Robert E. Lee's advance on Gettysburg in late June and early July 1863 threatened the bank itself. "Funds of the bank and books removed," the minutes noted tersely on July 1, "the rebels being in too close proximity." The minutes of the Dauphin Deposit recorded little, and whether its funds were removed is not known. By war's end both the Harrisburg Bank and Cameron's State Capitol Bank sought and obtained federal charters under the new National Banking Act. The former became the Harrisburg National Bank, the latter became the First National Bank of Harrisburg.[33]

So far as Pennsylvania's capital city was concerned, the war, far from retarding business and industry, acted as a stimulant. Some standards used for measuring the impact of the war on national economic development, such as the rate of new railroad construction, are inappropriate for the Harrisburg area.[34] Railroad construction before the war had been impressive in the region, but by 1861 Harrisburg's railroads were nearly all built. Except for a few relatively insignificant feeder lines, no new routes were built during or after the war. At the same time, the very considerable upgrading of lines by double-tracking both the Pennsylvania and the Northern Central during the war are not counted as new construction. When considering the impact of the war on railroads in the Harrisburg area, increased tonnage and number of passengers hauled and both the gross and net earnings of the lines are far more useful yardsticks.

The three major railroads serving the community prospered, although not all to the same degree. The Pennsylvania and the Northern Central railroads, given their critical role in the movement of goods to Washington and the eastern front, profited handsomely, as might be expected. During the first two years of the war, the gains for the Pennsylvania came chiefly from freight rather than passenger traffic; in 1863 and 1864, earnings in both categories increased; and in 1865, the year the war ended, passenger service increased but freight traffic declined. An unexpected bonus to the Pennsylvania was the immediate and permanent diversion to its line of grain traffic that had once gone down the Mississippi.[35] Payments for moving troops ranged from 9.4 percent of total passenger revenues in 1861 to 23.2 percent in 1863 and averaged 17 percent over the period 1861–65. Net revenues increased 53 percent in 1861 and another 37 percent in 1862. Average net earnings as a percentage of paid-in capital between 1857 and 1860 had averaged 18 percent a year, between 1861 and 1864 they swelled to 33 percent a year, and for the next two four-year blocks, 20 percent and 19 percent.[36]

The experiences of the Northern Central were similar. Both traffic and profits grew rapidly, as Cameron had anticipated. The line's net revenues in 1862 ($927,300) nearly equaled its gross revenues of 1859 ($929,500), and its net in 1865 ($1,326,000) exceeded its gross of 1860 ($1,018,000). Payments for moving troops and military supplies amounted to 18 percent of gross revenues for the years 1861–64. For the same years, its net earnings per year as a percentage of paid-in capital averaged 30 percent, in spite of very large expenditures for improvements in rolling stock and track charged off as expenses. Its annual dividends were 11 percent in 1860 and 1861, and 12 percent until 1866, when they dropped to 8 percent.[37]

The Cumberland Valley Railroad, a much smaller company and one that ran from Harrisburg into enemy territory, did well in 1861 when transport of soldiers and munitions increased net earnings from $117,000 to $159,500, a gain of more than 36 percent. However, Confederates invaded its right of way in 1862 and 1863, tearing up track, damaging facilities, and reducing traffic and earnings. The number of passengers carried by the line, including a large number of soldiers, increased by one-third between 1862 and 1865; freight tonnage declined 42 percent in the same period. Net earnings declined 15 percent in 1862 and another 13 percent in 1863, before rising a spectacular 64 percent (to nearly $194,000) the last full year of war. Net earnings as a percentage of paid-in capital from 1861 through 1864 averaged just under 16 percent, about half that earned by the Pennsylvania and the Northern Central.[38]

Other favorable impacts of the war on the railroads, though more difficult to measure, were important. For example, the excessive burdens imposed by the war led both the Pennsylvania and the Northern Central to double-track their lines and increase and upgrade their rolling stock. They also experimented with ways to reduce the need to replace rails. The Pennsylvania, finding that in some places rails had to be replaced every six months, in 1862 imported from Great Britain 100 tons of high-carbon crucible steel rails, at a cost of $150 gold per ton. Although twice the cost of iron rails, they lasted eight times longer and clearly were superior. The year the war ended, the Pennsylvania Railroad Company stood ready to move to the use of steel rails.[39]

Even less tangible were certain lessons taught by the war. Railroad companies that had previously been implacable rivals quickly saw the advantages of long-distance hauls without having to transfer freight from one company's cars to another's. Cooperation was not only important to the war effort, it often proved more profitable than competition.[40] This fostered postwar agreements among the trunk lines and accelerated the absorption of small independent lines into ever-larger systems. During the war, for example, the Pennsylvania leased both the Northern Central and the Cumberland Valley railroads in the Harrisburg area and began absorbing them as subsidiary lines.

A lack of comparable data on the earnings of Harrisburg's major industries makes the degree to which they suffered or benefited from the war less certain than for the railroads. It is clear, however, that the war affected the city's various firms differently. Its impact on the Harrisburg Cotton Company, for example, was almost wholly negative. Earnings were good only for the first year of the war, allowing the company to pay 6 percent in dividends—the best since the profitable year of 1856. Unfortunately, the supply of raw cotton soon dried up, and in July 1861 the directors voted to close the mill as soon as the present stock was gone. At the end of November 1863 they rented company lands for $500 a year to the army invalid corps for the erection of frame hospital barracks. They also turned over 150 feet of space in the mill for drilling soldiers. Dividends fell to 1 percent in both 1863 and 1864, and the value of the company's stock fell to 45 by September 1863.[41] In the first thirteen years of its existence, the firm paid a total of 19 percent in dividends. At the close of the war, the directors decided to put the firm up for sale, advertising in Baltimore, Philadelphia, New York, and Boston papers.

When no responses came, the stockholders decided to sell at a public auction for a minimum price of $100,000 in cash to be paid within thirty days. J. Donald Cameron apparently had hope that cotton cloth could be manufactured profitably after the war. On August 23, 1865, he purchased the mill on the terms advertised. The shareholders received an additional 27 percent dividend, the company books were closed, and the firm passed to its new owner.[42]

A lack of business records makes it impossible to measure directly the impact of the war on Harrisburg's iron industry. Certainly the war interfered with the usual patterns of trade and drained both laborers and managers from the factories. Both Henry McCormick and one of the smaller ironmasters, William W. Jennings (founder of the Franklin Iron Works in 1859), enlisted as officers, commanded regiments, and rose to the rank of colonel. At the same time, the market for iron increased dramatically. According to Jones Wister, operator of the furnace at Duncannon, north of the city, government contracts for iron products had by 1862 resulted in a doubling of prices for bars and nails. He and his three brothers made plans to build a new furnace within the city at war's end.[43] Most of Harrisburg's existing ironmasters, and a few newcomers to the business as well, followed the same path, accumulating capital and planning for postwar expansion.

In the second year of the war the Baileys renewed their partnership. They were able to reduce their dependence on Morris Patterson of Philadelphia, who had provided most of the initial capital for the firm ($20,000 in 1854 and an additional $25,000 in 1859). Patterson's share fell to $12,500, on which he received a guaranteed 6 percent annual return. Once the war ended, the Baileys began breaking ties with McCormick, reorganized the Central Iron Works as a corporation in 1866 (without Patterson), and the next year built their own Chesapeake Nail Works.[44]

Meanwhile, James McCormick added fourteen acres to the site of the West Fairview Nail Works and began expanding that plant's capacity. Between 1866 and 1874 the output doubled (it had been 60,000 kegs in 1860 and reached 84,000 kegs by 1870).[45] In 1865 Henry McCormick returned from war, took an extended overland trip to the Pacific Coast, and then assumed managerial control of the family mills. The McCormick iron empire would expand rapidly in the late 1860s and early 1870s. One further evidence of the profitability and prospects of the industry was the appearance of three large new mills between 1864 and 1867: a huge facility for rolling iron rails, an even larger plant for producing steel and steel rails, and yet another new anthracite blast furnace.

Immediately after the assault on Fort Sumter, President Calder of the Harrisburg Car Works fired off a letter to the War Department. Although addressed to Chief Clerk John P. Saunders, the letter ended up in the personal papers of Secretary Cameron. Knowing that Cameron was busy, Calder simply asked Saunders to call to his attention that the car works could supply the government with "army waggons, Gun Carriages, or anything in that line. Also to furnish cannon Balls, Bomb Shells and munition of that kind in large quantities." The car shops had facilities to "melt" up to forty tons of iron a day and had a large stock of seasoned lumber on hand. It could have 250 persons working within a week's time and had a capability of employing 300.[46] In November 1861 Calder told the congressional committee on war contracts that by that time his firm had contracted to supply 400 wagons to the military. He noted that "all the wagon-making community" resented his firm's ability to turn out inexpensive wagons: the first hundred had cost $120 each, the second hundred had cost $114 each, and the final 200 cost only $105.[47] Apparently the firm received no contracts for cannonballs, bomb shells, or other munitions.

Again, there is no direct information about the wartime profits or losses of the firm. In his history of the company, Hildrup discussed how in its early years the firm squeaked through the Panic of 1857 and then enjoyed a year or two of prosperity before the short depressed period at the outbreak of the war. In another passage, he declared that the business "had a steady and sure prosperity from its start." Whichever is correct, midway through the war, in 1863, Fleming steered a charter of incorporation for the firm through the Pennsylvania Assembly. By then, of the nine original partners, Haldeman and McKinley had died. Two others had sold their shares to the remaining partners, one because of misfortune, the other because of policy differences. In nine years, the firm had paid only one dividend; the earnings had instead gone to plant expansion, to paying off debts, and into reserves. The remaining parties at interest divided some $47,000 of the reserve fund among themselves and dissolved the partnership. That amounted to a return of 188 percent, or 18.8 percent a year, on the original investment, not including the expanded facilities and the one paid dividend. Moreover, the new corporation had a paid-up capital of $75,000, "the assets of the Company being the original $25,000, with earnings for the ten years of its existence."[48]

The war years saw a complete turnaround in the fortunes of Hickok and his Eagle Works. Employment figures for the firm support Hildrup's account of the impact of the war on business. Using the number of workers for the first two-week payday in June and November of each year, Hickok in November 1860 had 76 employees. The onset of the war cut that number to 37 in June 1861 and to 29 by November. Thereafter confidence returned gradually, and by November 1863 employment at Hickok's climbed to 72 workers. Until the end of the war the figure fluctuated between 54 and 69.

Hickok's reputation with Dun & Company followed a similar pattern. In June 1863 he was referred to as the agent for James McCormick, who owned the works. "The concern seems to be doing well," the reporter added. "Hickok is a decided mechanical genius. Can make anything but is too fond of ingenious contrivances to make it pay as well as he might. The Company, I believe, pays its debts." A year later the concern was "considered good." The workshops were worth at least $25,000, and Hickok was rumored to have settled his old debts and expected to "shortly take the concern into his own hands again." A report in March 1866 marked completion of the transformation: "Out of all his troubles now. President of City Council."[49]

Whether the Civil War stimulated or retarded economic development and industrialization generally, its effects on Harrisburg's economic growth were overwhelmingly positive. This was due partly to Harrisburg's strategic location vis-à-vis the eastern front and partly to timing. Whereas other cities of the Northeast had industrialized a decade or more earlier, Harrisburg was still in the first flush of its furnace- and factory-building stage. Its new industries were only three to eight years old when the Panic of 1857 slowed them, and that crisis had hardly passed when the uncertainties of 1861 overtook them. From 1862 onward, however, except for the cotton mill, the city's economy prospered. Its entrepreneurs and merchants profited from the war in a variety of ways; the assets and dividends of its banks grew; the railroads that served Harrisburg earned handsome profits from swollen wartime traffic even as they improved their tracks and rolling stock; ironmasters accumulated capital and made plans for expanding the capacities of their furnaces and rolling mills before the war had ended; the car works profited from supplying both the army and railroad companies with rolling stock; and Hickok and his Eagle Works emerged from a cloud of debt to respectable prosperity. Peacetime would bring a surge of industrial expansion.

Expansion and Consolidation

Harrisburg's entrepreneurs and industries stood poised at the end of the Civil War for a resurgence of economic growth. Over the next two decades the new industrial order achieved maturity. The community's largest and most successful enterprises continued to center around railroading and iron and steel manufacture, much of which related to railroading. The Pennsylvania Railroad Company expanded its large repair facilities at Harrisburg, the Harrisburg Car Manufacturing Company increased its capacity for producing railroad cars, and new facilities for rolling rails, both iron and steel, appeared in or near the city. At the same time, the McCormicks and Baileys continued to produce boiler plate and large quantities of nails. William O. Hickok's machine works, in spite of its foundry and casting work and dependence on the railroads for marketing its products, remained on the periphery of the city's main industrial thrust. Although Hickok thrived, his firm did not keep pace with the larger enterprises. The activities of the remaining major industrial firm—the cotton mill—had no direct relationship with the others. It limped along haltingly into the 1880s. Although these were by far Harrisburg's largest industrial firms, they were not the whole of the industrialized sector. Lumber mills, printing establishments, brick and fire-brick yards, and various producers of food and clothing also moved to large-scale production.

As had been true before the war, expansion and contraction at the local level closely paralleled the boom-and-bust pattern of the national economy. Prosperity in the late 1860s and early 1870s and through much of the 1880s alternated with depression, collapse, and consolidation during the severe Panic of 1873 and the brief downturn in 1884–85, which badly hurt railroads and railroad-related enterprises. Although all the major firms weathered the depression of the 1870s, two had been weakened seriously. The cotton mill failed in the 1880s and passed to outsiders, and the car works fell into bankruptcy during the preliminary stages of the Panic of 1893. By the end of the century, manufacturing in the city had reached its zenith and was in retreat.

The Pennsylvania Railroad

The increase in traffic occasioned by the Civil War soon overwhelmed the prewar facilities of the Pennsylvania Railroad at Harrisburg. Construction went on apace during and after the war. By 1870 the company had built a second roundhouse for engines, enlarged the repair shops, and expanded the yards for switching cars and assembling trains. The manufacturing census that year listed six operations for the company: a machine shop, a blacksmith shop, a carpenter repair shop, a boiler shop, a paint shop, and a tin shop. Capitalized at nearly $39,000, these steam-operated facilities utilized a total of 50 horsepower, employed 131 workers, and provided an annual payroll in excess of $75,000. Although the company periodically upgraded the equipment in the shops, the physical plant remained much the same until repair work was shifted to new yards across the Susquehanna early in the twentieth century.[1]

The Iron and Steel Industry

Iron and steel production led Harrisburg's post–Civil War industrial growth. New plants appeared, and old plants stepped up their output. Local interests founded a giant new firm for rolling iron rails, while the powerful Pennsylvania Railroad secretly helped finance an even larger plant to produce steel rails immediately south of Harrisburg, run by Philadelphia interests. Within the city, the McCormicks and the Baileys, now rivals, spurred each other into new expansions. Except for the Baileys, the other iron and steel producers integrated their operations, erecting blast furnaces near their rolling mills, or rolling mills near existing furnaces, to increase overall efficiency and ensure lower operating costs.

Fig. 11.   Pennsylvania Railroad Roundhouses and Repair Shops.   This aerial photograph was taken in the 1920s, but the buildings remained almost unchanged from the 1860s, when they were built, until they were torn down shortly after this picture was taken.

The Lochiel Rolling Mill

In 1864, with his public career for the moment in eclipse, Simon Cameron turned to new business ventures. He had helped make Harrisburg, where he now lived, a major railroad center with connections to all the important markets of the Northeast. He knew how close the city was to both coal fields and iron ore. Anticipating a great postwar expansion of railroads, Cameron joined with a number of local capitalists, including his son J. Donald, his banking partners William Calder and Jacob Eby, and David Fleming and William T. Hildrup of the Harrisburg Car Works, to establish a large plant for rolling new rails and rerolling used ones. Much of the capital came from outside: Sunbury, Philadelphia, and Baltimore interests pledged $100,000 of the $250,000 raised. The new facility, located in South Harrisburg between the river and the canal, consisted of two main buildings over 300 feet long, one for heating and the other for rolling iron. The organizers expected an estimated work force of 500 to produce seventy tons of railroad iron daily.[2]

Once construction began, Calder, president of the new Lochiel Iron Company, encountered needless delays and large cost overruns. He asked Hildrup, general superintendent of the Harrisburg Car Company, to take charge. Already fully employed, Hildrup agreed only to visit the site daily and spend at least an hour directing construction. After carefully examining all contracts and correcting prices, he notified the suppliers and builders that further failures to deliver on schedule, or unnecessary delays in construction, would result in forfeiture of contracts. The mill went up in record time. When it was finished, Hildrup turned the management over to others. The company struggled along, heavily in debt, until August 1870, when President Calder fell seriously ill. Hildrup, called back as acting president, persuaded the board to refloat the debts and provide additional funding to get the plant into full production. A mill for producing merchant iron was built for greater versatility. By the end of the year the firm had netted $28,000.

Within the year, Lochiel again ran into difficulties when employees unionized and resisted changes in work rules. The company met this challenge by closing, paying off the workers, selling its raw materials, paying its debts, and forming a new corporation. Labor troubles alone did not account for these actions. "The Lochiel Iron Co. expended $250,000," an informant for R. G. Dun & Company reported. "Their paid up capital was $400,000. The concern was badly managed and in July '71 the property was purchased [by new owners] for about $280,000." The new Lochiel Rolling Mill Company was capitalized at $300,000. Its Harrisburg stockholders included Donald Cameron ($39,200), Henry McCormick ($39,200), and the Harrisburg Car Company ($5,600). The great bulk of stock, however, was owned by outsiders. "Some of the most prominent Rail Road men in the Country" were interested, Thomas A. Scott of the Pennsylvania Railroad, who invested $39,200, being the best known. A new manager, Andrew J. Dull, replaced Hildrup, who returned to the car works. More important, Colonel Henry McCormick became the new president, bringing the firm for the next several years into the McCormick orbit.[3]

The Pennsylvania Steel Company

Cameron had been right about the need for rails, but increasingly few of them would be made of iron. The development of the Bessemer process in both England and the United States was about to make steelmaking easier and cheaper. The Pennsylvania Railroad's wartime test of steel rails had proved them superior to those made of iron. By 1865 the railroad company was prepared to invest in the manufacture of steel rails in the United States. Its hand in forming the Pennsylvania Steel Company, which arose just south of Harrisburg in what later would be called Steelton, long remained hidden. Also kept secret at first was the new firm's intention of rolling that steel into rails.

On June 26, 1865, a number of railroad executives and other steel-users met to form a new company to manufacture the metal in the United States. Included in the group were J. Edgar Thomson, president of the Pennsylvania Railroad; Samuel M. Felton, recently retired president of the Philadelphia, Wilmington & Baltimore Railroad; Nathaniel Thayer of the Baldwin Locomotive Works; and William Sellers, a Philadelphia inventor and machine-tool maker. All were "alert to the need for steel and its usefulness to the railroad industry" and were determined to end dependence on English and European suppliers. Felton became the firm's first president, a post he filled until his death in January 1889. The others, and Thomas Scott of the Pennsylvania Railroad, sat on the board of directors. Initially the firm incorporated with a capital of $200,000. Given the scope of their plans, that sum soon had to be increased nearly tenfold. Of the company's $1,834,000 in capital stock, the Pennsylvania Railroad subscribed to more than a third ($679,800 worth). The company meanwhile negotiated with the American holders of the patents for a license to produce steel using the Bessemer process.[4]

Ironically the owners of the firm, intending to free America from dependence on foreign steel, turned to England for an experienced steelmaster to design and manage their plant and for a converter and other machinery to equip it. In the end, circumstances forced the substitution of both an American planner-manager and American equipment. The expert they originally hired, William Butcher Jr., of Sheffield, had recently settled in Philadelphia and was building a steel mill there. When that undertaking proved unsuccessful, he departed. In his place the Pennsylvania Steel Company hired Alexander Holley, one of a handful of bold young American pioneers in steelmaking. Later the ship carrying the equipment purchased in England sank, and American companies, chiefly in Philadelphia, supplied replacements.[5]

In the meantime, the company decided to locate its works along the main line of the Pennsylvania Railroad immediately south of Harrisburg. So far as is known, no local residents were directly involved with the company or this decision. However, both William Calder Jr. and J. Donald Cameron were later credited with bringing the facility to the

Fig. 12. Harrisburg, 1850–1880. Map on facing page, key to map below. Basemap, including street grid, from Boyd's Harrisburg Directory, 1907.

Infra-Structure

$ ① Harrisburg Bank 1814
  ② Dauphin Deposit 1834

  ③ Water Works Pumping Station 1843
  ④ Water Works Reservoir 1843
  ⑤ Harrisburg Gas Works 1849

RR ⑥ Pa RR Station 1849-57-87
   ⑦ Phila. & Reading RR Station 1858

Factories/Mills

Pre-Civil War

Ⓐ Pratt Bros. Rolling Mill 1845
Ⓑ Cotton Mill 1849
Ⓒ Central Iron 1853 (Old Hot Pot)
Ⓓ Hickok's Eagle Works 1853
Ⓔ Harrisburg Car Works 1854

Ⓕ Porter Furnace 1850
Ⓖ Keystone/Paxton #1 1853

Post-Civil War

Ⓗ Pa RR Roundhouse 1861

Ⓘ Pa RR Repair Shops 1861
Ⓙ Lochiel Rolling Mill 1864
Ⓚ Chesapeake Nail Works 1866
Ⓛ Paxton Rolling Mill 1869
Ⓜ New Central Iron Works 1878
Ⓝ Harrisburg Car Works Foundry 1871

Ⓞ Pa RR Roundhouse #2 1871

Ⓟ Wister Furnace 1867
Ⓠ Lochiel Furnace 1872
Ⓡ Paxton Furnace #2 1872

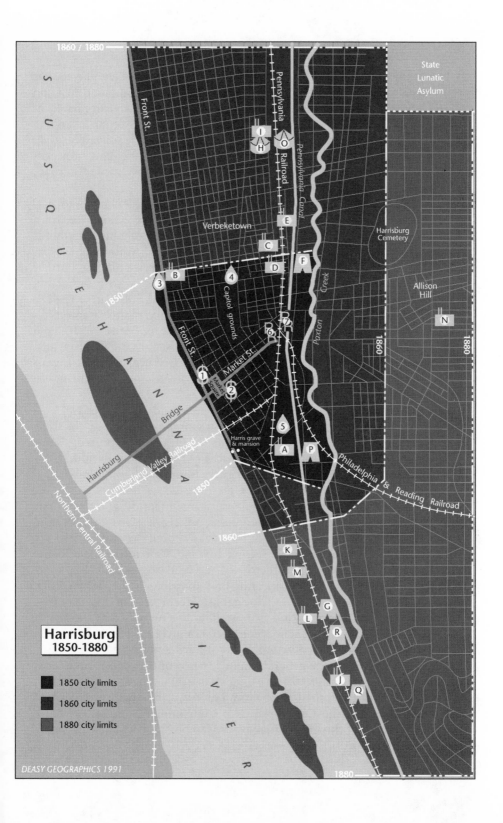

Harrisburg
1850-1880

■ 1850 city limits
■ 1860 city limits
■ 1880 city limits

DEASY GEOGRAPHICS 1991

Susquehanna

River

Harrisburg Bridge

Cumberland Valley Railroad

Northern Central Railroad

Front St.

Pennsylvania Railroad

Pennsylvania Canal

Paxton Creek

Verbeketown

Capitol grounds

Market St.

Harris grave & mansion

Philadelphia & Reading Railroad

State Lunatic Asylum

Harrisburg Cemetery

Allison Hill

Harrisburg vicinity. When, how, and what role, if any, they played, has not come to light. Certainly the site had much to recommend it. Land and water (the Susquehanna) were abundant, the extensive ore banks at Cornwall lay only twenty-six miles to the east, and limestone, charcoal, and coal—both bituminous and anthracite—were readily available. Moreover, the canals and railroads of the area offered adequate transportation, and a principal anticipated consumer, the Pennsylvania Railroad, operated on adjacent property.[6]

Eventually the Pennsylvania Steel Company would expand to 600 acres and extend four miles along the river. It began, however, on a ninety-seven-acre farm that Harrisburg hardware merchants Rudolph F. and Henry A. Kelker had inherited from their father. At first the Kelkers refused to part with their patrimony, at least at the price offered. Further negotiations succeeded, however, and the Kelkers, whose father acquired the land for $37 an acre in 1830, received $300 an acre. The brothers promptly bought other land in the area and platted it into fifty-foot lots for workers' housing, each of which sold for between $100 and $250.[7] To pay for the land, the promoters proposed (as the Pennsylvania Railroad had when it located its yards at Harrisburg) that the townspeople buy up the site and give it to the company as an "inducement" to build at the place already selected.

The general public first learned of the plan in mid-November 1865. As early as October 16, however, Cameron's organ in the city, the *Telegraph,* hinted at important developments. At present, it reported, "some of the largest manufactories and furnaces in the Commonwealth," employing between 1,000 and 1,500 workers to produce iron, lumber, and cotton, were located in the city or just outside its borders. A decade earlier the town had made little pretense at manufacturing; twenty years in the future it would be a leading industrial center. Given its location and resources, only the apathy of its citizens could prevent Harrisburg from becoming Pennsylvania's second city and "one of the great inland cities of the Nation." The directors of the Harrisburg Bank had more precise information. At their meeting on November 1, they learned that "a body of capitalists proposed to establish Steel works in this vicinity— on condition that the requisite ground should be furnished by the people of the place." The board, whose membership included Henry A. Kelker, authorized the bank to contribute $1,000 for the project.[8]

On November 17 the general public learned of the plan from a letter published in the *Telegraph* revealing that a giant new steel company intended to locate on the Kelker farm. However, the townspeople would have to purchase the farm, give it to the company, and subscribe to $50,000 worth of the firm's stock by November 22. The letter proposed a public meeting on November 20 to raise the money and suggested that the Harrisburg City Horse Passenger Railway Company, which had much to gain from the project, might want to subscribe $5,000. The value to the community as a whole would run to millions of dollars and elevate Harrisburg's reputation as a first-class manufacturing center. "Your action," the letter concluded, would reach "the prominent members of this enterprise, the Messrs. Butcher, of Sheffield, England," and Harrisburg would "henceforth be known in this country and the world as Sheffield is in England."

The *Telegraph* took up the theme, endorsed the project, and observed that a number of other communities were vying for the works. Some were offering greater and more valuable inducements than the company was asking of Harrisburg. In reporting the public meeting, it declared that "a large amount" had been subscribed, but apparently not enough. Although the November 22 deadline passed, agents were named to solicit additional subscriptions in the various wards of the city. A few days later the *Telegraph* cautioned that with foreign capital seeking investment Harrisburg must not be backward in displaying the "energy and liberality" needed to attract it. The jibes of a Pittsburgh paper that it did not know whether to wonder most at "the cheek of the company . . . or the vendacity [*sic*] of the Harrisburg folks in agreeing to their proposals" led the *Telegraph* to reply that Pittsburgh was jealous. The enterprise would "soon put at rest the pretensions of Pittsburgh of being the great center of iron and steel manufacture in this country." Harrisburg residents had "freely subscribed" to the project because they "believe in the profit of the investment, while the donation of land is a mere trifle to our liberal citizens." Harrisburg's other leading paper, the *Patriot,* favored the project but expressed concern over a rumor that half the employees would come from England.[9]

With the land provided and the money raised, construction of the new and entirely integrated plant began in May 1866. Although physically located only a few miles south of the city's boundaries, the new facility never fully identified itself with Harrisburg, nor was it ever wholly regarded as part of that community. It was owned and managed by outsiders. "They only have their works here," a local agent reported to R. G. Dun & Company. "Their office and officers are located in Philadelphia which is their headquarters and financial depot."[10]

On May 25, 1867, the company's Bessemer furnaces poured their first steel. Because the rolling mill would not be in operation for a year, the ingots were sent to the Cambria Iron Company at Johnstown, where they were rolled into the first commercial steel rails produced in the United States, and sold to the Pennsylvania Railroad. Improvements and additions were made to the mills in 1868–69. The manufacturing census of 1870 showed a complex plant, capitalized at $603,000, that included facilities for making steel, a foundry, a rolling mill and forge, and a machine shop. Steam engines, generating 2,550 horsepower, drove the machinery. About 275 workers produced 5,500 tons of steel rails and 450 tons of castings worth $770,000.

Although first in the field, the Pennsylvania Steel Company soon faced competition. American ironmasters recognized the inadequacy of iron rails but hesitated shifting to steel until questions over patent rights, financing, and technical problems, such as the high phosphorous content of American ore, were resolved. In the meantime, Abram Hewitt and Andrew Carnegie unsuccessfully experimented with a halfway measure. To strengthen rails without making them entirely of expensive imported steel, they fused steel heads onto iron rails. Once Michigan ore became available and the other problems were resolved, the rush to steel began. By 1876 eleven firms were producing steel, some exceeding Pennsylvania Steel's output. In 1875, officers of the leading firms, including Felton of Pennsylvania Steel, formed the Bessemer Steel Association to exchange information, maintain tight control over patents, stabilize prices, and lobby for tariff protection. The next year they entered into a tentative pooling agreement for dividing the market. Cambria at Johnstown, already the leading producer, was allotted 19 percent of the total, Pennsylvania Steel was allotted 15 percent, and so on. Carnegie, given only a 9 percent share, threatened to undercut the others and provided facts and figures showing he could probably do so. To win his temporary cooperation (pools rarely lasted longer than it took to set them up), the association made his share equal to Cambria's. By 1880 Carnegie dominated the field.[11]

Nonetheless, Pennsylvania Steel continued to prosper. Four large blast furnaces added between 1873 and 1884 enabled the company to produce its own pig iron for the Bessemer converters. By 1876 its work force had swelled to 1,656 employees, who produced 48,660 tons of steel worth $3,162,900. "This has become a mammoth operation," commented Dun's agent in August 1879. The company was doing a "very prosperous business," was enlarging its capacity, and had orders "for more than a year ahead." A second report, the next June, declared Pennsylvania Steel "one of the largest works of the kind in the country, well managed, doing a large business & it is thought a profitable one." When open-hearth furnaces proved more efficient than Bessemer converters, Pennsylvania Steel was quick to accept them, installing two in 1883, another in 1889, a fourth and fifth in 1890, a sixth in 1892, and six more in 1893. Meanwhile, in 1887 the Pennsylvania Steel Company established another large facility at Sparrows Point, south of Baltimore, to process foreign ore and to ship steel to Steelton for rolling. That year the firm's capital stood at $2 million and represented a total investment of $5 million.[12]

The company's share of the American market for all rails, both iron and steel, grew steadily from less than 1 percent in 1868 (when iron rails still dominated) to a high of nearly 19 percent in 1885. Between 1882 and 1886, with only steel rails being used, the company's share averaged 14.4 percent of the market total. Of all rails produced in the United States in the twenty years between 1867 and 1886, Pennsylvania Steel produced more than 1.1 million tons, or 5.8 percent of the whole. As a producer of basic steel, its share of the national output reached 21 percent in 1871. Between 1867 and 1886, however, Pennsylvania Steel's share averaged a little over 9 percent.[13]

Other New Firms

The Wister Brothers' new anthracite furnace, erected in 1867, was the third ferrous metal works constructed at Harrisburg after the war. Somewhat earlier, members of the Wister family who came from Germantown (and, on and off, continued to live there) had taken over the Duncannon Iron Works, located twelve miles above the city in Perry County. There they built and operated one of the earlier pre–Civil War anthracite furnaces in the region. Their postwar furnace in Harrisburg proper supplied pig iron to the Pennsylvania Steel Works. Between 1867 and 1878 its capacity grew from 7,200 to 10,000 tons a year.[14] One other iron furnace built in the greater Harrisburg area after the war never got into production. Its builder was Gilliard Dock, a local machinist, who believed that a small-scale, inexpensive, no-frills furnace could compete successfully with the larger, fancier, and costlier furnaces then in fashion. Unfortunately, the boom times of the early 1870s ended just as he began building his 5,000-ton furnace in 1873. The Panic halted construction for several years. Though he subsequently finished it, Dock was obliged to sell the furnace in 1880 to people who used it for other purposes.[15]

The McCormick and Bailey Iron Properties

Meanwhile, as soon as the war ended, Harrisburg's leading ironmasters, the McCormicks and the Baileys, for reasons unknown, severed their business ties with one another and began expanding their respective enterprises. At that point James McCormick owned Paxton Furnace No.

1 (with a capacity of 7,000 tons), the rolling mill (originally built by Hunt & Son), which he leased to the Baileys in 1862, and the nail works in West Fairview, which he ran in conjunction with Charles L. Bailey. In 1866 Bailey left the nail factory, and his firm, the Central Iron Works, declined to renew its lease on McCormick's rolling mill after 1869. McCormick's son, Colonel Henry, home from the war, resumed management of Paxton Furnace in 1866 and replaced Bailey as manager at West Fairview. In 1869 he moved the rolling machinery from the family's idled mill in Harrisburg to the nail factory to produce nail plate on site. A year or so later, when the empty building in Harrisburg burned, it was not rebuilt. It was no longer needed. In 1869 the McCormicks had integrated their works by building a new rolling mill at Paxton to convert much of that furnace's output into boiler plate. Meanwhile, the West Fairview Nail Works acquired not only the rolling equipment from Harrisburg but also other new and better machinery and a cooper's shop to turn out kegs for shipping the nails.[16]

At a slower pace dictated by their more limited means, the Baileys also expanded. A major restructuring of their iron interests began in 1866, the year Charles L. Bailey ended his association with McCormick. The Baileys formed a partnership, Charles L. Bailey & Brother, which built the Chesapeake Nail Works in Harrisburg proper. At the same time, the Central Iron Works, owned by the Baileys in partnership with Morris Patterson, gave way to a corporation that would manufacture boiler plate and tank iron. Under the change, the number of owners remained very limited. Alexander C. Durbin, William C. Wallace, and Dr. George Bailey each held $30,000 worth of the firm's $120,000 capital stock. Who held the balance is not evident, but Charles L. Bailey at this point apparently had no interest in the firm. Except for George Bailey, the stockholders (who were also the company's officers) were absentee owners. Durbin, the president, was not from Harrisburg, and Wallace, the vice president, was a New York businessman. They, in turn, brought in another outsider, Robert R. Crisman, to superintend the works.

For three years Charles Bailey managed Chesapeake, then left to run an ironworks at Pottstown. George Bailey remained to look after the family interests in Harrisburg. Between 1869 and 1874 the output of the nail works grew from 100,000 to 155,000 kegs a year. In the same period, Central Iron ("Old Hot Pot") increased its output of rolled iron from 2,230 to 3,000 tons. Annual dividends at Central Iron averaged nearly 22 percent for those five years, although 18 percent of the 30 percent paid in 1869 came from the sale of land to the Pennsylvania Railroad. Central Iron's best year in the period was 1872. The Panic that began the next year marked the start of seven years of famine.[17]

Between 1870 and 1874 the McCormick and Bailey interests underwent substantial changes that set the terms of rivalry between them for
the balance of the century. James McCormick, an eminent lawyer near-
ing sixty-nine years of age, died intestate in 1870. His principal heirs
were his widow, Eliza, their two sons, Henry and James, and a daughter,
Mary, the wife of Donald Cameron. Over the years, McCormick had
acquired a complex estate consisting of "farms, lands and factories" in
Dauphin, Cumberland, Lancaster, York, and Perry counties and personal property that included "goods and chattels, stock, bonds and
notes, contracts and other choses in action." Among his major properties were the Paxton Furnace and the new rolling mill adjoining it, the
nail works in West Fairview, the lot on which the Harrisburg Cotton
Mill stood, the Paxton Flour Mills in Harrisburg, Sheesley's (Rosebud)
Island in the Susquehanna River, the family home at 233 Market Street
(purchased from Simon Cameron in 1834), a 103-acre farm in Dauphin
County, seven farms totaling 1,053 acres in Cumberland County, two
farms totaling 192 acres in Lancaster County, and 756 acres of mountain land in Perry and Lancaster counties. For thirty years James Mc-
Cormick had been a principal stockholder and president of the Dauphin
Deposit Bank, a major stockholder and president of the Harrisburg
Bridge Company, and probably held substantial amounts in stocks,
bonds, and notes of other businesses.[18]

Instead of probating and distributing the estate, the heirs established
the McCormick Estate Trust. Who thought up the device is not known.
The old entrepreneur himself may have proposed it before he died, or
James Jr., also a lawyer, may have conjured it up. But the arrangement
also smacks of the Camerons. In any event, the purpose of the trust
clearly was not to protect incompetent or inexperienced persons. The
sons, thirty-nine and thirty-seven years old respectively, were men of
substantial affairs. Although women were not regarded in that era as
competent to handle money or business matters, Mary's interests would
have been looked after by her husband, and Mrs. McCormick's by her
sons.

The principal reason for the trust was recognition that the value of the
entire estate was greater than the sum of its parts. Breaking up the estate,
with an iron furnace to one heir, rolling mills or a nail factory to a
second, the Dauphin Deposit to a third, and miscellaneous real estate to
a fourth, would have diminished the total value. And so, thirteen days
after James McCormick's death, the four heirs entered into a written
agreement for the purpose of avoiding "for the present" any necessity of
distributing the estate and keeping it together "for their common benefit." The McCormick sons and Cameron in effect formed a "quasi-

partnership" to act as "agent and attorney in fact" for all the other heirs and in their "own name or otherwise," to transact all business of the estate, its assets, or management. The three were to hold title to all real estate left by James McCormick, with full power to "grant, bargain, sell and convey the same or any part thereof." The principal heirs confirmed and ratified whatever the trustees might do and pledged to "truly account to each other" for actions taken under the agreement. Each heir was entitled to one undivided fourth of the whole estate and its assets and earnings. Should a conflict of interest arise, the agreement was to be broken and each member "remitted to their original rights and remedy of partition and account."[19]

No available records indicate how the trust fared over the next seventeen years. In October 1887, however, the trustees organized the McCormick Company, probably to enjoy the legal benefits of incorporation. All but five shares in the new firm were held by the McCormick Trust (and voted by the trustees, Colonel Henry McCormick, James McCormick, and Donald Cameron). The five shares were individually held by the same three trustees, by Henry McCormick Jr., and by Cameron's son, James M. Cameron. For all the vast holdings of the McCormick Trust, only a minutes book of the McCormick Company appears to remain, and it reveals little of the firm's affairs beyond changing relationships among various properties held by the trust.[20]

In the absence of substantial records for the various firms and properties, tracing the growing capacity, output, and profitability of the McCormick empire must be done chiefly from outside sources. According to the manufacturing census for 1870, Paxton Furnace No. 1, with 40 employees, produced 6,000 tons of pig iron worth $170,000. The new rolling mill at Paxton, employing 120 operatives, turned out 5,000 tons of boiler plate worth $500,000, and the nail works at West Fairview, using 195 workers, shipped out 84,000 kegs of nails worth $350,000. The McCormicks promptly set about to enlarge the scale of their operations. In 1872 they doubled their output of pig iron by building a second furnace, Paxton Furnace No. 2. That same year, when the Lochiel rolling mill came under Colonel Henry McCormick's management, it integrated backward, adding a blast furnace. The rolling mill in 1870 had employed 510 laborers and produced more than $1.6 million worth of rails and merchant iron. A survey of Harrisburg's iron industry in 1875 showed measurable growth. The three furnaces then controlled by the McCormicks (Lochiel and the two at Paxton) turned out 21,200 tons of pig iron, the rolling mill at Lochiel produced 24,000 tons of rails and bar iron, the Paxton rolling mill turned out 10,000 tons of boiler, ship, and tank iron, and the West Fairfield plant produced 182,000 kegs of nails.

A decade later the same furnaces produced 60,500 tons, the mills rolled 28,000 tons, and the nailery manufactured 200,000 kegs of nails.[21]

Even as Henry McCormick became Harrisburg's preeminent iron-master, Charles L. Bailey returned to Harrisburg in 1874 to assume control of both the Central Iron Works and the Chesapeake Nail Factory. Bailey apparently bought the stock of President Durbin and replaced him as head of the firm. A longtime ironmaking associate of Bailey, Artemus Wilhelm, replaced Wallace as vice president, but Wallace did remain on the board until 1882. With the Panic at its worst in 1875, the board learned that the company could not run at a profit without expanding its productive capacity and securing a more talented work force. Lacking both the resources and the profit margin to make those improvements, Bailey, with Wilhelm's support, called for a "convention" of stockholders (an inflated description for a meeting of so small a group) to offer the company, or the stock of such stockholders as wished to sell, for sale.

Only Bailey, Samuel Matlack (the secretary of the company), and Wallace attended the session on August 3 in Wallace's New York City office. The absence of a majority prevented any business from being transacted, and so the company lived on. The next year, Bailey's brother-

Fig. 13.   Chesapeake Nail Works and Central Iron Works.   The nail works was built in South Harrisburg in 1866. In 1877–78 a completely new Central Iron Works arose nearby. The original "Old Hot Pot" mill was located northeast of the state capitol building.

in-law, Gilbert M. McCauley, began his long association with the company, apparently buying the stock of Dr. George Bailey and succeeding him as treasurer. Charles L. Bailey dominated the firm through the difficult mid-1870s and the eleven prosperous years that followed. At first, profits and dividends were dismal, averaging less than 4 percent a year from 1873 to 1877, with none at all in 1878. Even so, Bailey persuaded the directors in 1877 to authorize him to rebuild and update the works. Old Hot Pot became a puddling mill. The new mill in South Harrisburg, costing $72,000, began rolling in 1878, just in time to take advantage of the industry's return to prosperity. Oddly enough, a major addition that might have been expected was not built. The Baileys neither erected nor leased a blast furnace to supply iron to their rolling mill and nailery. This meant that they continued to rely on others for that essential raw material. Between 1881 and 1891, dividends ranged between 12 percent and 90 percent and averaged nearly 42.5 percent a year for the eleven years. In this period, Bailey's sons joined the firm one by one as they graduated from Yale. William E. became a member of the board in 1882 and secretary of the company in 1884, Edward became vice president in 1887, and James B. became a director in 1890.[22] The Baileys were firmly in control and prospering.

The production data available for the Bailey and McCormick iron properties suggest that the Baileys, though operating on a smaller scale, were the more successful producers of nails and rolled iron. Chesapeake, from its beginning in 1867, consistently put out more nails with fewer nail machines and fewer employees, except for the suspiciously large output reported for the McCormick's West Fairview works in the 1875 Dauphin County atlas (see Table 5.1).

Similarly, the Baileys' Central Iron Works more than kept pace with the McCormick equivalent, Paxton Rolling Mill. With three-fifths the output of Paxton in 1870, Central Iron by 1880 was producing almost the same tonnage, and by 1897, thanks in part to a new universal mill, half again as much (see Table 5.2). McCormick, on the other hand, also owned or managed furnaces that dominated the production of pig iron in the city for most years from 1855 until 1885, when Lochiel was sold to outside interests (Table 5.3). It should be noted that Pennsylvania Steel's pig-iron capacity grew rapidly from 7,500 tons in 1875 to 200,000 tons by 1894. Though not in Harrisburg proper, a portion of that output supplied rolling mills in the city.

In the production of rails, the McCormick-managed Lochiel plant, which rolled and rerolled iron rails, was no match for Pennsylvania Steel, which produced steel rails. Lochiel's capacity in 1873 amounted to 28,000 tons of rolled iron rails and merchant bars. That same year,

Table 5.1. Comparative Output of McCormick and Bailey Nail Works

	McCormick's Nail Works			Bailey's Nail Works		
	Output (kegs)	Workers	Machines	Output (kegs)	Workers	Machines
1860	60,000	130	n/a	—	—	—
1870	84,000	195	55	100,000	75	n/a
1871	n/a	n/a	n/a	125,000	200	56
1873	126,000	n/a	73	134,000	n/a	66
1875	182,000	n/a	73	155,000	175	66
1877	121,880	310	n/a	n/a	n/a	n/a
1880	161,817	418	75	n/a	n/a	67
1885	200,000	500	80	250,000	400	103
1894	200,000	n/a	83	260,000	n/a	103

SOURCES: For 1860 and 1870, computerized manufacturing census schedules; for 1871, 1873, 1877, and 1878, Pennsylvania Bureau of Statistics, *Annual Reports;* for 1875, *Dauphin County Atlas,* p. 30, and *Pennsylvania Manufacturers,* pp. 299–300; for 1880 and 1894, American Iron & Steel Assoc. Directory, pp. 30, 109–12, and 18, 103–10; for 1885, Harrisburg Board of Trade, *Industrial and Commercial Resources of Harrisburg,* pp. 29–30; for 1897, *Iron Age* 77:44.

n/a = not available.

Table 5.2. Comparative Output of McCormick and Bailey Rolling Mills

	McCormick's Paxton Mills		Bailey's Central Iron	
	Output (tons)	Workers	Output (tons)	Workers
1870	5,000	120	3,000	70
1873	8,750	n/a	3,000	n/a
1875	10,000	n/a	6,500	n/a
1878	8,750	n/a	3,500	n/a
1880	9,164	170	9,000	150
1885	10,000	275	14,000	300
1894	48,000	n/a	58,000	n/a
1897	60,000	n/a	90,000	n/a

SOURCES: For 1870 and 1880, computerized manufacturing census schedules; for 1873, 1878, 1880, Pennsylvania Bureau of Statistics, *Annual Reports;* for 1875, *Pennsylvania Manufacturers,* pp. 299–300, 437; for 1880 and 1894, American Iron & Steel Assoc. Directory, pp. 30, 109–12, and 18, 103–10; for 1885, Harrisburg Board of Trade, *Industrial and Commercial Resources of Harrisburg,* pp. 29–30; for 1897, *Iron Age* 77:44.

n/a = not available.

Table 5.3. Tons of Pig Iron Production, 1850–1885

Year	Harrisburg Total	McCormick Interests				Other Furnaces			
		Paxton	Lochiel	Total	% Harrisburg	Porter	Wister	Total	% Harrisburg
1850(c)	3,500	—	—	—	—	3,500	—	3,500	100
1855(p)	8,309	4,504	—	4,504	54	3,805	—	3,805	46
1860(p)	5,480	5,480	—	5,480	100	0	—	0	0
1870(p)	17,460	6,000	—	6,000	34	5,700	5,760	11,460	66
1875(c)	33,800	14,000[a]	7,200	21,200	63	5,400	7,200	12,600	37
1878(c)	51,000	25,000	10,000	35,000	69	6,000	10,000	16,000	31
1880(p)	38,921	21,415	6,828	28,243	73	926	9,752	10,678	27
1885(c)	60,500	43,000	17,500	60,500	100	—[b]	—[b]	—	—

SOURCES: For 1850, 1860, and 1870, computerized manuscript manufacturing census schedules; for 1855, Lesley, *Iron Manufacturer's Guide*, p. 16; for 1875, *Dauphin County Atlas*, p. 29; for 1878 and 1880, Pennsylvania Bureau of Statistics, *Annual Reports*; for 1885, Harrisburg Board of Trade, *Industrial and Commercial Resources of Harrisburg*, pp. 28–35.

NOTE: (c) = capacity; (p) = amount produced.
[a]Paxton Furnace No. 2 went into production in 1872.
[b]Porter and Wister furnaces were dismantled by 1885.

Pennsylvania Steel produced nearly 223,000 tons of steel rails. By 1880 Lochiel's actual output fell to 12,035 tons, while Pennsylvania Steel's rose to more than 1,397,000. In 1885, investors from Danville purchased Lochiel. By century's end, Pennsylvania Steel owned it.[23]

The Harrisburg Car Manufacturing Company

The incorporation of the Harrisburg Car Works in 1863 marked a new stage in that firm's development.[24] Three of the original nine co-partners, William Calder, David Fleming, and William T. Hildrup, acquired all the stock of the new corporation. To meet the law's requirements regarding the number of incorporators, they sold one share each to three other men, who served as token stockholders and directors. Calder's banking partner, Jacob R. Eby, also later joined the board. Calder presided over the company until his death in 1880 and was succeeded by Vice President Fleming. Hildrup, who was not a director, held the posts of superintendent and general manager of the works and secretary-treasurer of the corporation throughout its history.

In response to a rapid expansion of business in 1864 and the anticipated railroad boom at war's end, the firm raised its capital from $75,000 to $300,000. The market's initial reaction to peace was a slump in orders for new cars. Hildrup offered the directors a gloomy prognosis for the future: several of the larger railroad companies had started to build their own cars, dimming the prospects for private builders. Accordingly he proposed shifting from railroad cars to the manufacture of machinists' tools. Several consequences followed acceptance of that strategy: the firm permanently abandoned the building of passenger cars, hired new workers skilled in toolmaking, and acquired new machinery. It also raised its capital to $500,000 and acquired five Philadelphia businessmen as major stockholders. The Harrisburg entrepreneurs hoped that the important connections the new investors had would bring a flood of orders. Foremost among the five were John M. Kennedy, a member of the Pennsylvania Railroad's board of directors, and John A. Wright, an original member of the Pennsylvania's board and now president of the large Freedom Iron Works in Lewistown.

As matters turned out, when the railroads discovered that making their own cars cost more than expected, they again began buying from independent companies. The press of orders for freight cars soon restored the car company to its original business. At the same time, machinery and toolmaking assumed "moderate importance" and held promise for the future. Apparently still toying with the idea of making its own cars, the Pennsylvania Railroad in 1869 opened negotiations to purchase the car company. Hildrup, anticipating that the Pennsylvania would want more room for expansion than the company could provide, persuaded the board to buy as much adjacent real estate as the railroad might need. Unfortunately, the negotiations with the Pennsylvania fell through. Hildrup, easily disturbed at the sight of vacant land, persuaded the board to use it for a new and larger machine shop and foundry for the machinery and tool division. Both orders and earnings were extremely strong from 1868 through 1874. Total manufactures of the car company never fell below $500,000 a year, in 1869 and 1872 they exceeded $1,000,000, and in 1873 they reached a little more than $1,500,000. Earnings those years ranged between $54,500 and $105,000.

A period of trouble for the Harrisburg Car Company began in one of the good years, 1872. In April the entire facility except for the new machine shop and foundry burned to the ground. Insurance fell short of covering the loss by nearly $80,000. Hildrup immediately ran notices in the local newspapers assuring employees that there would be plenty of work for all. Some would continue building cars in the machine shop and foundry, the others would be used in reconstructing the car shops. That night Hildrup sketched out plans, and rebuilding began the next morning, even before the ashes cooled. Within three months the rebuilt plant was turning out ten cars a day. Hildrup went on vacation, only to be recalled to Harrisburg a few days later when the machine shop and foundry burned. They were rebuilt in thirteen and a half working days.

The car works now had new and greatly enlarged and improved facilities. The new plant consisted of nineteen buildings. Included were a large office building with a drafting room, a blacksmith shop where 125 persons worked at forty-six forges, a machine shop where about 100 employees made axles, bolts, and other car parts, a foundry employing 50 molders and 90 other workers to turn out 120 railroad-car wheels a day, a framing shop where another 100 workers prepared lumber for the construction shops, two car-building shops in each of which 100 employees turned out seven flatcars daily, a shop that made 6,000 bolts a day, and various warehouses, paint shops, car repair shops, and drying buildings. As a precaution against future fires, the company installed pipes, fireplugs, and hoses throughout the works and organized its employees into a volunteer fire company. A powerful pump was installed to bring water directly from the canal to fight any fires that might occur.[25] Although many of the new buildings allegedly were "fireproof," fires again destroyed the planing mill in 1880 and the machine shops and foundry in 1881. Both, however, were fully insured.

All of this, nonetheless, left the firm saddled with debts of $600,000 in spite of the booming business cycle then in progress. A crisis developed with the onset of the major business panic in 1873. The company could not collect what was owed it, as one railroad after another went into receivership. The failure of the Allegheny Valley line in 1874 cost the car company over $25,000 in losses. Not surprisingly, friction developed between Calder and Hildrup, who previously had worked in complete harmony. Calder accused Hildrup of being rash, buying up land in hopes of selling it at a profit to the Pennsylvania Railroad, then building the machine shop and foundry on it chiefly to fill the land he could not sell. Calder charged him with being more "an expansion-mad maniac than a conservator of the company's assets." After three hours of shouting at one another, the two decided to take the matter to the board of directors.

There Calder, with a banker's concern for finances over the next few years, clashed with Hildrup, who defended the firm's longer range production capabilities from a managerial perspective. Calder predicted a lengthy depression for the overcrowded and highly competitive car-building industry because of the many railroad failures and reorganizations. He proposed closing the works, suspending the work force, and cutting back expenses until the firm could be reopened certain of a profit. Hildrup argued that such a policy would be fatal because the company was dependent on the skills and knowledge of its workers. If they were lost, they could never be reassembled. The directors backed Hildrup and issued "shinplasters" (interest-bearing notes backed by the assets of the company) in denominations of from 25 cents to $5 to be used to pay its employees. Both local merchants and the city itself accepted them in lieu of lawful currency.[26]

The crisis was not yet over. In 1875 the machine shop and foundry, which had been separately incorporated, found itself in financial straits. The car company accordingly took over the firm and its debts and operated it at an annual loss of $10,000. Although Hildrup won the debate, his policy was abandoned in 1875 when the works for the most part stood idle. The next year, no dividends were paid and managerial salaries were slashed 25 percent. Finally, in 1877 and 1878, the clouds began to lift as orders trickled in. Even the losses of the machine shop and foundry fell to $8,000 in 1878.

A second postwar boom began for the car company in 1879 and lasted four years. Hildrup, moving quickly to take advantage of the new wave of prosperity, borrowed money to buy up materials in advance of orders. His move paid off handsomely as orders and prices advanced. The company again acquired more land and built new buildings, dividends reached double digits for the first time since 1873, and managerial salaries were restored. In 1881 the company turned out 3,402 cars, and at year's end it had orders for 2,630 more. That record production year was followed by its highest gross income (in excess of $2 million) in 1882. The company built a new repair shop to keep pace with growing business in that line.

The second boom lasted only through 1883. The failure of the banking house of Grant & Ward the next year touched off another panic for railroad and industrial securities. That same year the car company lost two lawsuits that it had fought all the way to the Pennsylvania Supreme Court. For the first time in thirty years, the Harrisburg Car Company earned less then it spent, resulting in a loss of $81,000. Although the panic was short-lived, recovery dragged out over the rest of the decade. In his summary of the firm's earnings for the twenty-eight years between 1863 and 1890, Hildrup noted that the company had done a total business in excess of $23 million, on which it had paid dividends of nearly $1,175,000, an average of 20.5 percent a year.[27]

The Cotton Factory

As with most enterprises controlled by the Camerons, once J. Donald bought the cotton mill its affairs were rarely open to public scrutiny. Shortly after the purchase in August 1865, the passing of bales of raw cotton through the streets of Harrisburg raised hopes that full-scale production would soon resume. Apparently Cameron reopened the works with a capital of about $25,000 but devoted to it only such time as he could spare from his other business and political affairs. After three months of haphazard operation, William Calder, with the "aid of a friend to the enterprise," raised about $300,000 and reorganized the works on a more solid footing. He and his friends continued to operate the business, a contemporary encyclopedia sketch declared, "with the generous and charitable object of keeping this class of hands [women and children] employed." Often the business earned no more than enough to pay its employees.[28] Another explanation might be that Calder, once he acquired the mill, had few other choices. Certainly, running at minimal profit was better than standing idle, earning nothing. Selling such a property or the machinery in it would not be easy, and in that era losses could not be converted into advantages when tax returns were filed.

Data collected in the manufacturing census for 1870 and by the state of Pennsylvania over the next several years give some indication of the work force, the annual wages paid, and the output of the mill (Table 5.4). Whatever it earned above costs in later years, in 1870 the cotton

Table 5.4. Data on Operations of the Harrisburg Cotton Works

| Year | No. of Workers | | | | Annual Wages | Cost of Raw Materials | Output (000 yards) | Output Value |
	Men	Women	Under 16	Total				
1870	60	135	85	280	$52,800	$246,895	2,600	$325,000
1872–73	40	109	131	280	67,000	n/a	3,500	n/a
1876	26	172	40	238	n/a	n/a	n/a	n/a
1876[a]	n/a	n/a	n/a	251	60,000	n/a	3,687	n/a
1877	37	124	89	250	48,000	121,219	n/a	n/a

SOURCES: Computerized manufacturing census schedules, 1870; Pennsylvania Bureau of Statistics, *Annual Reports*.

n/a = not available.

[a]Two reports on different aspects of the mill's operations were published in 1876.

works grossed only 10 percent on its capital above what it paid for raw materials and labor alone. For reasons unknown, the manufacturing census schedules for 1880 did not include an entry for the cotton works, and the Pennsylvania Bureau of Statistics stopped publishing data on individual firms.

Hickok's Eagle Works

Hickok's Eagle Works benefited considerably from the business boom that followed Appomattox. As firms grew in number, size, and complexity, so did calls for ruled paper on which to keep records. Accordingly, orders for Hickok's machines, which ruled sheets for account books and other business forms, increased. Hickok's postwar catalogs offered customers a choice of seven different ruling machines ranging in price from $200 to $465. His product had a worldwide reputation. As one writer remarked, "They have been sold and shipped to every civilized and half civilized country in the world. They rule the paper for the 'ukases' of the autocrat of all the Russias; the 'firmans' of the Turkish Sultan and the Khedive of Egypt; the 'edicts' of the Emperor of China; the 'mandates' of the 'Mikado' of Japan, and the 'yalmids' of the Rajahs of India. In short in whatever clime the art of printing has penetrated, followed by its handmaid, bookbinding, there will be found a ruling machine bearing the trade mark of the Harrisburg Eagle works."[29]

Competitors in 1860 and 1865 improved on the machine that Hickok first developed in the early 1850s. Still, a mechanism was needed that would enable the ruling pens to strike automatically in relation to the position of the paper. The "O-A Striker," perfected by Hickok in 1874, solved the synchronization problem by a method that was to remain the basis of ruling machines for most of the next century. In 1880 the firm introduced a machine that could simultaneously rule faint lines on both sides of a sheet of paper. A decade later, L-style ruling made it possible for two machines placed at right angles to be striking one half of the paper while faintly ruling the other half. The "Dual L" ruled one side of the sheet in both directions with a single operation, while the "Quad L," consisting of two double-decker machines at right angles, completely ruled both sides of the sheet at the same time.[30]

At least into the 1870s, Hickok also continued to manufacture and sell a complete line of bookbinding equipment and the portable cider press he invented twenty years before. The machine, with only slight improvements, was now called the Keystone Wine and Cider Mill.

Hickok touted it as especially useful for winegrowers. It had proved "very superior for grinding the grapes, just cutting the skin without breaking up the core." It could also be used for pressing "currants, cherries, berries, cheese, butter, lard, and tallow."[31]

Except for catalogs and payroll timebooks, the Eagle Works' own records for that period apparently no longer exist. Only census data and lists of employees suggest the scale of operations. According to the manufacturing census of 1870, the Eagle Works was capitalized at $102,000. It used $14,800 worth of raw materials and had a gross output that year valued at $50,000. The company employed seventy-five workers on average, operated all twelve months of the year, and paid out $31,000 in wages, or an average of about $413 each. Comparable data from the previous manufacturing census show that capital had increased more than 500 percent during the decade, that the value of raw materials used more than doubled, and that the value of output increased about 55 percent. Compared with 1860, some 78.5 percent more employees shared a payroll that had increased 135 percent.

Payroll records show that the 75 employees claimed in 1870 was a high number. The employees listed for the first two-week pay period each June from 1865 through 1874 ranged from 57 to 77 and averaged 68. In contrast to the 1860s, the 1870s brought setbacks instead of growth. By 1875, at the depth of the Panic, the number of employees had fallen to 45. Although recovery eventually came, it brought no great expansion. In 1879 Hickok reported 71 employees and a payroll of $22,000. The next year he told the census taker his firm's capital was $100,000 (slightly less than in 1870). On the other hand, the value of his output ($112,000) had more than doubled in the same decade, in spite of a smaller work force (65) that earned lower wages ($340 on average each per year).

Harrisburg's Banks

Although not part of the industrial sector, Harrisburg's banks continued to expand their intimate relationships with large-scale manufacturing interests. At the end of the Civil War the city boasted five banks. Two held federal charters (Harrisburg National and First National), two were state banks (Dauphin Deposit and Mechanics' founded in 1853), and one was private (City Bank, established with a capital of $100,000 in 1861). A decade later there were nine. The national banks were the same, but in 1874 the Dauphin Deposit sold off its assets and became an unchartered private bank belonging to the McCormick family. Meanwhile, two new state banks had come into being: the Real Estate Savings Bank in 1872 and the Farmers' Bank in 1873. Two additional private banks also opened: Dougherty Brothers & Company and the State Bank.[32] Of the principal banks, apparently only the records of the oldest and largest, the Harrisburg National, are extant after 1874.

In spite of considerable growth, certain characteristics of the Harrisburg National persisted from its founding in 1814 until the turn of the century. For example, three lists of stockholders in the bank's records show there were 235 shareholders in 1814, some 222 in 1861, and 261 in 1883. Twenty-four major shareholders in 1814 held blocks of stock with a par value of $5,000 or more. Their combined holdings amounted to 40 percent of the bank's total. In both 1861 and 1883 only eleven shareholders held stock worth more than $5,000 each, and the total holdings of the group amounted to about 30 percent of the bank's capital in those years.[33]

An era of frequent changes in the presidency of the bank had begun with Thomas Elder's death in 1853. Four short-term presidents followed: Jacob M. Haldeman (who died in 1856); William Kerr (who died in 1864); Haldeman's son, Jacob S. (elected in 1864 and 1865 but denied additional terms); a period with no president (1866–68); and finally Valentine Hummel, who served until his death in 1870. Stability returned to the office with the election of Dr. George W. Reily in 1870. Reily, the last of the entrepreneurs to be discussed, was the son of Dr. Luther Reily, a prosperous physician who in 1850, with $85,000 worth of real estate, ranked fifth among the town's landholders. After college, young George worked in a Pittsburgh bank for a year before following his father into medicine. Graduating with an M.D. degree from the University of Pennsylvania in 1859, he returned to practice medicine in his native city. After his father's death in 1854, Reily also became much involved with a wide variety of financial and business interests. Over the years, he served as president of the Harrisburg Gas Company and the Harrisburg Boiler Manufacturing Company and was a director of the Harrisburg City Passenger Railway Company, the Harrisburg Burial Case Company, the Kelker Street Market Company, and the Harrisburg Bridge Company, among others.

Only thirty-six years old when first elected president, Reily presided over the affairs of the bank until his death in 1892.[34] His son-in-law, Edward Bailey (son of ironmaster Charles L. Bailey), succeeded him and held the post until 1927. Similarly, the cashiership was stable, with only three men holding the post between 1865 and 1900. James W. Weir, first appointed in 1844, served until his death in 1878; James Uhler served until his resignation in 1892; and William L. Gorgas served until well into the twentieth century.

Under steadier leadership, the duties of the Harrisburg National's board of directors diminished, becoming more routine and sometimes only perfunctory. In 1876 the membership of the board was cut from thirteen to nine. Until 1841, directors had been eligible for reelection indefinitely, and many died in office during long stints of service. Between 1842 and 1884 the bylaws permitted only the president to sit on the board for more than three consecutive years. Many directors, however, following a single year's absence, won reelection. After 1884, continuous reelection was once again permitted and soon became the rule.[35] Of the thirty-two persons who served on the board between 1865 and 1900, twelve put in totals of between fifteen and thirty-six years. Throughout the nineteenth century, the great majority of the directors were merchants, businessmen, and professionals. Beginning in 1853 a few leading industrialists made it to the board. That year, William Calder and David Fleming of the Harrisburg Car Works became directors. Calder served only until 1856, when he became a partner in the banking firm of Cameron, Calder & Eby, and later, from 1874 and 1880, became president of its successor firm, the First National Bank. Fleming sat on the board for nine years between 1853 and 1873. Charles L. Bailey, president of Central Iron, put in nineteen years as a director of the Harrisburg National between 1865 and 1889.[36]

James W. Weir, the cashier, took effective charge of the bank during the period of transient presidents. He continued to dominate until his death in 1878, exercising greater authority than his official superiors. Weir sat on the board for nine of the ten last years of his life, something none of his predecessors had done. It was Weir who initiated ideas, generated reports, and made recommendations that were routinely approved by the board throughout the late 1860s and 1870s. For example, on January 27, 1864, he reported to the board that he had bought up enough U.S. bonds ($100,000 worth) on his own authority to meet the legal requirements needed to obtain a federal charter. The reason he gave was that it was the last week the bonds would be available for purchase. The board commended Weir for "his perseverance and efforts" and for being "always on the alert, watching and doing the best for the interests of the stockholders."[37]

Weir transacted much of the Harrisburg National's business with and through the Philadelphia banking house of Jay Cooke & Company. When that firm collapsed in the Panic of 1873, he persuaded the directors of the Harrisburg National to apply $25,000 from the contingency fund to reduce the indebtedness of Cooke & Company to the bank by 75 percent. As he pointed out, much of the Harrisburg bank's profitability since 1861 had çome from dealings with Cooke. For example, it had earned $57,507 in interest on balances deposited with the Philadelphia firm. During the war it earned more than $74,500 in commissions on government bonds obtained through Cooke & Company, and since the war it had earned $142,518 on a number of exchanges of its whole capital stock for various U.S. bond issues, again with guidance from Cooke's bank. These profits, in turn, had led to other income: profits on gold and premiums on government bonds.[38]

The last years of Weir's cashiership found the bank slowing down. Largely because of the Panic of 1873, its total assets in 1875 stood 17 percent below what they had been a decade earlier. President Reily was not yet providing much leadership. He may have lacked support, not being the enthusiastic choice for president at the time of his election. It had taken twenty-one ballots to elect him, and even then the vote had been only seven to five in his favor. Reily, however, quickly assumed command following Weir's death. He had the bylaws revised, making the president the chief executive officer and fixing responsibility in him for all moneys, funds, and valuables. The cashier's responsibilities were limited to the matters specifically entrusted to him.[39] Reily soon consolidated his position and by 1884 was, in effect, handpicking the directors. A loose tally sheet in the minutes shows that the directors elected in 1889 each received 1,814 votes. Reily cast 1,179 votes (including proxies), the other ten shareholders present casting a combined 635. In twenty-one years, Reily missed only five weekly meetings of the board.[40]

Although the Harrisburg National once more advanced under Reily, rival firms increased in number and moved forward even more rapidly. In 1885, for example, the Harrisburg National Bank remained first with assets of approximately $1,570,000. The assets of the Dauphin Deposit ($1,280,000) and the Mechanics' ($1,060,000), however, were not far behind. The city's other banks included the First National ($896,000), the Commonwealth Guarantee ($730,000), the Farmers' ($546,000), and the Merchants' ($200,000).[41]

The year 1892 brought another complete change of leadership. President Reily died in February, only a month after Cashier Uhler's resignation. In a single ballot, the board unanimously elected Reily's thirty-year-old son-in-law, Edward Bailey, as successor. The new president had been in office little more than a year when the Panic of 1893 struck.

It was a difficult time for banks, one calling for stringent measures. The Harrisburg National promptly cut back on long-standing loans, including loans to relatives of bank officers and directors. Notes or lines of credit in excess of $10,000 had to be approved in advance by the board. Further loans were denied to the financially troubled Harrisburg Foundry and Machine Shop, among others. In 1894 several customers were asked to begin reducing large debts to the bank. The property of one prominent debtor, a near relative of the former president, was taken by the bank and sold to make good his debt. Adding to the difficulties were rumors that the bank's condition was not healthy. In December 1894 the board hired a detective to run down the source of those stories, without success.[42] When the depression at last ended in the late 1890s, the bank returned to prosperity that carried it into the next century.

Throughout the post–Civil War period the Harrisburg National Bank continued to play a major role in the development of the community, mingling good business with good works. Not only did it provide financing for Harrisburg industries, it also supported the installation of new transportation facilities and periodic reconstruction of existing ones, loaned money to improve public utilities, and contributed to a variety of public enterprises. Just as its large loans to various ironworks had helped build the Paxton Furnace, the Central Iron Works, the Harrisburg Car Company, and the Foundry and Machine Works, so it contributed to the fund that brought the Pennsylvania Steel Company to the area in 1865. When the Harrisburg Bridge burned in 1867, the bank subscribed to 240 $4 shares for rebuilding it. Throughout the last three decades of the century, it bought city bonds to finance the paving of streets, the laying of sewers, and the building of schools and parks. In 1861 the City Passenger Railway Company formed to provide local transportation. The bank helped the company through numerous difficult periods until 1891, when it abandoned horse-drawn cars and sold its franchise to a firm that installed one of the nation's early electric trolley systems.[43]

The minutes book of the Dauphin Deposit Bank by 1865 provided only the barest record of that company's affairs. When the president, James McCormick, died in 1870, his eldest son, Colonel Henry, succeeded him; his son-in-law, J. Donald Cameron, replaced him on the board of directors; and his other son, James Jr., remained cashier. For reasons unknown, the McCormick Estate Trust, which controlled all McCormick interests after 1870, decided to allow the bank's state charter to expire in 1874. Accordingly, the bank was sold and closed, but then promptly reopened as a private bank belonging to the McCormick Trust, with James McCormick Jr. as president. Apparently the change was all but unnoticed by customers, who continued to bank as before in the same building.

The change was no doubt important to the McCormicks. Not being chartered meant, among other things, freedom from regulation. This provided greater "flexibility" at a time when Henry McCormick was vastly expanding his various iron properties and Donald Cameron was moving into his father's seat in the U.S. Senate. The nature of that flexibility is suggested by what happened when the Dauphin Deposit again became a state bank in 1905. The state banking commissioner, on checking the firm's first statement, observed that aggregate loans to directors exceeded the legal limit. That problem was solved by Cameron (who had borrowed a large sum) resigning from the board and being replaced by his son-in-law, J. Gardener Bradley.[44] That the bank also thrived in its institutional capacity is evidenced by the fact that in 1885 its total assets were 80 percent as great as those of the Harrisburg National Bank; in 1864 they had been but 53 percent.[45]

Along the way the leading banks became enmeshed in the rivalries among the leading industrialists. The Harrisburg National, once controlled by the Haldemans, in the post–Civil War era fell under control of the descendants of the wealthy physician Luther Reily and the Baileys, with whom they intermarried. The McCormicks and Donald Cameron had controlled, if not outright owned, the Dauphin Deposit since the 1840s. Simon Cameron, William Calder, and Jacob Eby founded, owned, and managed the First National. The precise relationships of the banks, the manufacturers, and their industries in the last half of the nineteenth century is not always clear. The banks obviously provided the industries with funding in periods when they were expanding, loans to carry them through bad years, and a variety of discounting, credit, and checking services. But they also provided the entrepreneurs with business information, investment opportunities, flexibility in handling their riches, a steady income from bank dividends, and power to thwart rivals, real or potential.

The Crisis of the 1890s

The stringencies of the 1880s and the Panic of 1893, which the Harrisburg banks survived, were somewhat harder on the city's industries. All were weakened, and two major firms passed to new owners even before the Panic got under way. It was perhaps appropriate that the cotton mill, the city's first manufacturing plant, was also the first to pass to outsiders for conversion to other work. So long as William Calder was in control, his personal wealth kept the firm's credit rating solid in spite of low earnings. "Not a very profitable bus[iness] but keep the factory running and are reg[arde]d perfectly relia[ble]," R. G. Dun & Company's informant reported in May 1875. Following Calder's death in 1880, control of the mill passed to one George Calder of Lancaster.[46] In the mid-1880s it was sold to the New York firm of Pelgram & Meyer. The new owners produced ribbons and dress silk and operated large mills in Paterson and Boonton, New Jersey. They promptly converted the Harrisburg mill to silk, which was shipped to New York City. At the time the plant reopened in January 1887, the owners anticipated needing a work force of nearly 1,000. That hope proved to be extravagant. The mill actually employed only half that number, or fewer.[47]

The Harrisburg Car Manufacturing Company went into bankruptcy. Calder had been its president as well as head of the cotton works. Fleming, who succeeded Calder at the car works, died in 1890. In addition to leadership problems, the firm never fully recovered from its financial difficulties of the mid-1880s. Conditions appeared to be improving by 1890, but in fact they were not. The company's assets consisted chiefly of orders for rolling stock amounting to $1.2 million from the Pennsylvania and other railroad companies. Those orders, plus others for a recently developed refrigerator car, seemed to justify expansion of the car shop to make the new product. Just as the testing of the refrigerator cars began, the final blow came. The sudden failure in 1891 of Baring Brothers in London, a major dealer in American railroad securities, touched off a new crisis in financial circles. Many of the car company's debtors, unable to meet their obligations, canceled unfilled orders and defaulted on the rest. Hildrup devoted every cent he had and all he could borrow to save the business. In the end the company collapsed, wiping out most of its shareholders' investments and completely breaking Hildrup, who was nearing seventy.[48]

Even as the car works was in the bankruptcy courts, a new firm began in its ruins. Faulty refrigerator cars with leaking pipes were coming back to the dying company for repairs. Three young men with nothing to lose—Hildrup's son, William T. Jr.; his classmate in engineering at the University of Pennsylvania, David E. Tracy; and J. Hervey Patton, a purchasing clerk in the old foundry—saw a way to turn a profit. They raised $300, founded the Harrisburg Pipe & Pipe Bending Company Ltd., and began repairing refrigerator cars. In the new century their enterprise would succeed admirably.[49]

The longtime rival Bailey and McCormick iron operations merged in 1897. The new firm, the Central Iron & Steel Company, had a capital of $1 million. Despite their history of emnity, union made sense. Both rolled iron and steel plate, and both depended on the same source for the slab steel they used—the Pennsylvania Steel Company.[50] Each brought particular assets to the combine. Central Iron contributed its new universal rolling mill, Paxton brought furnaces for making pig iron and an excellent shearing plant. Under single management, greater economy of operation and an easing of local competition might reasonably be expected. The Baileys, who had already bested their rivals at making nails and rolling iron, dominated the joint enterprise. Also, their guiding genius, Charles L. Bailey, for the moment remained vigorous.

McCormick and Cameron interests held 3,575 shares in the new firm and occupied four seats on the board of directors. The Baileys and their allies controlled 5,525 shares and held five seats. Charles L. Bailey assumed the presidency of the company; his brother-in-law, Gilbert M. McCauley, who had been general superintendent of the Central Iron Works, continued on as general manager and treasurer; and James M. Cameron, son of J. Donald Cameron, became vice president. These, however, were only the initial changes. Charles Bailey died suddenly in September 1899, Colonel Henry McCormick died ten months later, and Gilbert McCauley passed away in 1901. Control went on to the next generation, with Bailey's sons filling the important vacancies. Although there would be shifts in management from the Baileys to the McCormicks, the enterprise would last another half-century.[51]

Harrisburg's other ironworks had already disappeared. The city's first anthracite furnace, built by Porter & Burke, had gone out of blast in 1875 after twenty-five years of service. The Harrisburg Car Company purchased the building in 1881 and put it to other uses.[52] The Wister furnace had an even shorter history. Erected in 1867, it was demolished in the 1880s to make way for a viaduct over the Paxton Creek valley to carry the Philadelphia & Reading Railroad into the city.[53]

Even the Pennsylvania Steel Company at Steelton faced difficulties during the Panic. During the 1880s and early 1890s it had invested heavily in plant expansion and modernization: the replacement of the original Bessemer facility with a new one, the erection of two giant blast furnaces, and the installation of a dozen new open-hearth furnaces. To convert foreign ore into pig iron at tidewater, it had constructed the massive new Sparrows Point Works near Baltimore in 1887. Then deciding that it was cheaper to make steel directly from ore by continuous process, the company spent additional millions converting Sparrows Point into a full-fledged steel mill. Burdened with an unmanageable floating debt, the management in April 1893 asked to be placed in protective receivership.[54]

At Hickok's Eagle Works the deaths of the founder's sons before their father created a different type of crisis. William O. Hickok had brought the boys into the firm as they came of age, but Edwin, the elder, did not get along well with his father and left. William O. Jr. proved more satisfactory, but he contracted a fever while hunting in swamps along the

Susquehanna and died in October 1881. After Edwin's death five years later, Hickok, then seventy-one years old, incorporated the firm. Of 5,000 shares of stock, he held all but 32 shares, which went to five dummy incorporators, the minimum number required by law. When Hickok died in 1891, control of the stock passed to his wife and grandsons.[55] Management of the firm, however, went to Christian W. Lynch, who presided over the company until the young men completed educations at Yale and returned to administer the family enterprise early in the next century.

Although it was not recognized in the 1890s, industrialization had already crested at Harrisburg. Neither the industries that survived intact and passed to the next generation nor those reorganized and taken over by outsiders would ever again grow at rates comparable to the 1860s or 1870s. At the same time, no major new industries were arising to replace those that had stagnated. If anything, industrial manufacturing was fragmenting. Craft production, though continuing, provided ever-reduced incomes for its practitioners. The decisions of local factory managers and owners often fell short of providing for the long-term survival of their firms. However, by century's end, forces beyond their control were rendering the city's industries obsolete and noncompetitive. Harrisburg's economic base in the twentieth century would be in nonindustrial sectors.

The Process of Industrialization

Urbanization and industrialization at Harrisburg fit into a much larger regional pattern. Pennsylvania's capital city was only one of many communities in the northeastern United States that evolved into medium-size manufacturing cities between 1830 and 1870. The economic histories of four of them—Albany, Reading, Trenton, and Wilmington—at least in broad outline, had close resemblances to developments at Harrisburg. However, all were older communities, all began to industrialize from ten to twenty years earlier, and all were larger than their counterpart on the Susquehanna. Settlement came earlier because they were nearer the Atlantic and had adequate-to-excellent natural waterways for transportation. Wilmington stood at the point where the Christina and Brandywine rivers joined the Delaware. Trenton and Albany were, respectively, at the heads of navigation of the Delaware and the Hudson rivers. The somewhat shallower Schuylkill connected Reading with Philadelphia, some sixty miles to the southeast. Like Harrisburg, each began as a river entrepôt, peopled by merchants, artisans, a handful of professionals, and many casual laborers. At each, agricultural and forest products from their respective hinterlands were assembled, processed, and exchanged for locally crafted tools, furniture, and the like or for manufactured goods from Philadelphia, New York, or Europe.[1]

As happened at Harrisburg, the coming of canals and railroads contributed to the growth of each and touched off or at least accompanied the onset of industrialization. Albany became the eastern terminus of the Erie Canal, and in 1856 the New York Central located railroad yards, a roundhouse, and repair shops there. The land between the Hudson and the New York Central became Albany's principal industrial district.

Reading was served by two canals: the Schuylkill, which improved navigation to Philadelphia, and the Union, which linked the Schuylkill River with the Susquehanna at Middletown. The Philadelphia & Reading Railroad, which connected those two cities in 1839, located yards at Reading and there built steam locomotives and cars for its lines. Reading's industries, though scattered, were sited along the city's various canals and railroad lines. The arrival of the Philadelphia, Wilmington & Baltimore Railroad in the mid-1830s created an industrial corridor for Wilmington between the Delaware and the railroad. As at Harrisburg, the new mode of transportation created a need for rolling stock, the manufacture of which became one of the city's leading industries. Canals soon supplemented the Delaware River at Trenton, furnishing the city with transportation and raw materials critical to manufacturing. The opening of the Delaware & Raritan Canal in 1834 placed the city on an all-water inland route between New York City and Philadelphia. Two decades later a feeder line of the Lehigh Railroad initiated a flow of anthracite to and through the city.

Like Harrisburg, none of the four was a single-industry town. Although each developed specialties, iron products were common to all (see Table 6.1). Iron stove production began at Albany in 1830 and was the city's principal industry by 1850. Foundry and machine products, bricks, and pianos, chairs and cabinets developed as industries after the Civil War. Manufacturing arose at Trenton due to the nearness of iron, coal, silica, flint, and clay. Ironmasters Peter Cooper and Abram Hewitt had operations in the city before the Civil War, as did John A. Roebling, who manufactured iron-rope. In the 1850s and 1860s, English potters set up kilns and made pottery and earthenware Trenton's distinctive industry. Textile production, especially woolens, followed in the post–Civil War years. Reading moved from small craft shops to factories beginning in 1836. That year, naileries, foundries, and machine shops began to appear. In 1849 Reading acquired a cotton mill (as did Harrisburg the same year), and in 1853–54 it had its first anthracite furnace. By the closing decades of the century, wool hats, long a local craft specialty, were factory-made, and after 1880, when cycling became a craze, several large bicycle factories opened. At Wilmington, following the arrival of the Philadelphia, Wilmington & Baltimore Railroad, carbuilding joined shipbuilding as a leading industry. Foundry products used by both gave rise to yet another main venture, as did the tanning of leather.

The common feature of the economies of these and probably many other manufacturing cities of the region was that their industries were by-products of, or heavily dependent on, improved transportation. With

Table 6.1. Leading Industries of Harrisburg, Albany, Trenton, Reading, and Wilmington by Percentage of Total Output, 1880

Harrisburg		Albany		Trenton		Reading		Wilmington	
Iron/steel	(37%)	Foundry products	(12%)	Pottery	(18%)	Iron/steel	(30%)	Iron/steel	(15%)
Flour milling	(8%)	Malt liquor	(12%)	Iron/steel	(18%)	Wool hats	(9%)	Shipbuilding	(15%)
Printing	(6%)	Boots/shoes	(10%)	Woolens	(7%)	—		Leather	(14%)
Railroad cars	(?)	—		—		—		Railroad cars	(?)

SOURCE: *1880 Census, Social Statistics of Cities*, Part 1, pp. 458 (Albany), 731 (Trenton), 767 (Harrisburg), 880 (Reading), and 909 (Wilmington).

NOTE: Because railroad-car building was done by single firms in Harrisburg and Wilmington, output, though large, was included under "all other products."

few exceptions, they were typical of second-stage, derivative industrialization rather than primary innovative industrialization. Their factories were moderate in scale and locally owned and served local and regional markets. Much of their capital and most of their raw materials came from the immediate area. With a variety of products made in several moderate-scale firms, they represented the golden age of what has been called "civic capitalism."[2]

But to narrow our focus once more, what is to be said of the process of industrialization at Harrisburg? Certain things stand out. The new mode of production was slow in coming to Harrisburg. The simultaneous arrival of the Pennsylvania Railroad, and an apparent recognition by local entrepreneurs that manufacturing held more promise of profit than further commercial ventures, seem to have provided the initial impetus. Although large-scale factory production marked a radical departure, it was at the same time a direct outgrowth of the building of the infrastructure. Not only did improved transportation, especially major railroad lines, make large-scale manufacturing possible in the community, it also prompted the rise of firms that serviced railroads and produced their rolling stock. Further, several of the entrepreneurs responsible for bringing canals and railroads to Harrisburg provided much of the leadership and funding for the new industrial firms. Although local business leaders organized and financed these enterprises, the essential technology came almost entirely from outside. Whether early industrial plants succeeded or failed depended more on funding and timing than on the technological know-how of their founders. However, as throughout Harrisburg's history, geography fixed the ultimate limits of economic development. The area's location, topography, and natural resources invited the manufacturing that arose. But the same forces worked against those industries when the scale of manufacturing shifted to serving national and international rather than local or regional markets.

The decade or more delay in industrialization at Harrisburg stemmed in part from the area's lack of waterpower. Although the town was located between the Susquehanna River and Paxton Creek, aspiring industrialists in the 1820s and 1830s would have found the river far too wide to dam. On the other hand, Paxton Creek, meandering slowly through a swampy tract at the borough's eastern limit southward to the Susquehanna, did not even support millponds for gristmills or sawmills. It could never have provided the power industrial plants needed. Such plants became possible only with the advent of steam engines fueled by coal.[3]

Adequate power alone would not have been enough. Factory production also required certain other previous local developments: the founding and peopling of the community; the rise of agricultural surpluses sufficient to support nonagricultural activities; a flow of raw materials from the farms, forests, mines, and mills of the extensive hinterland to the north and west; improved transportation facilities to support growing commerce; an amassing of capital and a banking and credit system; the existence of a pool of workers either in the community or available for moving to the area; and the appearance of persons of talent and means willing to introduce and finance the new methods of production.

By 1849 all these factors were in place. Exactly what, beyond a quest for profitable investments, moved the town's entrepreneurs after that date to shift from banking and railroading to manufacturing is not entirely clear. Successful examples of manufacturing in nearby communities (cotton mills at Lancaster and anthracite iron furnaces at Columbia, Danville, Lancaster, Lebanon, and Reading) no doubt provided stimulation. For some, the arrival of the Pennsylvania and Northern Central railroads suggested the need for plants to produce rails and rolling stock. The Bailey brothers may have decided to set up their rolling mill and nail factory at Harrisburg in 1853 because Porter Furnace was available to supply the pig iron they would need. Only Hickok, who had no money to invest, simply expanded his small shop to industrial proportions as he found the means.

The advent of industrialization at Harrisburg was simultaneously revolutionary and evolutionary. On the one hand, it changed drastically the way goods were made. Prior to the coming of factories, production was small-scale and carried on in small shops with only a few employees using hand tools or simple machines powered by wind, water, or muscle. Local customers directly consumed most of what was produced. The 1850 manufacturing census shows that nearly 55 percent (by value) of Harrisburg's preindustrial manufacturing was devoted to consumer goods—food and drink, clothing, furniture, and the like—as opposed to producers' goods, such as lumber, iron, and leather (see Table 6.2). Furthermore, these firms employed three-quarters of the workers engaged in preindustrial manufacturing. The makers of producers' goods, on the other hand, employed nearly two-thirds of the capital invested in craft-type production. After mid-century, both old and new products were produced on a far greater scale than before, and increasingly by steam-driven machinery. In turn, the new and larger plants and sophisticated machinery required a greater magnitude of capital investment and an expanded labor force.

The 1850 manufacturing census points up the change at its very beginning. That year only one industrial-scale enterprise, Porter's & Burke's

Table 6.2. Industrial and Preindustrial Manufacturing in Harrisburg, 1850

	No. of Firms	Annual Product Value	Capital Invested	No. of Workers
All firms	46	$316,960	$251,450	324
Preindustrial firms	45	$236,460	$128,450	212
Percent of all manufacturing		74.6%	51.1%	65.4%
Consumer goods				
Bake/confection shops	7	$24,313	$5,400	26
Furniture shops	3	15,330	5,600	20
Shoemaking shops	6	14,000	3,350	31
Hatter shops	2	12,500	3,300	9
Tailor shops	4	12,500	4,900	21
Tin shops	2	11,800	3,200	8
Chandler shops	3	10,975	6,500	5
Breweries	2	9,075	10,000	6
Shoe "factory"	1	8,000	1,000	25
Cigarmaking shops	4	6,648	1,750	10
Saddlemaking shop	1	2,500	700	2
Total	35	$127,641	$45,700	163
Percent of preindustrial	—	54.0%	35.6%	76.9%

Table 6.2. *Continued*

Producers' goods			
Iron rolling mill	1	$45,000	$25,000	15
Sawmills	3	38,500	41,000	17
Tanneries	4	15,119	4,550	10
Burrmaking shop	1	6,000	10,500	4
Windowsash shop	1	4,200	1,700	3
Total	10	$108,819	$82,750	49
Percent of preindustrial	—	46.0%	64.4%	23.1%
Industrial firms	1	$80,500	$123,000	112
Percent of all manufacturing	—	25.4%	48.9%	34.6%

SOURCE: Computerized manufacturing census schedules, 1850.

blast furnace, was in operation. The cotton mill was still under construction. The contrast between the blast furnace and Harrisburg's other manufacturing firms clearly demonstrates the differences between the coming new industrial order and traditional manufacturing. That single firm accounted for almost half the town's total capital investment in manufacturing, a quarter, by value, of all goods produced, and more than a third of all workers engaged in manufacturing.

Industrialization was also part of an ongoing process in Harrisburg. A number of preindustrial firms were already moving to larger-scale production. If "industrial" firms are defined as enterprises employing at least twenty-five wage earners, using centralized power (in Harrisburg this meant steam engines), and having both substantial capitalization and production, only Porter's Furnace qualified.[4] Nonetheless, though far smaller than the furnace in these respects, four firms had between ten and twenty-five employees: a tailor's shop and a chairmaking shop each had ten, a rolling mill had fifteen, and a shoemaking shop had twenty-five. Of the four, only the rolling mill operated with steam. At the same time, the rolling mill, three sawmills, and a sashmaker used steam power, but each had only from four to fifteen employees. The rolling mill and two of the sawmills each turned out products valued at more than $13,000, and the rolling mill, two of the sawmills, and a burr-making shop were each capitalized at $10,000 or more.[5]

Even the large industrial plants that suddenly appeared after 1849 did not spring from thin air. They represented important continuities from the past. The improving of transportation, the forming of banks, the nurturing of entrepreneurs, and finally the building of factories all melded together. The process was not a neat progression of distinct steps or stages, each with its own special characteristics and personnel. The infrastructure was not completed in one stage and industry begun in a second. Rather, large-scale industries began when there were sufficient railroads and banks to support them. But railroads, banks, and utilities continued to evolve and change side by side with the new factories and, in effect, had a symbiotic relationship with them. As for personnel, the chief entrepreneurs of the two stages overlapped. Six of the eleven who built Harrisburg's infrastructure were among the fourteen leaders who later brought industries to the community; four industrial entrepreneurs were sons of builders of the infrastructure.[6]

That improved transportation had to precede industrialization at Harrisburg was confirmed when the entrepreneurs located nearly all their factories in the transportation corridor along the valley of Paxton Creek. In fact, as soon as the canal was completed there in 1834, many of the larger preindustrial manufacturing plants (sawmills, iron foundries, rolling mills, breweries, and distilleries) located or relocated along its course; later, so did brickyards, pottery furnaces, and warehouses. After 1849, with the coming of the Pennsylvania Railroad to that same corridor, all Harrisburg's anthracite-fueled blast furnaces were erected adjacent to the transportation systems that brought them their fuel from the hard-coal district.[7] The same transportation systems also brought them the ore and limestone they needed, and carried any iron not used locally to markets elsewhere.

Without the railroads, the Harrisburg Car Works, the Lochiel Iron Company, and the Pennsylvania Steel Company would never have been built. After all, it was the railroads' need for rolling stock and rails that called those firms into being. Similarly, without railroad rolling stock to be serviced, the extensive repair shops of the Pennsylvania Railroad would not have appeared in the industrial corridor during and after the Civil War. Hickok's Eagle Works would probably have remained a small firm turning out cider presses and bookbinding supplies for local customers were it not for the railroads that gave its ruling equipment a worldwide market.

Of the major firms of the area not located in Harrisburg's industrial corridor, one, the nailery across the river at West Fairview, was built years earlier to take advantage of the fall of the Conodoguinet Creek for waterpower and at first depended on the Susquehanna for transportation. By the early 1850s, however, the Northern Central Railroad ran adjacent to its property. Thus, of all of Harrisburg's major industries, only the cotton mill came into being with no apparent concern for direct access to improved transportation.

The capital that financed Harrisburg's industrialization came from the personal fortunes of its entrepreneurs, investments by outsiders, stock purchases by the general public, and extensive loans from Harrisburg's banks. The money invested by the major entrepreneurs consisted variously of accumulated individual and family wealth (often the savings of more than a single generation), professional and business earnings, profits from government contracts, borrowed money, and most often a combination of those sources. Few of the entrepreneurs limited themselves to single lines of business. As already seen, most were involved in numerous activities. Of those who built the infrastructure, Elder, Harris, and Forster were sons of the area's earliest major landowners and inherited at least part of their wealth. The elder James McCormick inherited lands from his father in Cumberland County. Haldeman began his iron furnace in that same county with a sizable loan from his father, a prosperous Lancaster County farmer. The Baileys came from a family of successful Chester County ironmasters. Four of the younger entrepreneurs,

William Calder Jr., Colonel Henry McCormick, the younger James Mc-Cormick, and J. Donald Cameron, were sons of older entrepreneurs. Reily started with capital inherited from his father, a physician with extensive real estate holdings. Government contracts added to the fortunes of several. William Calder Sr. carried the U.S. mail on his stagecoaches and canal packets, Simon Cameron profited from state contracts for printing and building sections of the Pennsylvania Canal, and Burke made his early fortune as a canalbuilder in both New York and Pennsylvania. Government contracts during the Civil War swelled the holdings of Calder Jr., the Camerons, Eby, and Hildrup, among others. Several were lawyers: Ayres, Elder, Fleming, Forster, Harris, and both James McCormicks. The Camerons, and to a lesser degree Harris, were active in national politics.

Harrisburg's infrastructure of banks, transportation companies, and utilities, because they were engaged in quasi-public functions, were incorporated. This meant that part of their capital came from stock bought by the general public. Not infrequently, state and local governments contributed to such firms, especially to transportation companies. Those that were financed from their own private resources included the senior Calder's stagecoach and canal packet lines and Cameron's and Burke's canal construction firms. The capital for those undertakings, in turn, came from earlier business successes, reinvested profits, and loans secured by public contracts.

Similarly, the capital used to start the principal industrial firms that appeared at mid-century came from a variety of sources. The cotton mill began with capital raised through the widespread sale of its stock in the community, supplemented by extensive bank loans and what might best be described as short-term credit from Senator Charles T. James of Rhode Island, who sold the idea of the mill to Harrisburg entrepreneurs and supervised its construction. The pioneers of Harrisburg's iron industry, Hunt & Son, Pratt & Son, Porter & Burke, and Bryan & Longenecker, relied on their own resources from earlier enterprises and on extensive bank loans. All soon failed and sold out to others. The elder James McCormick, with his enormous personal resources (a thriving legal practice, extensive real estate holdings, and easy access, as president, to the resources of the Dauphin Deposit Bank) bought up and expanded many of the properties that the earlier ironmasters lost. When the Baileys entered the iron business in Harrisburg, a third of the initial capital was money they had earned in previous enterprises; two-thirds was provided by a silent partner from Philadelphia. The Harrisburg Car Works began with $3,000 investments by each of eight Harrisburg businessmen. The one underfinanced enterprise to succeed was that of William O. Hickok. Wiped out first by a fire and later by business reverses, he used a series of partnerships and loans (some of which were "settled" rather than repaid) to get a toehold. Although twice sold out at sheriff's sales to satisfy creditors, Hickok's perseverance and the faith of others in his potential kept him going. About a decade after its founding, with support from James McCormick, his enterprise began to thrive.

Timing played an important part in the success or failure of most of Harrisburg's industries. The very first builders, those who promoted the cotton mill and who built the first anthracite furnaces, were victims of poor timing. The cost of raw cotton rose, and the price of cotton textiles fell just as the cotton mill began operations. A depression in 1857 further set the firm back, and the outbreak of civil war cut off its supply of raw materials. None of Harrisburg's supposed other advantages, such as a lower cost of living and being nearer to raw cotton, proved to be beneficial. Similarly, several of the major new ironworks failed during the Panic of 1857. The owners found themselves unable to service their construction loans, much less borrow additional funds to get into operation. Lochiel's appearance at the end of the Civil War to manufacture iron rails represented unusually poor timing. The shift to steel rails outmoded the firm even as it was starting.

By contrast, fortunate timing helped others. McCormick, for example, bought up the distressed furnaces and other iron properties during the panic years for low prices, just in time to take advantage of the wartime market for iron. The Baileys too, after a shaky start, benefited from wartime and postwar demands for iron. The Harrisburg Car Company survived the prewar panic to find that it had difficulty meeting the demands of the rapidly expanding railroad network for rolling stock, and, with the great expansion of industries nationwide, the unparalleled demand for record-keeping created a flood of orders for business forms that could be produced by Hickok's ruling machines.

The two major new enterprises that arose at the close of the Civil War, the Lochiel Iron Company and the Pennsylvania Steel Company, required immense investments of capital. Their founders tapped similar sources. The Harrisburg residents behind Lochiel combined large personal investments ($150,000) with outside funds ($100,000) and extensive bank loans to build the large facility. When reorganized with a capital of $300,000 in 1872, locals provided only $84,000, with $216,000 in capital coming from outside interests, including Thomas Scott of the Pennsylvania Railroad. Scott, Thomson, and other Philadelphia capitalists provided two-thirds of the Pennsylvania Steel Company's capital, while the Pennsylvania Railroad supplied the rest. Harrisburg residents also bought up and donated the land on which the plant was built and purchased a substantial block of stock.

The preferred business form of the first industrial establishments was the joint co-partnership. The Baileys' Central Iron Works and Chesapeake Nail Works began as partnerships, as did the Harrisburg Car Works. Apparently McCormick and Charles Bailey were partners in the West Fairview Nail Works before 1866, and the McCormick Estate Trust that controlled the various family properties after 1870 described itself as a joint co-partnership. Of Harrisburg's earliest industrial firms, only Hickok's Eagle Works was a single proprietorship, and the cotton mill was the only corporation. Over the years the major firms all incorporated: the Harrisburg Car Manufacturing Company in 1863, the Central Iron Company in 1866, the W. O. Hickok Manufacturing Company in 1886, and the McCormick Company (owned in turn by the McCormick Trust) in 1887. The two major post–Civil War firms, the Lochiel Iron and the Pennsylvania Steel companies, began as corporations. However, with the exception of the cotton mill, all these corporations began with and usually continued to have only a handful of stockholders each. Despite the corporate form and the legal advantages of incorporation, their close-knit operating relationships remained very like partnerships.

Apparently, adequate capital was of greater importance to industrial success in Harrisburg than technological know-how. Most of the major entrepreneurs themselves had no technical skills, and the technologies they employed were imported from elsewhere. The mechanical innocents who founded the cotton mill relied on outsiders who were familiar with textile manufacturing (Senator James and others) to run the business until locals could be trained. The founders of the car works knew only that they needed someone with carbuilding expertise, so they sought out Hildrup and made him a partner in the firm. Porter, who built the first blast furnace, did have considerable experience with charcoal furnaces but promptly failed because of financial difficulties. The elder James McCormick, a lawyer and banker, knew little about iron manufacturing; until his son Henry completed an apprenticeship in furnace operations, he relied on the Baileys, who had considerable experience in iron production. Later, capitalists, not ironmakers and steelmakers, founded the Lochiel Iron Works and The Pennsylvania Steel Company. They simply hired the experts they needed to operate their firms. The Baileys and Hickok were the only skilled technicians to found major industries in Harrisburg, and Hickok, despite his mechanical genius, came near to failing for lack of adequate funding.

Perhaps more important, with the possible exceptions of the Eagle

Works and the car works, the Harrisburg enterprises did not operate at the cutting edge of technology in their fields. Most simply followed the lead of others. Obviously Senator James was promoting a new technology for manufacturing textiles by steam-driven looms, but the Harrisburg Cotton Mill was not one of his early projects, nor was the technology notably successful until widely employed on a much greater scale in southeastern Massachusetts after the Civil War.[8]

A lack of technological innovation characterized even iron and steel where Harrisburg had pretensions of making its mark. Although the furnaces, rolling mills, naileries, and other iron facilities installed the latest equipment when built, they did little pioneering. For the most part they duplicated what had been or was being done by others in the region. The introduction of anthracite blast furnaces, for example, represented the adoption of an already widespread technology; twenty-six such furnaces had been erected in Pennsylvania before 1847, and by 1850 there were fifty-six, including Porter's.[9] Rarely did Harrisburg entrepreneurs attempt to expand the frontiers of iron technology. Their furnaces were of average size and capacity and introduced no new processes. To be sure, the Wister brothers experimented with mixing coke and anthracite to fuel their furnace in the 1860s, but only because others reported favorable results from the process.[10] Henry McCormick, in 1866, was one of only four Pennsylvania ironmasters to experiment, briefly and unsuccessfully, with the Clapp-Griffith converter for making steel. From time to time Harrisburg's mills updated the machinery in their plants, but usually only after they found themselves falling behind the competition, not because they sought to set the pace for the industry.[11]

None of this prevented Harrisburg's chief industries from growing rapidly for most of a generation. Continuing to use the standard of twenty-five or more employees, steam power, and substantial capitalization and value of output, only the big-five firms fully qualified as industries in 1860. The cotton mill, car works, Bailey's iron properties, and McCormick's Paxton facilities each produced goods valued in excess of $100,000, hired between 40 and 140 employees, were capitalized at $60,000 or more, and used steam power. The Eagle Works had forty or so employees and used steam power, but had much less capital ($20,000) and output ($32,000). Several other firms came close to the industrial standard. Seventeen operated with steam power (breweries, distilleries, a farm equipment facility, small iron and brass foundries, machine shops, planing mills, printing plants, and a tannery), but none had more than eighteen employees. Two firms, both brickyards, had sufficient employees but used horses rather than steam engines. Seven firms (three distilleries, two planing mills, a farm equipment firm, and a tannery) were capitalized at $20,000 or more, and four produced goods valued at no less than $20,000 (two distilleries, a planing mill, and a machine shop) but either lacked sufficient employees or did not use steam power.[12]

A decade later, the big five had expanded their own holdings and grown to seven in number with the addition of Lochiel Iron and the Pennsylvania Railroad Shops.[13] By then nine other firms also qualified as industries: the anthracite furnaces of Wister Brothers and Price & Sharp (formerly Porter & Burke), three printing establishments, a machine shop, a farm equipment manufacturing plant, a steamboiler manufactory, and a planing mill. Some of these had appeared in the 1860 census as small traditional shops but had meanwhile grown to industrial proportions; others were new enterprises. Although falling short of industrial status because of too few workers, there were another thirteen firms that employed steam power: three breweries, two sawmills, two flour mills, two foundries, a fire-brick yard, a bakery, a machine shop, and a coachmaking firm. Another five firms (three brickyards, a planing mill, and a shoemaking firm) had more than twenty-five employees but did not use steam.[14]

Over the two decades, the five (and then seven) major industrial establishments increased their hold on Harrisburg's total manufacturing. Porter & Burke in 1850 had produced a quarter of all manufactured goods. By 1860 the big five produced more than half, and by 1870 the big seven turned out nearly two-thirds of all goods manufactured in the city. This output was accomplished with just under half the total capital invested in manufacturing in 1850, a little over half in 1860, and almost exactly half in 1870. Porter & Burke in 1850 employed about a third of all workers in manufacturing. The share of the big five in 1860 approached half, and that of the big seven in 1870 was nearly two-thirds. The same firms in 1860 controlled more than two-thirds of the total horsepower generated by steam, and in 1870 they controlled more than three-quarters. If the nine other industries in 1870 are included, industrial-scale firms controlled more than three-quarters of both output and workers, nearly two-thirds of total capital invested in manufacturing, and over nine-tenths of the steam power. A lack of data on individual firms after 1870 makes similar breakdowns impossible.[15]

Iron and steel production dominated manufacturing in Harrisburg throughout the second half of the nineteenth century (see Table 6.3). However, its position in the manufacturing sector peaked in the 1870s, when the value of its products reached nearly half of the total—more than two-fifths of capital investments—and when it employed a third of the manufacturing work force. Although iron and steel's share of capital investment exceeded half of the total in 1880, its portions of total output and labor declined. A decade later the industry's share of the city's

Table 6.3. Iron and Steel Industry's Share of Harrisburg Manufacturing

Year		Output	Capital	No. of Workers
1850	Total manufacturing	$316,960	$251,450	324
	Percent iron/steel	39.6%	58.9%	39.2%
1860	Total manufacturing	$1,256,954	$1,020,430	1,199
	Percent iron/steel	20.8%	27.5%	7.6%
1870	Total manufacturing	$7,073,508	$3,246,849	2,950
	Percent iron/steel	46.7%	44.2%	33.1%
1880	Total manufacturing	$7,663,508	$4,026,457	3,660
	Percent iron/steel	37.1%	50.2%	30.8%
1890	Total manufacturing	$10,538,444	$6,716,074	6,898
	Percent iron/steel	27.4%	27.5%	14.1%
1900	Total manufacturing	$16,064,597	$8,749,516	7,766
	Percent iron/steel	47.3%	35.0%	22.9%

SOURCES: Computerized manufacturing census schedules for 1850, 1860, 1870; *1880 Census, Manufactures*, Part 2, p. 404; *1890 Census, Manufacturing*, Part 2, pp. 235–41; *1900 Census, Manufacturing*, Part 2, pp. 778–79.

manufacturing output and capitalization had fallen to only little more than a quarter, and it employed but one in seven workers. Moreover, it had become largely subsidiary to the mammoth steelworks at nearby Steelton. Pennsylvania Steel bought most of the pig iron produced by the city's furnaces and supplied Harrisburg's mills with the raw steel they rolled. Although by the turn of the century the industry improved its share in all three categories, the gain was temporary and the industry never returned to its standing of 1870.

From 1880 onward, three factors complicate determination of the industrial structure of Harrisburg: the omission of three of the city's larger firms from the manuscript manufacturing census schedules in 1880, the unavailability of manuscript manufacturing census information after that date, and the Census Bureau's policy of listing data on single-firm industries only under the heading "All Others." This lumping-together makes the category unacceptably large and distorts the proportions of all industries. It also frustrates efforts to single out four of Harrisburg's larger firms: the old cotton mill and car works, and the new shoe factory and manufacturer of typewriters.[16] Because the car works did appear in the manuscript manufacturing census schedules for 1880, carbuilding's ranking as Harrisburg's second industry can be established that year (see Table 6.4). But because the cotton works was not included (it may have been closed temporarily because of the death of President Calder that summer),

Table 6.4. Ranking of Harrisburg Industries, After Iron and Steel, by Output, Capital, and Labor Force, 1880, 1890, 1900

	Rank by Output		Rank by Capital		Rank by No. of Workers	
	Industry	%	Industry	%	Industry	%
1880						
2nd	carbuilding	24.0	carbuilding	17.6	carbuilding	25.1
3rd	flour milling	8.1	lumber-milling	5.1	printing	6.6
4th	printing	6.1	gas producing	5.0	brickmaking	4.6
5th	machine manufacturing	2.3	printing	4.3	machine manufacturing	3.6
1890						
2nd	printing	3.1	machine manufacturing	4.6	brickmaking	3.7
3rd	carpentering	2.2	printing	3.6	printing	3.3
4th	brewing	2.2	brewing	3.0	machine manufacturing	2.4
5th	machine manufacturing	2.0	brickmaking	2.4	carpentering	2.4
1900						
2nd	machine manufacturing	5.2	machine manufacturing	14.5	shoemaking	10.1
3rd	shoemaking	4.7	shoemaking	7.3	machine manufacturing	8.0
4th	cigarmaking	3.3	printing	7.1	cigarmaking	7.8
5th	printing	2.8	brickmaking	3.1	printing	4.9

SOURCES: Computerized manufacturing census schedules for 1880; 1880 Census, Manufactures, Part 2, p. 404; 1890 Census, Manufacturing, Part 2, pp. 235–41; 1900 Census, Manufacturing, Part 2, pp. 778–79.

the ranking of cotton manufacture is lost in the "All Others" category. Similarly, after the mill shifted to silk manufacture, its rankings in 1890 and 1900 cannot be determined.

Despite these difficulties, it is evident that no other industry became sufficiently significant to fill the gap left by the decline of iron and steel production. Even as it faded, iron and steel continued far and away to lead the city's other industries well into the twentieth century. In the absence of any new major industry, the older crafts, such as printing or making bricks, machinery, or shoes, tended to evolve into industries or to move to factory-size units. The manufacture of cigars included both traditional craft shops and small factories. Carpentry, although carried on in the old manner, sometimes involved construction firms with as much output and as many employees as factory industries.

In spite of the decline of iron and steel, at least prior to 1900, the percentage of people listed as working in manufacturing did not drop. However, probably no more than between 50 and 60 percent of them worked in factories employing twenty-five or more workers. The rest continued working in traditional craft shops or small new shops that used power-driven machinery, once again demonstrating how long craft production persists in an industrialized community.[17]

Several factors contributed to the decline of iron and steel. The Mc-Cormicks and Baileys blamed "cut throat" competition (first from Carnegie Steel and then from its successor, United States Steel) and unfair rates charged by the railroads.[18] In fact, the Harrisburg firms, by failing to grow as fast as their rivals, to secure control of the raw materials essential to production, and to expand into more lines of iron and steel production, put themselves at a disadvantage. Their rivals enjoyed economies of scale as they grew, lower costs as they acquired the sources of their raw materials, and wider markets because they could supply customers with a greater variety of products.

To a significant degree Harrisburg's geographic location turned from being an asset to being a liability. So long as anthracite was the chief fuel in iron and steel production, nearness to the hard-coal fields worked to the advantage of Harrisburg's firms. Once coked bituminous proved to be significantly more efficient, the advantage shifted to the area south and southeast of Pittsburgh. For many years the Harrisburg firms bought the coke they needed rather than acquiring bituminous mines and coking ovens of their own. Their primary source was the Frick Coking Company, the industry's principal producer. In 1881 Carnegie merged with Frick, thereafter getting his coke at cost and selling it, so long as it pleased him, to the Harrisburg firms at market price. Similarly, as Pennsylvania iron deposits were depleted Carnegie bought up and leased mining properties on the Mesabi Range in Minnesota to obtain the ore he needed at cost. The McCormick furnaces relied on local ores, which they owned or leased, and when those ran out ore was purchased at market price from abroad or from companies on the Mesabi.[19] As for transportation costs, Carnegie owned ore boats and exerted considerable pressure on the railroads to force advantageous rates from them. The Harrisburg companies had to rely on Great Lakes and Atlantic Ocean shippers and on railroad companies, over whom they had little influence. Keeping up technologically was also a factor. Throughout the nineteenth century the McCormicks and Baileys continued to make do with existing facilities, installing new equipment only when necessary to avoid losing customers. Over the years, content to go on rolling iron as they always had, they did not bother to diversify.

Just as Porter's blast furnace provided a useful contrast between industrial and preindustrial manufacturing in Harrisburg in 1850, the history of the nearby Pennsylvania Steel Company at Steelton points up the shortcomings of the iron and steel firms in Harrisburg after 1865. Pennsylvania Steel fought to remain at the fore of the industry by moving to serve the national market. The McCormick and Bailey firms seemed happy with earning returns on past investments, made little effort to keep up with their competitors, and continued to focus on largely local markets.

Although founded by railroaders and other users of steel, rather than by steelmasters, the new firm at Steelton pioneered in the production of steel rails. Its owners included some of the foremost railroad managers in the country, persons who recognized the importance of adequate financing to maintain thrust. Setting out with $200,000 capital, they quickly increased that to $603,000 by 1870 and to $2 million by the end of 1873. Though their capitalization remained unchanged in 1885, the firm's investment exceeded $5 million.[20] They hired the best available expert, Alexander Holley, to plan, build, equip, and superintend their plant. When he left, other top managers were brought in. Pennsylvania Steel kept abreast of the latest developments, often leading the field. It was among the first to produce Bessemer steel on a large scale and was the first to roll steel rails commercially. In 1876 it began installing blast furnaces so it would not be dependent on others for pig iron, and it subsequently added facilities for producing its own coke. It built the large Sparrows Point plant near Baltimore in 1887 to make pig iron for Steelton from foreign ore, but then converted it into a second steel-producing unit when making steel directly from molten pig was determined to be more efficient than reheating it later. When open-hearth furnaces proved to be superior to the Bessemer process in making steel, the Steelton firm began building and using the new furnaces in 1883—ahead of Carnegie, who installed open-hearth furnaces earlier but did not use them until 1888.[21]

The McCormick and Bailey enterprises fell far short of that standard. Although their combined capital exceeded that of Pennsylvania Steel through 1870, it steadily fell behind after that year. In 1860, five years before Pennsylvania Steel came into being, McCormick's iron properties, including the West Fairview nailery, were capitalized at $300,000; Bailey's were at $60,000. A decade later the McCormick iron interests (including Lochiel, then managed by Colonel Henry McCormick) represented a total capital of $1,069,000; the Bailey interests were capitalized at $470,000. When the two firms merged into the Central Iron & Steel Company in 1897, the capital of the new firm was only $1 million.[22] By then the McCormicks no longer managed Lochiel, and their other properties had settled into the routine production of pig iron, some of which was rolled into plate at Paxton or into nails at West Fairview; the rest was sold to Pennsylvania Steel. The Baileys restricted themselves to rolling plate and nails and depended on others for the pig iron they needed. In 1892, to acquire greater flexibility, they installed a universal mill that enabled them to produce a wider variety of plate sizes and to roll both iron and steel plate. The raw steel they used came from Pennsylvania Steel at Steelton. Not until 1904, following the merger of Bailey and McCormick interests, did Central Iron & Steel produce its first open-hearth steel, twenty-one years behind Pennsylvania Steel and sixteen years after Carnegie.[23]

For managerial talent and technological expertise, they relied on themselves, their immediate families, and a handful of hired managers. The McCormick interests depended chiefly on the skills and experience of Colonel Henry McCormick, who had learned the art of ironmaking in the 1850s. He was assisted by managers such as John Q. Denny and, from time to time, by sons and nephews who spent brief stints in the mills. Similarly, the Bailey operations were managed by Charles L. Bailey, who also acquired his expertise in the 1840s and 1850s, by his brother-in-law Gilbert M. McCauley, and, as they came of age, his sons. Although the next generation of McCormicks, Camerons, and Baileys took over from their fallen elders at century's end, they clearly lacked the zeal, drive, singleness of purpose, and resources needed to compete successfully with Carnegie's successors, Charles M. Schwab, Elbert H. Gary, and a score of other steel-producers. Worse, few of the younger Baileys or McCormicks had any interest in manufacturing, preferring instead

banking, politics, journalism, or real estate. Some avoided business altogether in favor of social, religious, and charitable activities; a few became gentleman farmers.[24]

For all their shortcomings, the industrial leaders of Harrisburg had
remarkable staying power. For a full century the entrepreneurial families
dominated the economic and social life of the city, long after manufacturing there faded. It had taken them less than the single decade of the
1850s to establish firmly the town's only major industries. Thereafter,
for the next fifty years, they expanded those enterprises while effectively
blocking outside industrialists from the city. The only important exception was the mighty Pennsylvania Railroad, over which they exercised
no control. Even if they wanted to, they could not have refused it the
land it needed for a right-of-way through the industrial corridor or
blocked its purchase of the canal and its right-of-way from the state.
Once the Pennsylvania Railroad dominated the corridor, it located its
extensive repair and maintenance shops in the heart of the district. And,
as noted, the railroad and its chief officers were behind the Pennsylvania
Steel Company a few miles outside the city proper.

Otherwise, to all intents and purposes, industrial Harrisburg consisted
of the big five (later big seven) firms. Several factors enabled this small,
close-knit band to retain its hold over the city and its industries. First
was their control of the principal banks. James McCormick acquired the
Dauphin Deposit Bank more than thirty years before he moved into
industry. The Baileys reversed the process, first establishing themselves
in ironmaking, then becoming managers of the Harrisburg National
Bank and intermarrying with the Reily family, who managed it during
the late 1870s. The Camerons, who already controlled the Middletown
Bank, acquired links to both the Harrisburg National and the Dauphin
Deposit by intermarriage with the Haldemans and McCormicks, and
then joined with Eby and the younger Calder to establish the First National Bank.

With banks to serve as financial balance wheels for their industries,
the entrepreneurs were firmly in control. For the most part, they financed the maintenance and expansion of their firms from reinvested
earnings. But bank loans were always available if needed. Rarely did the
industrialists bring in outside money, as when the Baileys sold stock in
Central Iron to New York investors. By self-financing, of course, they
preserved control of their firms. Their banks did far more than support
their factories with loans, however. They also gave the entrepreneurs
access to a steady flow of information about the state of business and
finance in the community and elsewhere, and advance notice of pending
real estate transactions.

How this benefited the entrepreneurs can be seen on detailed maps of Harrisburg showing ownership of undeveloped real estate in the city and on its outer reaches. Particularly important was ownership of major empty tracts in the industrial corridor and on the southern and eastern edges of the city, where any aspiring industrialists would have to set up their enterprises. Most of those tracts were tightly held by the entrepreneurial families and their allies. In 1850 the principal landholders in the industrial corridor were Thomas Elder, Henry Buehler (James McCormick's brother-in-law), Simon Cameron, John Forster, Michael Burke, and Jacob M. Haldeman. Haldeman and Forster also controlled the undeveloped property immediately above North Street. Many of the town's merchants and professional leaders owned lots in the residential sections of Harrisburg in addition to their own homes. Michael Burke was probably the most extensive landowner there, owning houses and lots in the center of the borough, and particularly in the district behind the Capitol, which was heavily inhabited by the town's African Americans.[25]

By 1871 undeveloped land in the corridor had largely passed to the Pennsylvania Railroad. However, the McCormick Trust, Simon Cameron, Jacob Eby, the Baileys, and the Reilys all owned tracts there. The 1860 additions to the city included considerable land to the north between Forster and Maclay streets and to the east from the canal (now Cameron Street) to Eighteenth Street. At the north, Jacob Eby held the northernmost thirty blocks (between Maclay and Muench streets). South of Eby's holdings, between the river and the industrial corridor, were some ninety-six city blocks belonging to the heirs of Dr. Luther Reily (George W. Reily Sr. and his sisters), a tract that included both vacant land and rowhouses under construction. The Forster family and Richard Haldeman owned less-extensive holdings in the area immediately north of the Capitol. In the additions east of the industrial corridor, large tracts were held by William Calder Jr.; his brother, the Rev. James Calder; and William T. Hildrup, Simon Cameron, the Bailey brothers, and Dr. George Reily.[26]

A detailed real estate atlas of the city for 1901 shows that the most important areas for development remained largely in the same hands. The Pennsylvania Railroad controlled rights-of-way for both the railroad and the canal—and virtually all the land between, from the northern limit of the city to Hanna Street in the far south. Below that point the corridor belonged to the McCormick Trust. Immediately to the west of the railroad, lands still undeveloped belonged to David Fleming, R. J. Haldeman, members of the Alrick family (in-laws of James McCormick Jr.), and the McCormick Trust. East of the industrial corridor the principal landholders included the Alricks, the Simon Cameron Estate, various members of the Calder family, J. Donald Cameron, and the McCormick Trust.[27] With ownership of most of the real estate that might be used for manufacturing in the city, and control of the banks, the entrepreneurs held a tight rein on the community's industrial future.

By the close of the nineteenth century, industrialization had swept through and drastically altered the processes of production in Harrisburg. Several large mechanized factories, with hundreds of employees producing massive quantities of producers' goods for regional markets, had replaced the small shops that formerly made goods almost entirely for local consumption. Having examined that process at length, the impact of industrialization on the work and the lives of the people of Harrisburg and their families—the entrepreneurs, the artisans, the factory hands, and the various minority groups—remains to be addressed.

II

The Impact of
Industrialization

Fig. 14. Harrisburg After Thirty Years of Industrialization, 1879. Thomas Hunter's "Harrisburg, Pa., 1879. View from Fort Washington."

A Generation Later

THE MOST OBVIOUS CHANGES brought by industrialization can be seen by again turning to sketches and maps of the city and to federal census schedules. Thomas Hunter's "Harrisburg, Pa., 1879: The View from Fort Washington" (Figure 14) once more portrayed the community from across the Susquehanna.[1] Because both the population and the area had grown remarkably, Hunter's perspective was broader and deeper than Whitefield's of 1846. Census returns in 1870 reported 23,105 residents, a 195 percent increase in twenty years. In another decade, nearly 31,000 persons would live in the city. To accommodate them, Harrisburg in that same period had expanded its area sixfold. It annexed tracts equal to the whole town of 1850 at both the north and the south, tripling frontage along the Susquehanna and, to the east, lands at least as extensive as the total area bordering the river.[2]

The grave of the area's first settler, John Harris, still a shrine at water's edge, was now part of a small park. The nearby stone mansion built by his son, the founder of the city, had been purchased, remodeled, and enlarged by Senator Simon Cameron in 1863. Tall monuments commemorating the Mexican War and the Civil War occupied places of honor on the grounds of the State Capitol and at the center of State Street at Second, midway between the river and the Capitol. Near the heart of the old town stood an imposing new courthouse with a classical portico and tall cupola across from the county prison that was new in 1850.[3]

Church cupolas no longer dominated the skyline south of the Capitol. Elegant steeples of a half-dozen newly built or remodeled churches had taken their places. Among the older churches, only Salem Reformed and Bethel Church of God still had cupolas. Zion Lutheran Church now boasted a bell tower with steeple. When the Presbyterian church burned in 1858, the congregation split and built two large, steepled churches, one at Market Square, the other at Pine Street and Third opposite the Capitol. Although the fracture was said to have involved theological differences between "Old Lighters" and "New Lighters," it also reflected business rivalries among Harrisburg Presbyterians. Central Iron's president, Charles L. Bailey; the Harrisburg National Bank's president, George W. Reily; its cashier, James A. Weir; and David Fleming of the Harrisburg Car Works were prominent in affairs at the Market Street church. Henry and James McCormick Jr., and Simon and J. Donald Cameron, owners of the various McCormick enterprises and Lochiel Iron, and officer-owners of the Dauphin Deposit and First National banks, were active in the breakaway Pine Street congregation.

The years also brought prosperity to the Methodists; for example, William Calder Jr., president of the Harrisburg Car Works, was among their members. In 1873 they erected Grace Church on State Street, equaling the two new Presbyterian churches in size and steepled grandeur. Next door, the Roman Catholics completely rebuilt St. Patrick's, considerably enlarging it and providing it with a central dome and two towers befitting its stature (since 1868) as pro-cathedral of the diocese of Harrisburg. A number of smaller congregations added less-imposing churches to the scene. Between 1850 and 1875 the older established churches sponsored no fewer than seventeen missions or offshoot congregations throughout the city. New religious groups included Free Baptists, who by 1873 had three churches, and a Jewish congregation, Olev Sholom, established in 1858.[4]

Another notable new structure near the center of the city was the towered Italianate depot erected by the Pennsylvania Railroad in 1857 to house its divisional headquarters. To the north, the State Capitol and the arsenal remained much as before, but seemed smaller than in 1850 because of the many imposing new buildings nearby. Also unchanged in number and appearance were the two bridges leading into the city. On the rises beyond the transportation and industrial corridor that ran unseen (in the vista) behind the Capitol were the main building of the Pennsylvania Asylum for the Insane, built in 1851, and clusters of houses and other structures on the hills in the central background that were now part of the city.

Other than the greater span of the city and the substitution of steeples for cupolas, most notable in the 1879 sketch were many tall chimneys and stacks belching black coal smoke. Inclusion of the smoke by the artist was probably designed to portray the community's bustle and prosperity, not to comment on industrial pollution. Turning to the 1871 map, the smoking chimneys at the right, or southern, end of the vista can be identified as belonging to the Lochiel, Paxton, and Central Iron furnaces and rolling mills. On the hills beyond the industrial corridor appeared the stacks of a foundry and machine shops owned by the Harrisburg Car Works. The columns of smoke arising from the corridor itself came from Hickok's Eagle Works and from the old iron furnace built by Porter, both directly east of the Capitol. In the northern portion of the corridor, smoke poured from the stacks of the car works and the two huge roundhouses and extensive repair shops of the Pennsylvania Railroad. At the center of the riverfront stood the tall, smoking chimneys of the city's water-pumping station and the cotton mill. By contrast, the sky in the mid-century lithographs had been clear, though certainly the widespread burning of wood and charcoal at the time would have produced clouds of blue-gray smoke, especially in winter.

A careful examination of the city map of 1871 reveals features of Harrisburg's growing infrastructure and industries not obvious in Hunter's portrait of the city. In 1850 two sets of track had run through the corridor behind the Capitol. By 1871 the area was filled with tracks and spurs for switching cars, providing access to repair facilities, and servicing industrial plants. The Pennsylvania Railroad dominated the area. If the same traveler who passed through the corridor twenty-one years earlier repeated the trip in 1871, it would seem a quite different place. Thanks to the growth of the city, the corridor now extended as far both north and south of the old borough limits as the entire original trip. Activity on the canal (which now belonged to the Pennsylvania Railroad) had subsided, as first passengers and then much of the freight once borne by packet boats shifted to the faster and more comfortable railroads. Most of the farmland immediately south of town in 1850 had given way to heavy industry. Just below Harrisburg stood the extensive rail mills of the Pennsylvania Steel Company. Once within the city, in quick succession came the Lochiel Rolling Mills, the Paxton Furnace and Rolling Mill, the Paxton flour mills (all owned or managed by the McCormicks), and the Central Iron Company and Chesapeake Nail Works, which belonged to the Baileys. Together these industries provided work for more than 1,200 people. Next appeared a large distillery, brickyards, small iron rolling mills, and the city's largest sawmill. Above Paxton Street (the town's southern boundary in 1850) were the Harrisburg Firebrick Works, another McCormick-owned rolling mill, the large and odoriferous Gas Works, the new furnace of the Wister brothers, and assorted warehouses, coal sheds, and storage facilities. The trip would be interrupted by a stop at the impressive new depot stretching between Market and Chestnut streets.

Resuming its run, the train would ease by the less-imposing depot of the Philadelphia & Reading, one block to the north. As it picked up speed, the traveler, straining to see the Capitol on a rise six blocks to the west, perhaps would not notice either Hickok's Eagle Works beside the track or, out the opposite window, Harrisburg's first anthracite furnace, which had gone into production about the time of the earlier trip. More small planing mills and brickyards came next, followed by the Harrisburg Car Works, which gave employment to 442 workers in 1870. Next the train would pass the Pennsylvania Railroad's own extensive maintenance facilities: switching yards where trains were made up or broken apart; roundhouses (one on either side of the track), where engines were inspected, repaired, or turned around; and a variety of repair shops housed in a single large building. Finally, just as the train left the city to the north, the traveler would have seen and smelled stockyards that held livestock for local butchers.

The double-track Lebanon Valley Railroad (by then part of the Philadelphia & Reading system) entered the city from the southeast. Generally paralleling Paxton Street, which lay to the south, it passed through its own small yards with engine house, crossed the old State Works canal immediately north of the Pennsylvania Railroad station, and terminated at its own depot. A single track belonging to the Cumberland Valley Railroad still crossed the railroad bridge over the Susquehanna, carrying that line's passengers and freight into Harrisburg. Its tracks ran in the center of Mulberry Street until they joined the Pennsylvania line and ended at that company's station house. Across the Susquehanna, the Northern Central Railroad, now double-tracked, still paralleled the river, stopping to deposit or pick up passengers to or from Harrisburg, and servicing the large McCormick-owned nail works at West Fairview.

During the Civil War a street railway had been built from Camp Curtin to provide transportation for soldiers who wanted to go to or from the center of the city. That line continued in operation after the war. Starting at Camp Curtin, it ran south along Sixth Street, turned west onto "Broad or Verbeke Street" (as the map identified it), and skirted the new sheds of Harrisburg's second market for farm produce. Turning south on Third, it passed in front of the Capitol before reaching Walnut Street. Running west for one block, it turned onto Second Street, continued past the produce sheds at Market Square, ran eastward along Market Street for three and a half blocks, and finally terminated at the Pennsylvania Railroad station. In 1875 a branch line extended to Steelton. At first the cars were open and drawn by horses or mules. During the 1880s the lines were electrified and ran larger, more comfortable, closed cars. Once the eastern portions of the city developed, a second company constructed a line to bring passengers from that section to center city. Not until the final years of the century did the two consolidate into a single firm with sufficient passengers to turn a regular profit.[5]

Through most of the nineteenth century, the chief obstacle to a profitable street railway system was that Harrisburg, like similar communities, remained a walking city.[6] This was possible because the community was still relatively compact, extending north and south about two and a half miles and no more than a mile and a quarter from the river to its eastern limit. Moreover, people who could afford them usually had horses and carriages of their own, while the poor could not afford to hire transportation. In his memoir of growing up in Harrisburg, Charles R. Boak wrote of his grandfather:

> He got a job in an iron foundry about three miles from his home on Broad Street. Rising early in the morning, he had to walk to the mill in West Fairview—down Front Street to the old Camel Back Bridge, which must have been almost a mile, and about a mile up the other side. There he worked twelve hours in the foundry, and then walked home. Sometimes in the winter when the river was frozen hard and deep, Grandfather could walk across the ice to West Fairview and saved about two miles each way. But airholes could be dangerous in the darkness of both morning and evening.[7]

Other physical changes that occurred during the first two decades or so of Harrisburg's industrialization also were evident on the maps. Although the borough in 1850 extended north to Forster Avenue, little building had taken place above South Street save for two blocks along the river. Except for industrial and commercial buildings, the area east, southeast, and south of the canal was unsettled swampland beyond the borough limits. By 1871 the city extended north to Maclay Street, east beyond Eighteenth Street, and south to Poplar Street. Large tracts of this newly annexed land remained vacant. Except for Front Street mansions belonging to George W. Reily and his sisters, and a twenty-square-block district the Reilys had platted and were developing to the east, no building had taken place above Reily Street. In six scattered districts east of the industrial-transportation corridor, developers were similarly platting and building, much of the housing intended for workers. Finally, on both sides of Paxton Creek in South Harrisburg immediately east of the Bailey and McCormick ironworks, contractors were erecting rowhouses for mill hands and their families. Six large schools, some with spacious playgrounds, stood in the more heavily settled parts of the city.

Harrisburg's principal community cemetery, chartered in 1845, overlooked the city from a hill east of Paxton Creek. Now fully laid out, the
large burial grounds were gradually filling up. Deceased members of the
elite entrepreneurial, merchant, and professional families occupied choice
lots from which the living could view the city while visiting the graves of
family members and friends. Ordinary folk were buried in the less scenic
central and eastern portions of the cemetery.[8] A few blocks to the north
and east, the "Free Cemetery" provided burial sites for African Americans. The Catholic church, many of its congregation being Irish workers,
had acquired land for a cemetery in southeastern Harrisburg along the
tracks of the Lebanon Valley Railroad, which some of them had helped
construct. The church also had a erected a convent on a large tract on
Derry Street, between Crescent and Summit, in East Harrisburg.

Federal census returns show that the first twenty years of industrialization produced the highest rates of growth in Harrisburg's history—more
than 70 percent each decade (see Appendix B). Only during its first
decade as a borough, 1790 to 1800, had there been a comparable rate
(68.2 percent). After the 1860s the next highest population growth (33.1
percent) would occur between 1870 and 1880 as industrialization
crested in the community.

The influx of newcomers brought surprisingly few changes in the
nativity, race, and ethnic patterns of the community, and those changes
were small (see Appendix H). In all three censuses the proportion of
residents native to Pennsylvania ranged between 79 and 85 percent.
Only between 4 and 9 percent were born in other states. Although the
African American community continued to hover around one-tenth of
the total population, this apparent overall stability disguised a remarkable change in birthplace among Harrisburg blacks—from Pennsylvania
in 1850 and 1860 to the former slave states after the Civil War (see
Chapter 10). Even with the sponge of industrialization, the city attracted
far fewer immigrants than most large industrial centers. The percentage
of foreign-born persons reached highs in the 1850s, 1860s and 1870s yet
amounted to little more than 12 percent of all residents at the peak in
1870. Three nationalities predominated: Irish, German, and British (the
latter English, Welsh, and Scots combined). The Irish, who had been
most numerous in 1850, were by 1870 second and made up only a third
of the foreign-born. German-born residents, the second largest group in
1850, became dominant by 1860 and remained so in 1870 and after.
Even so, German immigrants never amounted to more than 6 percent of
the total population. Only a handful of British immigrants appeared in
the censuses of 1850 and 1860. By 1870 they made up a fifth of all
immigrants but were only 2.5 percent of the city's population. After

1870 the proportion of foreign-born steadily declined, the most numerous groups coming from southern and eastern Europe rather than northern and western Europe after 1910.

The computerized census data for 1870 reveal that Harrisburg's residents lived in 4,575 dwellings—more than three times as many as in 1850. The percentage of persons living apart from families had fallen to 13 percent since 1850, when nearly a quarter appeared as roomers or lived in their employer's household. The gender imbalance of 1850 (505 females per 1,000) increased notably to 520 females per 1,000 by 1870. Whether this was related to the war, to industrialization, or to both is not clear, but the war did bring about a geographic unsettling of people, particularly blacks.

Real estate ownership patterns also changed during the first two decades of industrialization (see Appendix F). The total value of land owned by Harrisburg residents rose at a faster rate than the number of adults living there. While values swelled 310 percent (from $3.6 million in 1850 to $8.5 million in 1860 to $14.6 million in 1870), adults increased only 195 percent (from 4,000 to 7,000 to 12,500).[9] Meanwhile, the distribution of those holdings broadened considerably during the 1850s, only to narrow again in the next decade. One adult in ten owned land in 1850, one in five by 1860, one in seven in 1870. Among the propertied, holdings continued to be concentrated at the very top, though less so than in 1850. The portions held by the top 1, 5, 10, and 20 percent all declined, while the shares held by each of the four lower fifths of landowners increased. Most dramatic was the drop in the share of those at the very top; in 1850 the wealthiest 1 percent held a third of all land by value; two decades later they owned only a fifth of the total. Although this suggests less concentration of wealth, it must be remembered that by 1870 many of the wealthy were investing in stocks, bonds, and other securities, and proportionately less in real estate.

In sum, two decades of industrialization had altered Harrisburg's skyline and physical appearance. It had also contributed to the city's increased population and expanded boundaries and had shifted its economic base from commerce to manufacturing, thereby adding greatly to the community's productivity and wealth. It may even have contributed to broader holding of real estate. More important, however, was the impact that industrialization had on the occupational patterns of the city's residents. On the one hand, the coming of railroads and factories created a number of new jobs that had not existed previously. With the railroads came baggage masters, brakemen, conductors, engine builders, and locomotive engineers and firemen, to name but a few. The iron industry brought, among others, heaters, puddlers, and rollers. The Harrisburg

Car Works needed car inspectors, and it and Hickok's Eagle Works employed patternmakers. Meanwhile, the more general use of steam power in all production offered such new jobs as boiler maker and stationary engineer. On the other hand, most occupations, such as baker, barber, blacksmith, carpenter, cooper, lawyer, painter, physician, shoemaker, waiter, and whitewasher, continued from the preindustrial period.

Measuring changes in the occupations of residents is more complex and less precise than might be expected. Even detailed data from census schedules provide only impressionistic answers at best. For example, the manufacturing and population census takers gathered their data in different ways. The former contacted firms that had produced at least $500 worth of goods the previous year and asked the number of employees, the average payroll, and in 1870 the number of months out of twelve they had been in operation. "Total employees" did not differentiate between residents of Harrisburg proper and persons from outside the city. The manufacturing census provided no information on such matters as degree of skill, ethnic background, or tasks performed by employees, but the population census takers asked questions at each household about all persons living there. This provided information on individual backgrounds, including occupation, but usually did not indicate employer, place of employment, or whether the person was currently employed in his or her trade or profession.

Because some of the new industrial plants were located on the outer edges of the town, at least part of their work force probably came from the surrounding countryside. Similarly, some residents of Harrisburg, such as Boak's grandfather, found employment beyond the borough limits. If the data base were expanded to include adjacent townships in hopes of including everyone who worked for Harrisburg firms and all outside employers of Harrisburg residents, the problem would only be extended, not solved. Some residents and businesses in those townships, in turn, found employment in, or hired residents of, still more distant jurisdictions.

The spotty reliability of population census schedule data on occupations further complicates the problem. Census takers listed whatever occupations their informants reported. The quality of that information depended on whether it was provided by the persons involved or by others, and on whether the respondents in fact knew and reported accurately or exaggerated or understated occupations. Even when accurate, "merchant" could mean anything from itinerant peddler to large-scale wholesaler, "carpenter" could mean anything from nail driver to contractor, and "foundryman" anything from foundry worker to ironmaster. And did "loafer" refer to a specialty of bakers or ironworkers, or was it the census taker's characterization of a respondent? Most people reported a particular trade (blacksmith, driver, machinist), but some gave employer (Pennsylvania Railroad, the car works, the cotton mill) or type of employer (railroad, machine works, factory).[10] Notwithstanding these difficulties, the various census schedules provide the best available data on Harrisburg's workers.

The overall proportion of the population with listed occupations increased nearly 2.5 percent over the twenty years, but this was due to the inclusion of women's occupations in the census rather than to industrialization drawing or forcing more people into the work force (see Appendix C). Otherwise, when only the most general occupational groupings are considered (major manufacturers, executives, and managers; merchants, merchant-manufacturers, and other small manufacturers; professionals; clerks, salespersons, and semi-professionals; workers; and miscellaneous), two decades of factory production apparently had little impact on the way Harrisburg residents made their livings. Over that span the percentage of persons in each grouping except workers declined very slightly, the most being professionals who declined nearly 1.5 percent. The gain in the proportion of workers amounted to less than 3 percent.

When industrial workers are separated from other laborers, however, the impact of industrialization becomes more evident (see Appendix D). The percentage of factory and railroad workers (whether data are from the population census or the manufacturing census) increased much faster than either the overall population or the total number of workers of all types. Over the twenty years, while the city's total population nearly trebled, the total number with listed occupations increased more than three times. At the same time, if population schedules are used, the number of factory and railroad workers increased seven times, while manufacturing schedules show an increase of nearly twenty-fold.[11]

A further breakdown of workers, distinguishing craft workers, industrial workers, and general workers (unspecified laborers, servants, and other unskilled workers), further indicates the impact of the factory system on the occupations of Harrisburg residents (see Appendix E). Before the factories arrived, nearly all workers in town labored in their own craft shops or were hired by skilled artisans to work in their shops. Although the number of craft workers doubled between 1850 and 1870, their proportion in the total work force declined from almost half of all workers to a little more than a quarter. Meanwhile, the proportion of industrial workers grew almost to that of craft workers; two decades earlier they had made up little more than a tenth of the work force. General workers remained relatively steady, growing from less than 40 percent in 1850 to only a little more than 45 percent by 1870.

These figures considerably understate the extent to which workers shifted to factory employment. Many skilled and semi-skilled workers who reported their trade but not their employer to census takers were counted as craft workers when many of them actually pursued their trades in factory settings. Similarly, by 1870, many persons listed as laborers would have been classified as industrial workers if their employers had been known. In 1850, when Harrisburg's only important industrial plant was the new blast furnace of Porter & Burke, the overwhelming majority of those calling themselves "laborers" worked at craft-type or casual jobs. Two decades later many if not most of them probably worked in the new industrial factories, although the number is not known.

A substantial increase in the number of reported female servants further distorts the picture. In part this was a function of changing instructions to census takers. The 1850 forms called only for the occupations of males over the age of fifteen, but census takers at Harrisburg included some younger males and more than 100 females. The 1860 forms called for the occupations of both males and females over fifteen years of age, but local census takers also listed occupations for 176 children under sixteen. In 1870, census forms asked the occupations of all persons without regard to age or gender. Although the increase in the number of female servants may reflect a growing prosperity that allowed middle- and upper-class families to employ more household help, it is more likely that the listing of females with occupations after 1850 increased the number reported without greatly changing the percentage actually so employed.

Given the problems of the data, what can be said with reasonable certainty about the shift of Harrisburg workers to industrial positions between 1850 and 1870? At the earlier date, about one worker in ten had jobs in the new factories, and nearly all of those were employed at the new Porter & Burke furnace. Twenty years later more than one in four workers were certainly factory employees. If those listed as craft workers and as laborers who actually worked in factories are included, by 1870 as many as half of the work force, or more than a tenth of the total population, may have been employed in industry. This will be discussed at greater length in Chapter 9.

The meaning of technological and economic changes to those who lived through them is not always easy to determine. Much hinges on whether one considers the views of those who benefited from the changes or those whose fortunes were affected negatively. The former probably saw much progress and promise in the new order; the latter probably saw decline and the passing of a kindlier, better way of life. The speed of the changes and the degree to which they altered living patterns no doubt were also major factors. Where transformations were rapid and extensive, social disruption tended to be greater, and alienation more common. Where transformations were slow and partial, people had time to adjust and consequently suffered less.

Certainly this was true of transportation changes at Harrisburg during the first half of the nineteenth century. At that time, most people hailed the improvements as opening the community to the world and promising new opportunities. However, drivers of packhorses and mules, the original carriers of freight in the region, regarded roads and wagons as intruding on their rights. Forty years later the teamsters, in turn, objected to the coming of canal packetboats and railroads as dooming them.[12] In all likelihood, reactions to the rise of the factory system followed the same pattern. The entrepreneurial families no doubt saw their projects as desirable and beneficial to the public. Those displaced or lowered in status by the changes probably decried what happened and regretted that their children were being deprived of what had been meaningful to them. Adults who came of age before factories and who lived through the process of industrialization experienced its consequences most fully. Less so their children. What may have seemed to the parents a period of dramatic and often chaotic and disruptive change, their children would have perceived as part of the natural order.

By examining individual careers—first of residents who already were adults in 1850 and then of their sons—it is possible to gain insights into the effects of the changes on both generations. Because the entrepreneurial families left the best records, changes in their life-styles can be most fully documented. Detailed information is lacking for those in the middle and lower social ranks. Federal census data, however, make it possible to determine increases or shrinkages in the number employed in the various occupations and such changes in overall employee characteristics as ethnicity, family status, and property ownership. The same sources can also be used to trace, at ten-year intervals, the career histories of individuals who remained in Harrisburg. It is possible to determine whether they persisted at the same occupations or moved to others, whether they acquired real estate over the years, and whether their children followed in their footsteps or struck out in new directions. The chapters that follow deal with those and related matters.

The Entrepreneurs and Other Elites:
Fathers and Sons

The impact of early industrialization on Harrisburg illustrates the interplay of change and continuity in human affairs. It is not surprising that, soon after 1850, the leading manufacturers elbowed their way into Harrisburg's elite. Crowded aside (but not out) were the established merchants, landlords, and professionals. Their wealth, though no less than before, seemed somehow diminished compared with that of the newcomers. From the mid-1850s onward, the entrepreneurs, and later their children and grandchildren, would intermarry with the older elite and with one another and dominate Harrisburg society until the Great Depression and World War II. Despite the changes of personnel at the top, the new elite began by modeling their style of living on that of their predecessors. Over the years, however, they also altered it.[1]

The Entrepreneurial Families

Most of Harrisburg's entrepreneurs were born and grew up within thirty miles of the town. Most were Scots-Irish in origin and Presbyterian in religion (see Table 7.1). Thirteen of the nineteen were natives of Harrisburg or of Dauphin County or an adjoining county. Another, David Fleming, born in southwestern Pennsylvania, lived all but the first few months of his life in Harrisburg. Four others came from Chester County and the nearby states of Maryland, New York, and Connecticut. The only true outsider was Irish-born Michael Burke. Although the Harrisburg area was about evenly divided between Germans and Scots-Irish, only Jacob

Table 7.1. Backgrounds of Entrepreneurs

Name	Born	Birthplace	Ethnicity	Religion	Education	Occupation
Elder, Thomas	1767	Dauphin Co.	ScIr	Pr	Academy	Lawyer-banker
Harris, Robert	1768	Harrisburg	Engl	Pr	n/a	Landowner-politico
Forster, John	1777	Dauphin Co.	ScIr	n/a	Princeton	Banker-merchant
Haldeman, Jacob M.	1781	Lancaster Co.	Swiss	Pr	Private	Ironmaster-banker
Calder, Wm. Sr.	1788	Maryland	Sc	n/a	Self-educated	Stageline owner
Ayres, William	1788	Dauphin Co.	ScIr	n/a	Self-educated	Lawyer
Burke, Michael	1797	Ireland	Irish	Cath	Self-educated	Contractor
Cameron, Simon	1799	Lancaster Co.	ScIr/Ger	Pr-P	Limited	Banker-politico
McCormick, Jas. Sr.	1801	Cumberland Co.	ScIr	Pr-P	Princeton	Lawyer-banker
Fleming, David	1812	Washington Co.	ScIr	Pr-M	Academy	Lawyer
Hickok, Wm. O.	1815	New York	Engl	Pr-M	Academy	Inventor
Eby, Jacob	1816	Lancaster Co.	Ger	Luth	Common school	Merchant-banker
Bailey, Chas. L.	1821	Chester Co.	Engl	Pr-M	Academy	Ironmaster
Calder, Wm. Jr.	1821	Harrisburg	ScIr	Meth	Limited	Manufacturer-banker
Hildrup, Wm. T.	1822	Connecticut	Engl	Epis	Limited	Car builder
McCormick, Henry	1831	Harrisburg	ScIr/Ger	Pr-P	Yale	Ironmaster
McCormick, Jas. Jr.	1832	Harrisburg	ScIr/Ger	Pr-P	Yale	Lawyer-banker
Cameron, J. Donald	1834	Dauphin Co.	ScIr/Ger	Pr-P	Princeton	Banker-politico
Reily, George W.	1834	Harrisburg	ScIr/Ger	Pr-M	Yale	Banker

Note: Engl = English; Ger = German; Sc = Scots; ScIr = Scots-Irish; Cath = Catholic; Epis = Episcopalian; Luth = Lutheran; Meth = Methodist; Pr = Presbyterian; Pr-P = Presbyterian, Pine Street; Pr-M = Presbyterian, Market Square. n/a = not available.

Eby came of German stock, while eleven were Scots-Irish (although five of those had German mothers). The disparity lent credibility to the proverbial distinction between the cautious, hardworking Germans, who allegedly lived frugally and saved money, and the more adventuresome Scots-Irish, who made money. Simon Cameron was particularly proud that his heritage combined German doggedness and determination with Scots-Irish aggressiveness.[2]

Sixteen of the nineteen entrepreneurs left records indicating religious persuasion. As night follows day, all but one of the Scots-Irish who claimed church membership were Presbyterian, and so were two of English ancestry (Bailey had been a Quaker until dismissed from Meeting in Chester County when he married a Presbyterian) and one of Swiss ancestry. When that denomination split in 1857, four of the entrepreneurs continued to worship with the original congregation at Market Square, five with the new group on Pine Street. The remaining entrepreneurs included an Episcopalian, a Lutheran, and a Methodist. Burke was Catholic.[3]

The entrepreneurs differed in other important ways. The first was a matter of generation. The nine born between 1767 and 1801 did most of their work before the Civil War, built the town's infrastructure, and passed from the scene by 1870. The ten born between 1812 and 1834 became active during the 1850s or after and were industrialists, except for James McCormick Jr. and George W. Reily, who were primarily bankers. There was some overlap. Simon Cameron, for instance, lived to the age of ninety and worked in both groups. Five of the builders of the infrastructure (Haldeman, the elder Calder, Burke, Simon Cameron, and James McCormick Sr.) also helped launch Harrisburg's first industries. Three of the younger group (Eby, the junior Calder, and J. Donald Cameron) also built railroads and started banks.

The economic bases from which the entrepreneurs sprang also varied. Four (Calder, Ayres, Burke, and Cameron) were self-made men. Born in poverty and unschooled, they made their way in life by scratching together careers and fortunes as they went along. Others (Fleming, Hickok, Eby, and Hildrup), though from families of modest means, got ahead chiefly through their own efforts. They had the benefit of some schooling, and Fleming and Hickok attended academies. Still others (Elder, Harris, Forster, Haldeman, and the senior McCormick) were sons of wealthy farmers and landowners, and Bailey was the son and grandson of prosperous ironmasters. These sons of the well-off were well schooled, with two (Forster and McCormick) attending the College of New Jersey at Princeton. Among the younger group, four (Calder, the two McCormicks, and Cameron) were sons of the earlier entrepreneurs.

Another, Reily, was the son of Harrisburg's wealthiest physician. The McCormicks and Reily graduated from Yale; Cameron graduated from the College of New Jersey at Princeton. Instead of going to college, Calder left school at twelve to work for his father. The choice may have been his own, however, as his younger brother graduated from Wesleyan College and became a clergyman.[4]

Most of the entrepreneurs do not appear to have sought out wives whose wealth would raise their status, but none married substantially below his social rank either. Only one, Ayres, who chose the daughter of a prominent Harrisburg merchant and officeholder as his wife, married above his station. Several of the others (both James McCormicks, Forster, William Calder Jr., Eby, Fleming, Reily, and J. Donald Cameron) married into well-off and distinguished local families. Their social and economic rankings at marriage, however, were about the same or higher than those of their brides.

As the earlier group of entrepreneurs acquired wealth and power, they adopted features of the life-style of the town's older elite, the wealthy landowners and merchants. One feature was housing. As if to symbolize their standing in the community, two of the premier entrepreneurs moved into John Harris's old mansion on Front Street overlooking the river. Thomas Elder lived there until his death in 1853; in the 1860s Simon Cameron moved in and expanded and refurbished the place to meet his needs and tastes. Robert Harris had to be content with a house on Second Street separated by an alley from his family's former seat. The rest of the early entrepreneurs, in the manner of the older merchant elite, lived at or near their places of business in the vicinity of Market Square, the commercial hub of the preindustrial town. Jacob M. Haldeman's riverfront house was but three blocks from the Harrisburg Bank he headed. Lawyer-banker James McCormick Sr. lived on Market Street, across from the courthouse and within a block of his bank, the Dauphin Deposit. Calder was located on Market Square at his stage-line depot, Attorney Ayres had quarters on Market Street a block east of the courthouse, and Burke, the contractor, lived on Walnut Street near Front.

The younger group of entrepreneurs were able to live more opulently because the factories they owned provided them with greater incomes. After the Civil War old homes and business establishments along the river were torn down to make way for the larger and grander houses of the town's newest elite. Front Street became Harrisburg's most fashionable address. Along an eight-block stretch, anchored on the south by Simon Cameron in the Harris Mansion and on the north by J. Donald Cameron in a palatial residence at Front and State, lived most of the entrepreneurs: William O. Hickok, Charles L. Bailey, Jacob M. Halde-

Fig. 15.   FIRST-GENERATION ENTREPRENEURS.

James McCormick (1801–1870).
Lawyer; president of Dauphin Deposit,
1840–70; president of Harrisburg Cotton
Co., 1849–65; owner of Paxton Iron Co.,
1857–70, and West Fairview Nail Works,
1859–70

Jacob M. Haldeman (1781–1856).
Director (1814–53) and president (1853–
56) of Harrisburg Bank; director (1814–
56) and president (1846–56) of
Harrisburg Bridge Co.; founding partner
of Harrisburg Car Works, 1853–56.

Simon Cameron (1799–1889).   Printer;
canal and railroad promoter-builder;
banker; founder of Lochiel Rolling Mill
Co., 1864; U.S. Senator; U.S. Secretary of
War.

Michael Burke (1797–1864).   Canal and
railroad contractor; co-owner and builder
of the first anthracite furnace at
Harrisburg.

man (and, after 1856, his widow Eliza), James McCormick Jr., William
Calder Jr., and Colonel Henry McCormick. Interspersed among them
were the homes of Haldeman's sons Jacob S. and Richard; his widowed
daughters Sara Haly, Mary Ross, and Susan O'Connor; and Maria, the
widow of his son John. By 1871 Dr. George W. Reily and his three sisters
(one the wife of Dr. George W. Porter, the other two as yet unmarried)
lived on Front Street north of the developed section of the city, on land
inherited from their father. There, with a full city block for each, they
built three enormous mansions. Only two of the entrepreneurs failed to
follow the pattern. After the war, Eby built a home at Market and Fifth
near his wholesale grocery business. Later he moved to a farm along the
river just north of the city limits. Hildrup too lived on a farm, in
Susquehanna Township, less than two miles from Harrisburg. At the
same time, he maintained quarters at the Bolton House, a hotel in town.[5]

Several of the wealthy, once sons of poor-to-moderately well-off farmers, also owned country estates where they summered away from the
cares and heat of the city. These summer estates may also have been
token penances for having deserted the hard but virtuous examples of
their parents, or they might have been a fashionable form of homage to
the region's earlier agricultural ideal. The hallmarks of the successful
Pennsylvania farmer had long been well-tended fields, choice livestock,
substantial barns, and ample, solid farmhouses, preferably made of
stone. In addition to Eby and Hildrup, with their acreage and elegant
farmhouses north of the city, the McCormicks continued to own farmland in Cumberland County. Colonel Henry refurbished the old family
farm near Dillsburg, called "Rosegarden," and each summer moved
there with his family for the season. Simon Cameron, and later J. Donald
Cameron, lived part of each year at "Lochiel," their country estate between Harrisburg and Middletown. The Calders, father and son, owned
several model farms in Dauphin County where they raised prize horses
and other livestock. The Hickoks had a modest farm three miles east of
the city. Haldeman identified most completely with his agrarian roots.
He continued to own his farm on the Yellow Breeches Creek and most of
the other farms he had acquired in his early years. Although by 1850 he
had lived in Harrisburg for two decades, engaged chiefly in banking and
in business, he told the census taker that year that he was a farmer.

Another facet of the old aristocratic ideal included acceptance of political responsibilities. The active pursuit of public office was frowned on,
but men of substance were expected to serve brief stints in local or
county office, perhaps a session or two in the state legislature, and, for a
very few, a term or so in the United States Congress. Most of Harris-

burg's entrepreneurs complied with these standards. Except for the Camerons, who violated the rule in their constant quest for high office, they limited themselves to brief periods of public service. Elder held the office of Pennsylvania attorney general for two years; Harris was twice assistant burgess of Harrisburg and sat in Congress for two terms; Ayres was a justice of the peace for several years before serving on the borough council for six years and in the legislature for two; Burke and James McCormick Sr. each spent six years on the borough council; Hickok presided over the city council during the Civil War; Bailey too was a council member for two terms, before serving two years in the Pennsylvania Assembly; the younger Calder sat for one term on the city council; and Reily was a member of the city planning commission. Most were partisans. Before the Civil War, Elder, Haldeman, Fleming, and the younger Calder had been Whigs, while Cameron, the elder McCormick, and Burke were Democrats. After the split over slavery, Bailey, Calder, the Camerons, and Fleming became Republicans; the McCormicks and Reily remained Democrats. Partisanship seems rarely to have interfered with their business, personal, or social relationships.

The same men and their families dominated Harrisburg's religious, charitable, and social institutions well into the twentieth century.[6] The majority sat on the governing boards of their respective churches or, if less inclined to piety, at least were generous donors. Beginning in the late 1850s the entrepreneurs contributed large sums for building or rebuilding the town's principal churches. So intimate were the ties of Market Square Presbyterian Church to the officers of the Harrisburg Bank that it seemed only natural for Sunday collections to be taken there directly after services to be counted and deposited.[7]

The industrial entrepreneurs overall followed the lead of their more enlightened counterparts elsewhere in philanthropy and community service. Although only James McCormick Sr. served on the committee for chartering the new town cemetery in 1845 when the old churchyards were full, representatives of the entrepreneurial families thereafter regularly sat on its governing body.[8] William Calder Jr., J. Donald Cameron, David Fleming, William O. Hickok, Colonel Henry McCormick, and James McCormick Jr. were all at the meeting in December 1872 that called for the establishment of a city hospital. All contributed, but James McCormick Jr. provided the leadership. During the hospital's first fifty years, with the exception of one year, one McCormick or another served as president of its board. During the same years and after, male McCormicks, Baileys, Calders, Reilys, Haldemans, Hickoks, Hildrups, and collateral relatives all served on the board; their wives, sisters, and daugh-

ters, under the presidency of Mrs. Charles L. Bailey, Mrs. Henry Mc-
Cormick, or Miss Anne McCormick, among others, belonged to the
Women's Aid Society of the Hospital.[9]

The entrepreneurial families played major roles in establishing the
present Harrisburg city library. Again the younger James McCormick led
the movement, calling meetings of prominent citizens in 1889. A charter
was obtained later that year, and McCormick, Charles L. Bailey, and
Maurice Eby, of the entrepreneurial families, sat on the initial board of
trustees. Bailey served as president until 1894, McCormick as vice president until 1893. In 1892 McCormick had a building constructed on
Locust Street that housed the institution from 1892 until 1914. Upon
her death in 1896, Sara Haldeman Haly willed the Library $60,000. Her
executors made additional bequests of money and in 1900 gave the lot
on the southeast corner of Front and Walnut. Formerly the garden of the
Haldeman mansion, it became the site of the present library. Early in the
new century, the five bachelor sons and married daughter of James Mc-
Cormick Jr. became major contributors. When Donald, the last of the
sons, died, he devised to it the family home at Front and Walnut.[10]

The various entrepreneurial families all had favorite charities or
causes they supported. Baileys, Reilys, and Calders regularly sat on the
governing board of the State Lunatic Asylum, opened in 1851 just outside the boundaries of the borough. William Calder Jr. established the
Harrisburg Benevolent Association and was manager of the Children's
Industrial Home. He, his associate David Fleming, and James Mc-
Cormick Jr. actively supported the Home of the Friendless. William
Hildrup promoted a Mechanics High School for Harrisburg, but even
though he succeeded in getting a charter from the legislature, the onset of
the Panic of 1873 cut off funding for the project. James McCormick Jr.
and George W. Reily Jr. supported the YMCA; Anne McCormick and
Mary Calder Mains (daughter of William Calder Sr.) were active in the
YWCA.

The way the entrepreneurs reared their children reveals much about
their values. Information on the families of the older group who built the
infrastructure is incomplete.[11] Of thirty-five sons who reached adult-
hood, adequate sketches were found for only thirteen. Little more than
the names of seven others are known, and for the remaining fifteen only
bits of information, such as occupation, birth and death dates, or place
of residence, have come to light. If the eleven whose educations were
recorded were typical, the children of the early entrepreneurs were well
schooled, however lacking their fathers' formal educations may have
been. Only one was self-educated, eight attended the Harrisburg Academy, and two were tutored privately. Nine of the eleven went on to one

Fig. 16. SECOND-GENERATION ENTREPRENEURS.

Col. Henry McCormick (1831–1900). President of Dauphin Deposit Bank, 1870–74; president of Paxton Iron and West Fairview Nail Works, 1870–97; president & general manager of Lochiel Iron Co., 1871–85.

William Calder Jr. (1821–80). Co-founder and president of the Harrisburg Car Works, 1863–80; president of First National Bank, 1874–80; reorganizer, president and director of Harrisburg Cotton Co.

William O. Hickok (1815–91). Bookbinder-printer; inventor; founder-owner of the Eagle Works, 1853–91.

Charles L. Bailey (1821–99). Co-founder (1853) and president (1874–99) of Central Iron Co.; co-founder (1866) and president (1874–99) of Chesapeake Nail Co.

or more colleges: one each to Wesleyan, Bucknell, and the University of Pennsylvania; two each to Dickinson College and the College of New Jersey at Princeton; and four to Yale. One did postgraduate work at Heidelberg in Germany.

Fourteen are known to have left Harrisburg when they became adults, including all the sons of Elder, Burke (three each), and Ayres (four). Seventeen remained in the community, and the residences of four could not be determined. The occupations of twenty of the thirty-five are known. Four (one of whom was educated in law) became industrial entrepreneurs; nine took up professions (two civil engineers, three clergymen, two lawyers, and two physicians); two became merchants, two went into the military, and one each became an artist, a mechanical engineer, and a clerk. At least two of Harris's sons became dabblers. One, after a short stint as a merchant, retired to a farm; the other was briefly a navy midshipman but did not like that life so returned to Harrisburg. After "betaking himself to civil employments," in his later years he retired "generally to piscatorial enjoyment," reported his obituary in the *Patriot*. The *Telegraph* also put his life of leisure delicately: in recent years he had become "a student of, though not altogether a disinterested observer of the actions of men."[12] None of Haldeman's three sons pursued regular careers: John died young; Jacob S. served briefly as president of the Harrisburg National Bank and later as U.S. minister to Sweden, then gave himself over to the life of a gentleman farmer; and Richard J. briefly edited the *Patriot* and served two terms in Congress before retiring from active pursuits to a farm in Cumberland County.[13]

Not one of the eldest sons of the older group of entrepreneurs slipped easily into place behind their fathers. The first sons of Ayres, Burke, Calder, Elder, Forster, and Harris all promptly left home for other communities. Two died young: John Calder at the age of thirty-five of unknown cause, and Brua Cameron in 1864 of illnesses contracted as a Union army officer. Even Calder's second son, William Jr., threatened to strike out on his own until his father turned over most of his business enterprises for the son to manage. Eldest sons Henry McCormick and John Haldeman, apparently restless after serving as officers in the Union Army, headed west with a third young man in June 1865. Their objective was to look firsthand into gold-mine investments. In Denver, Haldeman fell ill and died suddenly, leaving in his estate a Wells Fargo gold note worth $10,000, which he was carrying with him. McCormick continued the trip but learned that gold mines were an expensive and risky investment.[14] When he returned, he resumed his prewar role as manager of his father's iron properties.

Information on the sons of the later industrial entrepreneurs, twenty-three of whom reached maturity, is much more complete. Of the twenty for whom sketches have been found, at least nineteen attended the same or similar schools as the sons of the earlier group. Eighteen attended one or more private academies. A dozen went to the Harrisburg Academy, and seven went farther afield to more-prestigious institutions: three each to the Hill School in Pottstown and to Phillips-Andover in Massachusetts, and one each to the academy in Westtown, Pennsylvania, and to Phillips-Exeter and St. Paul's schools in New Hampshire. A much larger portion (at least twenty of twenty-three, and perhaps all) attended college. Twelve went to Yale, three went to Princeton, two to the University of Pennsylvania, and one each to Harvard, the Pennsylvania State College, and Lafayette College. Three took postgraduate work—one at Heidelberg, one at Yale, and one at the University of Pennsylvania.[15]

The marriage patterns of the sons of all the entrepreneurs closely paralleled those of their fathers. All nineteen of the fathers had married and had one or more sons. Of the thirty-five sons of the pre–Civil War entrepreneurs, the marital status of twenty-nine is known. Although all but four married, only two marriages interlinked entrepreneurial families. Simon Cameron's daughter Margaretta married Richard Haldeman, and his son J. Donald married Mary McCormick. The others wed into local families who were not among the entrepreneurial elite, and a few wed women from outside the community. Of the twenty-three sons of the postwar industrial entrepreneurs, the marital status of twenty-two are known. Of those, one-third (including all five sons of James McCormick Jr. and two others who died before age thirty) remained bachelors. Again, intermarriage within the group was rare. All but two of the sons of the entrepreneurs married outside the charmed circle. The exceptions were sons of Charles L. Bailey, who married daughters of George Reily.

The careers the sons of the industrial entrepreneurs chose, and where they lived, were markedly different from those of the earlier group. Only about half the sons of the older entrepreneurs remained in Harrisburg, and no more than a third went into business. Several factors may have been involved. The older entrepreneurs tended to be owners of real estate, and few had businesses as such to pass on to their sons. Several had very large families and lived long lives, in a sense forcing their sons to strike out on careers of their own. All but four of the twenty-three sons of the industrial entrepreneurs returned to Harrisburg after college and remained there for the rest of their lives, fashioning careers in or on the fringes of their fathers' businesses.[16] Of the four who left, two again were elder sons. Charles Bailey's oldest dealt in real estate in Seattle a few years before returning home to develop land north of Harrisburg,

THREE GENERATIONS OF BUSINESS AND FAMILY TIES AMONG THE McCORMICKS, THE CAMERONS, AND THE HALDEMANS OF HARRISBURG*

(Showing only those active in business in Harrisburg)

JAMES McCORMICK (1801-70)

TRANSPORTATION
Hbg Bridge Co. pres 1856-70
BANKING
Dauphin Deposit pres 1840-70
INDUSTRIES
Hbg Cotton Co. pres 1849-65
Paxton Iron owner 1857-70
W Fairview Nail Works owner 1859-70

HENRY McCORMICK (1831-1900)

Dauphin Deposit pres 1870-74
McCormick Trust trustee 1870-97
General mgr and pres:
Lochiel Iron 1871-85
Paxton Iron 1857-61, 1866-97
W Fairview Nail Works 1866-97

MARY McCORMICK (1834-74)——————

JAMES McCORMICK (1832-1917)

Dauphin Deposit pres 1874-1908
McCormick Trust trustee 1870-1917

HENRY B. McCORMICK (1869-1941)

Dauphin Deposit dir
Hbg Bridge Co. dir

DONALD McCORMICK (1868-1945)
Dauphin Deposit pres 1908-45
Hbg Bridge Co. dir
JAMES McCORMICK JR (1863-1943)
Paxton Flour Mills mgr
McCormick Farms mgr
HENRY McCORMICK JR (1862-1939)
Paxton Iron & Steel mgr
ROBERT McCORMICK (1878-1925)
Dauphin Deposit dir
Harrisburg Bridge Co. dir

VANCE C. McCORMICK (1872-1946)

McCormick Trust trustee
Dauphin Deposit dir
Central Iron & Steel dir 1897-8,
1916-41 & pres 1939-41
Mayor of Hbg 1902-5

NO FOURTH GENERATION OF McCORMICKS

*Spouses, if any, omitted unless significant.

.SIMON CAMERON (1799-1889)

TRANSPORTATION
State Works contractor 1830s
Northern Central RR investor 1854-?
BANKING
Middletown Bank cashier 1832-50
Dauphin Deposit investor 1830s-40s
Cameron, Calder & Eby 1860-65
1st National Bank investor 1865-89
INDUSTRIES
Lochiel Iron co-founder 1864-71

OTHER
US Senator 1845-49, 1857-61, 1867-77
US Secretary of War 1861-62
Hbg Telegraph investor 1850s-70s

——— *m.* J. DONALD CAMERON (1833-1918)

Northern Central RR dir 1861-74 & pres 1863-74
Middletown Bank cashier 1850-62 & pres 1882-94
1st National Bank dir 1865-?
Dauphin Deposit dir & investor
McCormick Trust trustee 1870-1918
Lochiel Iron investor
Hbg Cotton Mill owner/investor
US Senator 1877-97

MARGARETTA CAMERON (1837-1915) *m.* ———⟶

cont'd next page

JAMES McCORMICK CAMERON (1865-1949)

Central Iron & Steel dir
Dauphin Deposit dir

NO FOURTH GENERATION OF CAMERONS

JACOB M. HALDEMAN (1781-1856)

TRANSPORTATION
Hbg Bridge Co. dir 1814-56 & pres 1846-56
BANKING
Hbg Bank dir 1814-53 & pres 1853-56
Dauphin Deposit investor 1840s-50s
INDUSTRIES
Hbg Cotton Mill major investor 1850-56
Hbg Car Works, co-partner 1853-56

MARY EWING HALDEMAN (1814-73)
m. ROBERT J. ROSS (1807-61)
(positions held by Ross)

Dauphin Deposit cashier 1839-61
Paxton Iron co-owner 1857-61
Hbg Cotton Mill investor 1850-61

MARGARETTA CAMERON
(from previous page)

m. RICHARD HALDEMAN (1831-85)

Hbg Patriot owner-editor 1857-60
Hbg Bank investor
Hbg Bridge Co. investor
US Congressman 1869-73

OTHER SONS:

JACOB S. HALDEMAN (1823-89)
Hbg Bank dir 1852-66 & pres 1864-65

JOHN HALDEMAN (1821-65)
Dauphin Deposit dir 1849-64

NO HALDEMANS IN THE THIRD GENERATION HELD
MAJOR BUSINESS POSITIONS AT HARRISBURG

much of it owned by his father.[17] Hickok's elder son, like his father, was an inventor. Apparently their natures were too similar to work together in harmony. After a brief stint at the Eagle Works, he left to try his luck with a steam-heated chicken hatchery he invented, then settled in Berks County.[18] Of the two younger sons who left home, James Bailey went into the iron business in Philadelphia, then in 1902 returned to Harrisburg as general manager of Central Iron & Steel. He boldly attempted to modernize and integrate the plant, but in the process the firm fell into receivership. Bailey moved permanently to Bryn Mawr.[19] James McCormick Jr.'s son William went into journalism, working for Boston, Philadelphia, and Bethlehem papers before founding first the *Allentown Leader,* and later the *Reading Herald.*[20]

All the others moved with apparent ease into family businesses or related enterprises. Most of the industrial entrepreneurs, after all, could offer their sons a variety of choices: factories to manage, banks to run, partnerships in law firms, or farms to supervise. If none of these appealed, they were assured incomes adequate to permit them to devote their lives to charitable, religious, and other good works.[21] For example, neither of Colonel Henry McCormick's sons was especially interested in business. Both did their duty, serving in managerial positions and on the boards of the various McCormick interests, but they turned with greater enthusiasm to their real interests. The elder, Henry Buehler, devoted himself to Pine Street Presbyterian Church, the Harrisburg Hospital, the Harrisburg Cemetery, Lincoln University, and the Harrisburg Country Club. Vance, who had been an All-American football star at Yale, became active in journalism and politics. For many years he was editor and publisher of the *Patriot,* sat on the city council, was mayor of Harrisburg (1902–5), made an unsuccessful bid for the governorship of Pennsylvania in 1914, and headed the National Democratic party during the presidential campaign of 1916. Both Henry and Vance married late in life and had no children.[22]

Although many of the second generation were dedicated and hardworking, most found far more time for social, charitable, and community work than their fathers had. Few if any of the sons exceeded the accomplishments of their elders. Having begun at the top, they apparently were less driven to succeed. Simon Cameron, when someone pointed out the "fine advantages" his son had received, replied, "Yes, he has had more than his father, but there is one supreme advantage that he has never enjoyed—the stimulus of poverty and hardship."[23]

Front Street remained the preferred address of seventeen of the nineteen sons of industrial entrepreneurs who remained in Harrisburg. They either moved into the houses of their parents or acquired homes of their own along the river. The two sons of Jacob Eby were exceptions. They had homes on Third Street and on State. The Haldeman and McCormick children continued the tradition of spending summers on their family farms in Cumberland County. Improved transportation and changing fashions inclined some of the others to go farther afield for their summer vacations. A favorite watering place by the turn of the century was Eagles Mere in Sullivan County, originally adopted by the Baileys and the Reilys.

The "Front Street Set," as they and their allies and neighbors were collectively known, determined the business and social tone of the city. At one time or another, most of the entrepreneurs actively supported the Harrisburg Board of Trade, or its successor, the Harrisburg Chamber of Commerce. They were also instrumental in founding and supporting the Harrisburg Club in 1897. Located at Front and Market and open only to males who could afford its high dues, the club featured waiters in livery who served drinks, luncheons, and dinners to those wanting a quiet place to meet and discuss business. Early in the next century, similar though less-exclusive imitations (the University Club, the Engineers Club, and the Civic Club) appeared. Eventually all gave way to the more socially oriented Harrisburg Country Club, located in the suburbs and open to both men and women.[24]

By the turn of the century the entrepreneurial families had become more exclusive and rigid. They had three social "assemblies" each season, which were "very formal," and "the line was drawn closely of those invited." The leading families sponsored "germans" and cotillions, and coming-out parties for their daughters and frequently held large, formal afternoon receptions. Mrs. Margaretta Haldeman, the widowed daughter of Simon Cameron, excelled at these functions. "Her house was large and service perfect and she a gracious hostess." Each Fourth of July the Henry McCormicks gave a large party at Rosegarden, ending the day with a fireworks display.[25]

The grandsons of the entrepreneurs, coming of age in the 1890s and afterward, received the same schooling as their fathers. They too started at the Harrisburg Academy, where they received excellent grades. "I was a poor student," Ross A. Hickok confessed, "but my marks were good as, I learned later, they marked well to please the parents." With few exceptions the boys went on to the Hill School, St. Paul's, or Phillips-Andover, and then to an Ivy League college—Yale remaining the favorite. Daughters attended either the Harrisburg Female Seminary (which did not last many years) or private finishing schools elsewhere. At least one, Sarah Fleming, graduated from Vassar.

Who their children married had become too important a matter for

Fig. 17. Two North Front Street Scenes, 1900. *Upper:* Looking north from Walnut Street. The mansion was that of the second James McCormick. *Lower:* Looking north from Pine Street. The principal building in view served as the Commonwealth's Executive Mansion from 1864 to 1891.

the entrepreneurial families to leave to chance, so they managed the social lives of their children carefully. During the 1890s they sponsored formal cotillions for the young, usually monitored by two widows with children of marriageable age, Mrs. Richard Haldeman and Mrs. William O. Hickok Jr. As "patronesses" of these affairs, they screened all invitations and served as chaperones. Each fall, in the same years, the sons of the elite gave a house party at Inglenook, a clubhouse on the Susquehanna several miles north of the city. About ten young men would each invite a young woman, usually "his best girl," and two chaperones, one of whom was always Mrs. Hickok. According to Ross Hickok, "a girl was considered very lucky to receive an invitation." The young couples would spend a week taking long walks, climbing the mountains, or swimming by day. Often they planked fish on an open fire on the beach. In the evenings each man would take one of the young women canoeing, gathering the boats together at the end of the island at ten o'clock for singing and setting off fireworks. No liquor was allowed, and at night the boys slept on cots in a large room on the ground floor while the girls slept in rooms on the second floor.[26]

Early in the twentieth century the grandsons of the entrepreneurs, as their fathers before them, took up positions in the family banks and business houses. The Camerons, the Ebys, the Hildrups, and the Mc-Cormicks, however, were at the end of their male lines and were gone by the end of World War II. The male line of Baileys died out soon after. Of the other families, today only the Hickoks continue to operate their original family business. Some of the other entrepreneurial families remain socially prominent in the community, others apparently moved away. None heads or manages any of Harrisburg's larger industries or banks.

Other Elites

In discussing Harrisburg's industrial interlude, it is easy to think of the community as consisting chiefly of the major entrepreneurs and the masses of people they employed. But many residents fell into neither category. Immediately below the great industrialists in wealth and influence were the town's older aristocratic families: the established wholesale and retail merchants and merchant-manufacturers, and the professionals, both the long-resident physicians, bankers, and lawyers and the more transient newspaper editors, clergy, and chief officials of state government. There were also new dignitaries created by the rise of factories: the upper-level managers hired to run them, and the owners of middle-size manufacturing enterprises. People from these groups plus those from the entrepreneurial families made up Harrisburg's upper classes. Below them in stature were the remaining white-collar groups: the dentists, the schoolteachers, and other semi-professionals; the growing army of clerks (government, industrial, and commercial); salespeople; and performers of various nonmanual services.

A close look at all these groups is not feasible, but the merchants and merchant-manufacturers, the professionals, and the lesser industrialists deserve consideration because of their particular ties both to those above them and to those below. The relationships of each of these groups with the entrepreneurial families were complex, combining a mild animosity with respect and deference. For example, the leading merchant and professional families who once dominated the town no doubt resented their loss of stature as the industrialists suddenly acquired great wealth and influence. At the same time, most of the entrepreneurs had sprung from these very groups, especially from the banking and legal fraternities. Eby had been a merchant. Outsider George Bailey, co-founder of the Central Iron Works and later its manager, held a medical degree, as did native-son George W. Reily. Elder, both Camerons, Forster, Haldeman, and the two James McCormicks had been and continued to be bankers. Ayres, Elder, Fleming, Forster, and the two James McCormicks were all members of the Dauphin County bar, and Donald Cameron was reading law at the time he entered his father's bank. Although none had been a clergyman or newspaper editor, several sons of the entrepreneurs—even those destined for careers in family firms—studied for and entered each of the various professions. Similarly, some sons of entrepreneurs became merchants.

The links between lawyers and entrepreneurs were especially close. By mid-century the legal profession had become increasingly involved in commercial and business law. Railroads and industries depended on lawyers to promote their interests by conducting negotiations for them and shaping legislation to meet their particular needs. Lawyers also assisted by advising them on the shadings between what was lawful and what was not and by defending them in court. Given their access to information and the growing dependence of business on legal expertise and services, it is not surprising that some lawyers took the extra step and themselves became business, banking, or industrial leaders.[27]

Other professionals, especially the more transient, may well have resented being "owned" by entrepreneurs or being overly dependent on them. Harrisburg's editors, for example, though almost always outsiders brought in by some political faction to run its newspaper, had of course always served the interests of those supporters. Isaac McKinley, a native of Virginia who long edited the *Democratic Union,* and Theophilus Fenn of Connecticut, editor of the Whig *Telegraph,* relied on political patrons for government printing contracts to finance their operations. Any limited independence they may have enjoyed, however, was lost once former editor Simon Cameron became both a state political operator and a major entrepreneur. His combined political savvy and enormous financial resources enabled him to buy and sell newspapers and editors as needed to serve his personal political ends. In the late 1840s and early 1850s, when he was a Democrat, his control of the *Democratic Union* was so blatant that Buchanan Democrats established the rival *Patriot*. In the mid-1850s, when he courted the Know-Nothings, he wrangled control of the *Telegraph* and installed the rabid nativist Stephen Miller of Philadelphia as its editor; then, a year and a half later, upon becoming a Republican, he helped German-born George Bergner purchase the *Telegraph* and become its editor and the mouthpiece of the Cameron machine. The *Patriot* too from time to time fell into the hands of entrepreneurial family members—Richard J. Haldeman, for example—buying into and editing it briefly in the late 1850s. At the turn of the century, Vance McCormick bought the *Patriot,* edited it for a few years, and controlled it well into the twentieth century.[28]

The clergy as well were nearly always dependent on the generosity of the wealthier members of their congregations. With the entrepreneurial families contributing heavily to church budgets and sitting on their governing bodies, it was the brave pastor indeed who preached against cruel masters or dwelt on the difficulties of the rich entering heaven.

Transient high officials of state, especially governors, were actually sought out by the entrepreneurial families. Simon Cameron, for example, was related distantly by marriage to Governor John A. Schulze and served as his adjutant general; the Haldemans entertained and were intimate with Schulze and his family, and a grandson of William O. Hickok married a daughter of Governor Daniel H. Hastings.[29] In a few instances, governors or their families remained in Harrisburg after leaving office— for example, David R. Porter, who established the town's first iron furnace; William F. Johnston, who became a local merchant; and the families of Francis Shunk and John F. Hartranft.

In tracing the impact of industrialization on the careers of individual members of these groups and the prospects of their sons between 1850 and 1870, it was not possible to distinguish merchants from merchant-manufacturers. The distinction was not clear in their own minds. They usually told population census takers that they were merchants even as they listed themselves as manufacturers in the corresponding manufacturing census schedules. Also, only merchants and merchant-manufacturers who owned real estate were included, in order to rule out those who exaggerated when they told census takers they were merchants or manufacturers. This group, nonetheless, increased much more rapidly between 1850 and 1870 than lawyers and physicians, including those who were landless (see Table 7.2). The procedure increased the percentage of merchants and merchant-manufacturers who were heads of family (com-

Table 7.2. Number, Heads of Household, and Ethnicity of Merchants and Merchant-Manufacturers, Lawyers, and Physicians

Year	No.	Increase	Heads of Household	White, American-born	Foreign-born	African American
Merchants & Merchant-Manufacturers						
1850	40	—	95.0%	92.5%	7.5%	0.0%
1860	89	123%	93.3	83.1	16.9	0.0
1870	171	92	97.7	76.0	23.3	0.6
Lawyers						
1850	40	—	50.0	100.0	0.0	0.0
1860	39	−3	69.2	100.0	0.0	0.0
1870	50	28	70.0	96.0	2.0	2.0
Physicians						
1850	20	—	85.0	80.0	10.0	10.0
1860	22	10	54.5	90.9	4.5	4.5
1870	40	82	85.0	92.5	5.0	2.5

Source: Computerized population census schedules, 1850, 1860, 1870.

pared with the two professional groups) by excluding sons who were in the same category but did not yet own real estate.

Foreign-born and African Americans were consistently among the merchants and merchant-manufacturers, and much more so than in the professions. This was not only because of professional barriers against minorities but also because minorities constituted a market for groceries, clothing, and other supplies that members of those groups rushed to meet. That was less true for lawyers and physicians. A lawyer could not practice or appear in court without being admitted to the bar, so until after the Civil War the law was generally closed to minorities. Harrisburg's one black lawyer in 1870, T. Morris Chester, though admitted to the bar, was in Harrisburg only temporarily. Ordinarily he practiced in other states. The lone immigrant lawyer was also a short-term resident of the community. As for physicians, neither of the black "doctors" included in the study was a bona fide trained physician (whether contemporary or modern medical standards are used), but they were included because they served people of their own race in this capacity.

By definition for this study, the merchants and merchant-manufacturers all owned real estate (see Table 7.3). So did half the lawyers and nearly as many of the physicians. The average value of land held by the professionals was twice that of merchants and merchant-manufacturers. When personal property (including inventories of goods in their stores and factories) was added in 1860, the average total holdings of the merchants group exceeded that of the two professions. When, by 1870, lawyers had invested heavily in the securities of the industries and banks they advised, they suddenly had total holdings more than twice as valuable as those of either the merchants or the physicians.

These same three groups had high rates of persistence in the community, which is not surprising (see Table 7.4). Although it is not evident how many died or left town between censuses, or simply had been skipped, nearly half or more of each group remained in Harrisburg between 1850 and 1860. Of the 1850 contingent, between a quarter and two-fifths were still there after two decades. As might be expected, no lawyer or physician changed occupation, though by 1870 more than a third of the remaining merchants and merchant-manufacturers had. Of those who changed, it appears that a few either failed as merchants or went into semi-retirement—in most instances returning to a craft or trade practiced earlier. One, however, was reduced to clerking in a store; another rose to a managerial position in industry.

Fewer sons of Harrisburg's merchants and merchant-manufacturers, lawyers, and physicians remained in the community than sons of the later industrial entrepreneurs (see Table 7.5). Of the fifty-five sons of

Table 7.3. Real Estate (RE) and Personal Property (PP) Holdings of Merchants and Merchant-Manufacturers, Lawyers, and Physicians

Year	No.	RE Holders	Avg. Holding	PP Holders	Avg. Holding	Total	Avg. Holding
			Merchants & Merchant-Manufacturers				
1850	40	100.0%	$ 6,914	n/a	n/a	100.0%	$ 6,914
1860	89	100.0	8,687	89.9%	$ 5,216	100.0	13,375
1870	171	100.0	12,319	92.4	5,437	100.0	17,343
			Lawyers				
1850	40	37.5	$12,020	n/a	n/a	37.5%	$12,020
1860	39	51.3	11,215	76.9	6,033	79.5	13,074
1870	50	50.0	25,272	54.0	34,078	60.0	47,187
			Physicians				
1850	20	45.0	$13,722	n/a	n/a	45.0%	$13,722
1860	22	45.5	12,100	63.6	2,693	68.2	10,580
1870	40	42.5	27,006	62.5	5,644	67.5	22,230

SOURCE: Computerized population census schedules, 1850, 1860, 1870.

n/a = not available.

Table 7.4. Merchants and Merchant-Manufacturers, Lawyers, and Physicians: Persistence in Harrisburg and in the Same Occupation

Year	No.	% In Harrisburg	% In Same Occupation	% Retired	% In Another Occupation
			Merchants & Merchant-Manufacturers		
1850	40				
by 1860	26	65.0%	69.2%	15.4%	15.4%
by 1870	17	42.5	29.4	35.3	35.3
			Lawyers		
1850	40				
by 1860	18	45.0	94.4	5.5	0.0
by 1870	11	27.5	100.0	0.0	0.0
			Physicians		
1850	20				
by 1860	10	50.0	100.0	0.0	0.0
by 1870	6	30.0	100.0	0.0	0.0

SOURCE: Computerized population census schedules, 1850, 1860, 1870.

Table 7.5. Occupations of the Sons of Harrisburg Merchants and Merchant-Manufacturers, Lawyers, and Physicians, 1850–1870

Sons of	No.	Executives	Professionals	Merchants	Clerks	Artisans	Industrial Workers	Laborers
Merchants & merchant-manufacturers	32	0 0.0%	6 18.8%	10 31.3%	12 37.5%	2 6.3%	2 6.3%	0 0.0%
Lawyers	13	1 7.7%	6 46.2%	1 7.7%	4 30.8%	0 0.0%	1 7.7%	0 0.0%
Physicians	15	0 0.0%	8 53.3%	0 0.0%	2 13.3%	0 0.0%	0 0.0%	5 33.3%

SOURCE: Computerized population census schedules, 1850, 1860, 1870.

merchants and merchant-manufacturers listed in the 1850 census, thirty-two (58 percent) were shown with occupations in the censuses of 1850, 1860, or 1870; the same was true of thirteen of twenty-nine lawyers' sons (45 percent) and fifteen of twenty-four sons of physicians (62.5 percent). With rare exceptions they followed their fathers into the same or closely related occupations. The younger ones began in clerical or other white-collar jobs, but it is safe to assume that they eventually moved into more prestigious positions. The largest departure was that of sons of merchants and merchant-manufacturers, nearly one-fifth of whom went into the professions. Five of six sons of lawyers who went into the professions were lawyers; the other became a civil engineer. Seven sons of physicians adopted the profession of their father; one read for the law. Only three sons from the combined groups went into industry in managerial or white-collar positions. That five sons of physicians became laborers is misleading. All were sons of Harrisburg's two black physicians, and the healing skills their fathers possessed apparently had not been passed on to them.

Another elite group was the lesser industrialists, a new and relatively small body of men who appeared almost entirely after 1855. Not as rich or powerful as the entrepreneurs, they also had less influence. Other than the mills of the entrepreneurs, the number of firms with twenty-five or more employees grew from one (a large shoe-shop) in 1850, to two in 1860, to fourteen in 1870 and thirteen a decade later. The number of firms with between ten and twenty-four employees was similarly low: three in 1850, twenty in 1860, fourteen in 1870, and fourteen in 1880. Together these manufacturers produced only a fraction of the goods made in Harrisburg: 21 percent in 1850, 19 percent in 1860, 24 percent in 1870, and 13 percent in 1880. They rarely provided jobs for more than a quarter of the industrial work force: 19 percent in 1850, 29 percent in 1860, 25 percent in 1870, and 16 percent in 1880.

Their relationships with their more powerful counterparts again were mixed. They did share a common endeavor and faced essentially similar problems. Still, some owned much older firms that had evolved into industries from craft shops or merchant-manufactories and along the way had been taken over by the major firms. Others, such as the owners of the Wister Furnace and a number of small rolling mills and machine shops, were quasi-rivals of the giant firms. At the same time, they and the firebrick manufacturers, limestone quarriers, and lumber mill operators, to name the more obvious, were also largely dependent on the giant firms to buy up and use the output of their firms. Given the lateness of their appearance, it is not possible to trace their careers or those of their sons through the censuses.

The resentments and antagonisms of all the middling groups aside, many of their members craved social contact with the wealthy and emulated their life-style as far as means allowed. The more successful among them constructed large homes on Front Street or within a block of two of it. Many of them sat on the boards of directors of the city's banks. They contributed to and were members of the governing bodies of the same churches, the library, the hospital, and other charitable institutions. Some rubbed elbows with the wealthier classes at the Harrisburg Club, more joined with them as members of the Harrisburg Board of Trade, and many attended the same social functions and took their summer outings too at Inglenook and Eagles Mere. Fortune smiled from time to time, when a son or daughter married into an entrepreneurial family.

Taken together, these lesser elites might well be thought of as the mediating center between the handful of industrialists at the top and their many hundreds of employees below. Because clashes between the extremes were infrequent at Harrisburg, and usually ended peacefully, instances of overt mediation were rare. There can be little doubt, however, that these groups served as a moderating force throughout the period. This role was perhaps less essential in Harrisburg than in single-industry communities or where absentee owners controlled the major businesses. Harrisburg, after all, had no single industry, firm, or family that completely dominated economically or socially. Just as the rivalry between the Baileys and the McCormicks had a healthy impact on their respective firms, so its reverberations in the city's banks, churches, politics, and social and charitable institutions worked against absolutist behavior by either family or by any of the city's other industrialists.[30]

None of the ties of these groups with the entrepreneurs put them totally on the side of the major industrialists, however. Their interactions may have predisposed them in times of trouble to defend the interests of the entrepreneurs, but, in any event, their own holdings would have inoculated them against rash assaults on property rights. Nearly as compelling, and sometimes more important, however, were their economic relationships with the working classes. The fortunes of the professionals and merchants to varying degrees depended on and fluctuated with those of laboring families. Prosperity or adversity in the mills was reflected directly in the sales of merchants and the fees of professionals. Factory payrolls, after all, provided the workers with the means to buy groceries, rent homes, pay medical and legal fees, subscribe to newspapers, and fill church pews and honor pledges. Thus the mediating role. In times of trouble in many communities such mediation was direct and open; to the degree that it occurred at Harrisburg, as will be seen, it was a quiet and steady call for reason and toleration, the avoidance of trouble, the urging of moderation, and the preaching of social harmony.

Recruiting and Persistence
of Industrial Workers

I n the early stages of industrialization, factories were not manned by armies of semi-skilled and unskilled machine tenders directed by a few managers and foremen. For most industries, machinery was still relatively unsophisticated, and production was carried on by a combination of skilled artisans, unskilled helpers, and semi-skilled machine tenders. Textile manufacturing was an exception, and Harrisburg's cotton mill employed only a few artisans. All the town's other factories, however, hired a mix of trained journeymen and unskilled laborers. The skilled artisans performed essential functions that required their special talents and served as mentors to the unskilled employees or as foremen and supervisors. These "factory artisans" were transitional figures who continued to practice their trade much as they had in craft shops, but with an important difference: most no longer completed all stages of production, but employed only part of their skill, executing one or two steps in the process and leaving the rest to other artisans or unskilled workers and machines to complete.[1] Factory artisans will be discussed in greater detail in Chapter 9. Here, although the recruiting of all industrial employees is involved, the focus will be on the unskilled and semi-skilled factory hands.

The promoters of industrialization in Harrisburg had declared, among other things, that factories would create more and better jobs for the town's residents, especially the poor.[2] In fact, the coming of industry added quite substantially to the number of employment opportunities. Exact numbers cannot be determined, but data from manufacturing and population schedules indicate that the railroads and new industries directly created approximately 2,600 positions by 1870 (Table 8.1). That amounted to slightly more than the total number of males in Harrisburg

Table 8.1.   Creation of Jobs by New Industries

Item	Census Schedule	1850	1860	1870
Persons with listed occupations	Pop Sched	2,336	4,269	7,518
Industrial workers				
Cotton Works	Mfg Sched	n/a	300	280
Eagle Works	Mfg Sched	n/a	42	75
Car Works	Mfg Sched	n/a	140	442
Iron & steel industry	Mfg Sched	112	91	1,092
Other industries	Mfg Sched	n/a	n/a	331
Pa. Railroad shops	Mfg Sched	n/a	n/a	131
Other railroad workers	Pop Sched	20	159	250
Total		132	732	2,601
Percent of listed occupations		5.7	17.1	34.5

SOURCE: Computerized population and manufacturing census schedules, 1850, 1860, 1870.

n/a = not available (the firms did not exist at that time).

between the ages of sixteen and sixty with listed occupations in the 1850 census. In effect, industrialization had within twenty years created the equivalent of a new job for every member of mid-century Harrisburg's male work force—at a time when the number of positions outside of industry was also increasing substantially. Many of those new nonindustrial jobs, it should be added, were by-products of the industrializing process.

The rising industries clearly created jobs faster than existing residents or their progeny could fill them. The remaining positions acted as a powerful magnet attracting swarms of newcomers to the community.[3] Still, the question remains, did long-term residents of the community get many or most of the new jobs or did they go to recent immigrants? Approximately half the employees of Harrisburg's industrial and railroad firms in 1860 could be found in the population census schedules for that year (see Table 8.2). Although a fifth of those identified had lived in Harrisburg for ten or more years, they made up little more than a tenth of the entire industrial force that year. A decade later, only slightly more than a third of industrial workers listed in the manufacturing census schedules could be found in the population schedules. Of that group, only one in seven (about 4 percent of all industrial workers in 1870) could be traced to the 1860 census. No doubt a portion of the factory workers lived in the area immediately outside the city, and others had moved to Harrisburg during the decade since the 1850 census, but even

Table 8.2. Employment of Established Harrisburg Residents by New Industries as of 1860 and 1870

1860 Employees	Manufacturing Schedule	Identified in 1860 Population Schedule		Previously in 1850 Population Schedule	
Industry	Total Employed	No.	% of Total	No.	% of Total
Cotton Works	300	161	53.7	31	10.3
Car Works	140	16	11.4	4	0.2
Eagle Works	58ª	46	79.3	19	32.8
Iron industry	91	87	95.6	11	12.1
Subtotal	589	310	52.6	65	11.0
Railroads	n/a	159	n/a	32	n/a
Total	n/a	469	n/a	97	n/a

1870 Employees	Manufacturing Schedule	Identified in 1870 Population Schedule		Previously in 1860 Population Schedule	
Industry					
Cotton Works	280	98	35.0	17	6.1
Car Works	442	86	19.5	9	0.2
Eagle Works	78ª	45	57.7	20	25.6
Iron industry	992	408	41.1	30	3.0
Subtotal	1,792	637	35.5	76	4.2
Railroads	—	663	n/a	116	n/a
Total employed	n/a	1,300	n/a	192	n/a

SOURCE: Computerized population census schedules, 1860, 1870.

n/a = not available.

ªCompany payroll, first week of July. The manufacturing census reported 42 employees in 1860, 75 in 1870.

if they could be identified and added in, that probably would not alter the likelihood that the great majority of new factory jobs were filled by relative newcomers.[4]

From where, then, did Harrisburg's new industrial workers come? The possibilities were limited. The great numbers of foreign-born immigrants, particularly natives of Ireland and Germany, who poured into America after the mid-1840s immediately come to mind. But relatively few of either nationality settled at Harrisburg. The city's foreign-born residents did little more than keep pace with general population growth before 1870, and after that date their proportion in the population declined steadily (see Appendix A). Unlike many industrial centers, Harrisburg's industrial work force was not foreign in origin.[5]

There were several possible sources of American-born laborers. In that era, for example, large numbers of floaters, rural and urban, drifted from place to place and from job to job in quest of ways to make a living. Youngsters too, from all classes and backgrounds, entered their first paying jobs. Farmers, farm laborers, and sons of farmers caught between lessening opportunities in agriculture and expanding urban job markets were yet another source. So were such craft workers, fully skilled or otherwise, who had been displaced from their usual occupations in the shift from artisan to factory production.

It is unlikely that many owners of farms moved to factory jobs. For them, farming was usually a deeply rooted way of life not easily abandoned. Moreover, in south-central Pennsylvania even people with small acreages and poor soil could earn a decent living. The few for whom that was not enough gave up farming, moved to town, and took factory jobs. Whatever the course taken by the farmers themselves in that era, however, their sons and daughters, and farm laborers, were flocking to industrial centers in search of what they hoped would be better livelihoods.[6]

When communities industrialized, necessity sometimes forced artisans to accept factory employment. The new modes of production simply eliminated their previous occupations and left them no other choice. But that did not happen in Harrisburg. Few if any of Harrisburg's skilled masters who owned shops had to give them up for factory jobs. As with the farmers, however, children of artisans, and their apprentices and journeymen, often sought at factories opportunities that craft shops could no longer provide. In the end, Harrisburg did not depend on immigrants, discontented farmers, or displaced master artisans for its factory hands. Those jobs went chiefly to floating casual workers; to youngsters seeking their first jobs or leaving the farms or the crafts of their fathers (particularly where the fathers were in declining crafts); or to propertyless farm laborers and journeymen artisans who lacked means to set up shops of their own. Most skilled journeymen who went to the factories, however, worked as factory artisans rather than as machine tenders.

Whether drawn from the city's own population or from elsewhere, what induced workers to leave their existing jobs as farm, artisan, or casual laborers to enter the factories? For many women and teenage youngsters the opportunity for gainful employment was inducement enough. Preindustrial society offered plenty of work for these groups, especially for females, but provided little or no remuneration in cash. Housework or child tending were nearly the only paying jobs for young girls not needed at home, where they would have done the same work without pay. To attract them, the cotton mill (the one major factory in Harrisburg that employed girls) needed only to run an announcement in the local paper: "Employment for girls. Eight or ten stout, active girls, sixteen to twenty years of age, are wanted at the Cotton Factory, where they will find permanent employment."[7]

Boys living in urban areas often had nothing to do once they completed or dropped out of school, if contemporary newspaper complaints were accurate. They tended to gather in the streets, where they became nuisances—fighting, drinking, and smoking; harassing women, drunks, and blacks; disrupting church meetings; and otherwise disturbing the peace. Those who found paying jobs were not always happy with their work. In a letter to the editor of the *Telegraph,* a young clerk employed by a Harrisburg merchant complained of having to work from five in the morning until the store closed at ten at night. Mechanics, he pointed out, put in but ten hours a day. Why should merchants demand seventeen hours of their clerks? The editor corrected the boy: generally stores were open only fourteen hours, from six in the morning until eight at night, leaving clerks "10 hours for rest and improvement." Even so, the editor and "many merchants" favored more time off, but only if "not spent in street lounging, corner carousing, idleness or dissipation." Perhaps the lads, especially those who had no parents to direct or control their "habits and tastes," could fit out a place with "books, maps, apparatus, lecture room, and lectures," where they could improve themselves. Once they demonstrated that they would use their leisure time well, shorter hours would follow.[8] The factories did not run ads for boys. Notice of factory openings apparently passed from friends or family members already employed.

Throughout the first two decades of industrialization, the chief inducement appears to have been higher wages (see Table 8.3). On average, factory employees, excluding those of the cotton mill, in 1860 received an average of nearly 35 percent more than employees of craft shops. Although on average the factory premium had sunk to 2 percent by 1870, iron workers that year still earned from 16 to 33 percent more than the average craft worker.[9]

Because the cotton mill hired more females than males and relied very heavily on teenage employees, average earnings there amounted to half (or less) those at other industrial firms. Only the manufacturing census of 1860 provided separate data on the pay of females, but again women who worked in industry received higher wages than those who worked in craft

Table 8.3. Comparative Average Wages of Industrial and Craft Shop Workers, Harrisburg, 1850, 1860, 1870

Industry	1850		1860		1870	
	Avg. No. Employees	Avg. Wage per Mo.	Avg. No. Employees	Avg. Wage per Mo.	Avg. No. Employees	Avg. Wage per Mo.
Cotton Works	—	—	300	$13.00	280	$18.86
Eagle Works	—	—	42	26.14	75	34.44
Car Works	—	—	(140)	—	442	43.04
Porter Furnace	112	$21.96	—	—	—	—
McCormick Iron Works	—	—	40	25.00	160	59.38
Bailey Iron Works	—	—	40	35.00	180[a]	56.60
Lochiel Iron Works	—	—	—	—	510	51.78
Pa. Railroad repair shops	—	—	—	—	131	48.14
Other industrial firms	—	—	—	—	429	42.98
All industrial workers	112	$21.96	422	$17.53	2,197	$44.48
Without cotton mill	—	—	122	28.67	1,917	45.66
All craft shop workers	167	19.73	355	21.03	386	44.70
Factory premium		+11.3%		+34.7%		+2.1%

SOURCE: Computerized manufacturing census schedules, 1850, 1860, 1870.

[a]Chesapeake Nail Works only.

shops. At the cotton mill, female employees averaged $10.50 a month (males received $18), while the two women employed at Hickok's Eagle Works (the only other female industrial workers) averaged $15.00 a month (males averaged $26.70). The 112 women working in craft shops, almost all at sewing, earned an average of $9.27.

That factories paid wages in cash was also of considerable importance. Writing of this early period, William Hildrup of the Harrisburg Car Company noted that "any regular pay day to mechanics, as a rule, did not exist, payments being in provisions of various kinds, and at irregular times, and but little money." By contrast, his company "never speculated" off employees' wages, always paying "in cash or its equivalent." The company did buy coal and flour in quantity and made those items available to the workers at better prices than otherwise available.[10]

One other attraction of the factories was the implication that they would provide "permanent work," as the cotton mill advertised. After all, the size of the plants, the number of employees they required, the expensive machines they housed, and the reputations of their owners as substantial people of affairs all bespoke a steadiness exceeding that of small craft shops, merchant houses, and farms. For these advantages the new factory hands gave up the more informal work arrangements and perhaps greater independence they had enjoyed on the farm or in the craft shop.

The Harrisburg Cotton Manufacturing Company

The Harrisburg Cotton Company was designed as an integrated facility. Bales of raw cotton entered the plant, all steps in the manufacture were completed there, and the product left the mill ready for use. Because the managers intended to produce only coarse materials, such as shirting, sheeting, and bagging, operations were simple and required little skill. The primary tasks were carding raw cotton, spinning it into yarn, weaving it into coarse cloth, dressing or finishing it, and packing it for shipment. Almost all of this was done by steam-driven machinery that could easily be tended by unskilled women and youngsters. Once under way, the mill would provide between 250 and 300 jobs that, as the *Telegraph* put it on December 19, 1849, would "be a source of comfort and competency to the industrious poor." Labor, in other words, was to be cheap.

The American textile industry provided at least three major models of labor-recruiting to guide Harrisburg's entrepreneurs. That of the extensive Philadelphia textile mills was inappropriate. Mill owners there either hired skilled European immigrants to turn out a variety of high-quality textiles or subcontracted work to smaller independent weaving shops. The Harrisburg mill did not need skilled workers, nor were they available in the community. Neither was it necessary for the Harrisburg entrepreneurs to erect dormitories and recruit farm girls from the countryside to tend the looms, as was done at Lowell and other textile centers north of Boston. Their mill would require only a modest number of employees. Because they had relied heavily on Senator Charles T. James of Rhode Island in setting up the mill, they might have adopted the "South of Boston" labor system—which would have involved erecting company houses near the factories, renting them to workers with large numbers of children ages six to sixteen, and employing whole families. Again, that model did not fit the local situation.[11]

Like the New England models, the Harrisburg mill would need a few adult males to do a handful of skilled jobs, maintain the machinery, and supervise the rest of the workers. Given the wide variety of better-paying jobs for men in the community, few adult males would have been available for unskilled work at the wage offered to women and youngsters. The town, however, did have an abundance of unmarried young women and large numbers of boys between thirteen and nineteen who were eager for employment. Because such youngsters ordinarily lived with their parents, no special housing would be required. The company had only to open its doors and hire those willing to work at the wages offered.

The general makeup of the cotton mill's work force can be reconstructed from the manufacturing schedules of the federal censuses and from reports to Pennsylvania's secretary of internal affairs (see Table 8.4). Apparently the proportions of men, women, and youngsters under age sixteen working at the mill varied considerably between 1860 and 1878, the last year for which a report breaking down the labor force was published. At different times, from as few as a tenth to as many as a fifth of the workers were adult males—that is, males sixteen or over. Between four-tenths and half were females over fifteen. Youngsters constituted from just under a third to nearly half of all employees. Put another way, well over half of all cotton mill employees were females under twenty-one. The entrepreneurs who founded the mill had done as they promised: they were providing jobs for those classes usually of no use in the production of wealth—women and children.[12]

Population schedules for 1860 and 1870 provide more detailed information about cotton mill employees, but unfortunately, any conclusions must be accepted with caution because large numbers could not be individually identified in that source. For instance, of 100 male and 200

Table 8.4. Cotton Works Employees by Gender and Age, 1860–1878

Category	1860	1870	1873	1877	1878
Males	100(33.3%)	n/a	n/a	n/a	n/a
Adult		60(21.4%)	40(14.3%)	24(10.1%)	36(14.5%)
Females	200(66.7%)	n/a	n/a	173(72.7%)	n/a
Adult		135(48.2%)	109(38.9%)	n/a	124(49.8%)
Children	n/a	n/a	n/a	n/a	n/a
Under 16	n/a	85(30.4%)	131(46.8%)	n/a	89(35.7%)
Boys	n/a	n/a	n/a	41(17.2%)	n/a
Total	300	280	280	238	249

SOURCES: For 1860 and 1870, computerized manufacturing census schedules. For 1873, 1877, 1878, Pennsylvania Secretary of Internal Affairs, *Annual Reports*.

n/a = not available (various reports classified workers differently).

female employees reported in the manufacturing schedules for 1860, fewer than two-thirds of the males and half of the females were found in the population schedules. Even smaller percentages could be identified in 1870. Of 180 employees (60 males and 135 females over fifteen years of age, and 85 youngsters, male and female, age fifteen or younger), just over half the men, well under half the women, and between a fifth and a fourth of the children could be identified as cotton mill employees in the population schedules.[13]

Of those identified in the 1860 census, more than four-fifths were natives of Pennsylvania (see Appendix I). The proportion of immigrants (16 percent), for the most part Irish, was more than half again as great as the proportion of foreign-born living in the city (10.5 percent). On the other hand, although a tenth of Harrisburg's population was African American, no blacks were listed as cotton mill employees in 1860. Cotton mill employees were young and predominantly female. More than three-quarters were less than twenty-one years old, and nearly a third were under sixteen. Only one in twenty was over thirty. Of those older than fifteen, women outnumbered men by more than two to one.

With regard to the status of cotton mill employees in the households in which they lived, fewer than a tenth were adults with children to support. Of that group, two were widows and one was a married woman—and all had young children. At the other extreme, a quarter of the employees were tenants (or perhaps family members with different surnames) living in Harrisburg households or boardinghouses. Many appear to have been relatives from nearby rural areas. Nearly two-thirds of the identified workers were unmarried sons or daughters of Harrisburg residents.

Although the mill rarely hired parents and children from the same household, many employees were closely related (see Table 8.5). Seventy were brothers or sisters of other employees, and five children worked

Table 8.5. Gender and Family Relationships of Cotton Works Employees Identified in 1860 and 1870 Censuses

	Gender		Relationship to Another Employee	
	Male	Female	Sibling	Child
1860 (n = 161)	38.5%	61.5%	43.5%	3.1%
1870 (n = 98)	44.9	55.1	55.1	6.1

SOURCE: Computerized population census schedules, 1860, 1870.

with a parent in the mill, two with their mother, and three with their father.

Within a decade the percentage of foreign-born employees dropped sharply (see Appendix I). Three blacks, a female age thirty-four, and two males age twenty-eight and thirty-eight appeared on the payroll. The work force was also a little older and included more males (especially those over the age of sixteen). On the other hand, more members of the same families worked together: more than half of the work force were brothers or sisters of one another (see Table 8.5). In three instances parents worked with their children in the mill: a mother worked with a daughter and two sons, a father worked with a daughter, and another father worked with two sons. In addition to being older, a larger portion of the workers were adults with children to support. Fewer workers were tenants, and a slightly higher percentage were children of local residents.

The high percentage of unmarried youngsters who lived at home justifies further examination (see Table 8.6). One-quarter of the 103 identified sons and daughters in 1860 helped support widowed mothers. The fathers of the others were chiefly common laborers and craft workers. Only a few were factory or railroad employees. The fathers of a tenth of the employees were clerks, merchants, professionals, or other white-

Table 8.6. Selected Characteristics of Cotton Works Employees Living with Parents, 1860, 1870

| | | Employee Place of Birth | | | Parents |
| | | Pa. | Other State | Foreign | Foreign-Born |
Year	No.				
1860	103	82.5%	1.0%	16.5%	29.1%
1870	67	94.0	6.0	0.0	19.4

| | | Status or Occupation of Parent | | | |
| | | | | Industrial | | |
		Widow	Laborer	Worker	Craft Worker	Other
1860	103	25.2%	32.0%	7.8%	24.3%	10.7%
1870	67	25.4	26.9	23.9	19.4	4.5

		Parents Owning Real Estate
1860	103	68.0%
1870	67	23.9

SOURCE: Computerized population census schedules, 1860, 1870.

collar workers. In 1860 some 45 percent were foreign-born or children
of immigrants. Although a sprinkling of cotton mill employees themselves owned real estate in 1860, the parents of more than two-thirds of
the unmarried employees living at home were property holders.

Ten years later a quarter of the unmarried employees still lived with
widowed mothers. Smaller percentages of the fathers were artisans or
unspecified laborers, but the percentages who were factory or railroad
hands had increased substantially. Although not one of the identified
young people was foreign-born, nearly a fifth had foreign-born parents.
As for land ownership, a slightly greater proportion of all employees
themselves owned land, but the percentage of landholders among the
parents of young employees living at home was sharply lower.

No breakdown of workers by task, gender, age-group, or rate of pay
is available for the cotton mill before 1873, so determination of wages
before that date can only be very general. The manufacturing schedules
of the 1860 census, for example, report that male cotton mill employees received on average about 69 cents a day without regard for age.
Female employees, a very large percentage of whom were under the age
of sixteen, received approximately 40 percent less, or 40 cents a day.
All males working in craft shops that same year—for the most part
adults—were paid an average of $21.36 a month, or about 82 cents a
day. Eighty females working in clothing and shoemaking shops received
an average wage of $9.10 a month, or 35 cents a day. Twenty-eight
others, employed in milliner shops, made an average of $14 a month,
or 54 cents a day. Given the number of youngsters under twenty-one
whose wages enter into the cotton mill averages, the wages the mill
paid adults probably equaled or bettered those paid by small-scale
manufactories.[14]

In 1873 the cotton mill reported a work force of 280 (see Table 8.7).
Forty were males over the age of fifteen, half of whom operated the
plant's steam engine, kept the machinery in repair, or supervised the
other employees. The mill paid this elite corps between $2.00 and $2.66
a day. The remainder were unskilled workers earning between $1.57 and
$1.85 a day. Taken together, the adult males made up less than 15
percent of the payroll but took home almost a third of the total daily
wages paid by the mill. A second contingent of employees consisted of
109 young women, most of whom were between sixteen and twenty-one
years of age. Nine (probably the older and more experienced) received
more than $1.00 a day. The rest were machine tenders or packers, whose
daily wages ranged from 66 to 92 cents. This group constituted 39
percent of the workers and received 39 percent of the wages paid each
day. The balance of the operatives consisted of 131 youngsters, male and

Table 8.7.   Cotton Works Employees by Task, Age-Group, Gender, and Daily Wage in 1873

Task, Age, and Gender	No.	Daily Wage
Males 16 and Over		
Engineers	2	$2.66
Dresser foremen	2	2.45
Spinner foremen	3	2.41
Machinists	6	2.40
Weaver overseers	7	2.00
Packers	2	1.85
Watchmen and laborers	5	1.58
Carders	13	1.57
Subtotal males 16 and over	40	1.95 (avg.)
Females 16 and Over		
Dresser	1	1.16
Drawer frame spinners	8	1.05
Weavers	67	0.92
Carders	25	0.75
Packers	8	0.66
Subtotal females 16 and over	109	0.87 (avg.)
Youngsters Under 16		
Carders	27	0.62
Weavers	7	0.50
Spinners	91	0.50
Dressers	6	0.50
Subtotal youngsters under 16	131	0.52 (avg.)

SOURCE: Pennsylvania Secretary of Internal Affairs, *Annual Report, 1872–1873*, p. 385.

female, under the age of sixteen. Their wages ranged between 50 and 62 cents a day, and though they made up nearly half the total work force, they received only 28 percent of the daily wages paid out.

How many employees remained at the cotton mill and how many used it as a stepping stone to other work is difficult to determine. In part this is because apparently fewer than a third of them remained in Harrisburg. That the majority were young females added to the problem, because many of them married, took their husbands' family names, and so, even if they remained in town, could no longer be identified. Of 161 cotton mill employees of 1860 found in the census that year, only 30 could be located in the census schedules a decade later. Twenty were males, four of whom still worked at the cotton mill. Of the ten females, none was married, and the three with listed occupations were no longer cotton mill employees. Six of the females continued living with their parents, one lived with her brother.

In summary, the cotton mill provided most of its workers, who were youngsters, with low-paying, entry-level jobs comparable to those that fast-food chains would give similar groups nearly a century and a half later. The great majority of jobs in 1860 went to children of foreign- and native-born residents of Harrisburg who themselves had already acquired property, not to children of the town's more numerous propertyless families. Those property-holding parents apparently pushed their children to acquire good work habits, to begin accumulating money, and perhaps to contribute to the family income. For the most part, however, jobs at the cotton mill offered no future. Only four employees remained with the firm as long as ten years. Of the twenty-six others who can be traced, eleven males moved to better jobs with the railroads or the post office or moved into a craft. Five became laborers. Female employees, except for those who married and cannot be traced, became domestic servants or housekeepers or remained with their parents without gainful employment.

Hickok's Eagle Works

The characteristics of employees of Hickok's Eagle Works are known only because the firm has preserved all its payroll books since 1851 with few gaps. Without the names found there, Hickok employees would be unidentifiable, because they gave their trades, not their employer, in responding to census takers in 1850, 1860, and 1870. Only the manufacturing census schedules for those same years, and an 1878 report to the state on the firm's foundry operations, provide additional information on its work force.

The 1850 population schedule showed William O. Hickok as a bookbinder who owned no real estate. Deeply in debt and hard-pressed for cash, he had not yet constructed the Eagle Works. However, his 1851 payroll books show that he already had twenty-three employees, thirteen of whom could be identified in the 1850 census. The data suggest a young and largely untrained group of workers. Only three were married heads of family, and only one owned real estate. Of the others, six were sons of Harrisburg residents, one lived with relatives, and three were tenants. Of the thirteen, six were under the age of twenty-one and only one was over forty. Most—nine of thirteen—were natives of Pennsylvania, three (like Hickok himself) came from out of state, and one was foreign-born. One was an African American. Five listed themselves as cabinetmakers or machinists, one was (or had been) a druggist, and the others were either laborers or gave no occupation. Hickok and a handful of older skilled artisans apparently had to train a group of raw youngsters in the skills the firm needed to produce a variety of machines.

During the next two decades the work force changed (see Appendix I). For one, a higher percentage were natives of Pennsylvania, and on average they were also older. Where more than 45 percent had been under twenty-one and less than 8 percent had been older than thirty-nine, by 1870 approximately a quarter were under twenty-one and a fifth were forty or older. It is not surprising that this resulted in more than half being heads of household in 1870, whereas fewer than a quarter had that status two decades earlier. The proportions of tenants and sons decreased accordingly. Ownership of real estate also increased in the same period, from less than 8 percent to more than 20 percent. By 1870 well over half the employees claimed a skill.

Fig. 18. Employees of Hickok's Eagle Works, 1878. Hickok hired a wide variety of workers representing several different ethnic groups. The tall African American on the left holding a whip was probably the firm's wagon driver.

The average wage per Hickok employee in 1860—reported in the manufacturing census schedules—was $26.14 a month, or about $1.01 a day. This exceeded the average pay of craft shop employees by more than 20 percent. A decade later, the $1.32 a day paid by Hickok lagged behind what craft-type employees were paid by more than 20 percent (see Table 8.3 above). But such averages are somewhat misleading. The payroll books reveal a wide range of rates for the same specialty. Because his work force was small, Hickok knew his employees and their degrees of skill personally. He carefully graduated their wages, apparently basing them on each worker's value to the overall operation. For example, in July 1860 Hickok employed four cabinetmakers: G. W. Brown, age thirty-one, received $11.00 a week; T. C. Sample, age forty-six, received $8.50; J. Ritner, age twenty-seven, $8.00; and Christian Boyer, age twenty-two, $7.50 a week. A decade later Brown earned $20.00 a week while two new cabinetmakers earned $13.50 and another earned $12.00. The rates of other occupational groups in the plant were similarly differentiated. What Hickok paid skilled adult workers compared quite favorably with what they could earn in other factories. His average in 1870, however, was below that of other factories (excepting the cotton mill) and craft shops because he continued to hire a higher percentage of youths at low pay and train them on the job.

More than 40 percent of identified workers in 1860 and 1870 had lived in Harrisburg for at least a decade. The same trend continued through 1880. Of fifty-six persons on the payroll that year, at least thirty-two (57 percent) had been residents of Harrisburg in 1870. An increasing percentage of workers also persisted at the plant. Four of the identified workers in 1860 (nearly 7 percent) had been on the 1851 payroll. In 1870 eight (over a tenth) and in 1880 twenty (over a third) had been on the company payrolls a decade earlier.[15] Of the persisters, two of the 1860 group acquired real estate by 1870, whereas one of the 1870 group added to previous holdings and two others acquired land for the first time.[16]

In addition to those who remained with the firm, three of the 1851 work force and seventeen from that of 1860 continued to live in Harrisburg though otherwise employed. All the 1851 contingent appeared to be doing well. George Baily had acquired $10,000 worth of real estate and continued as a machinist, George Forbes had become a superintendent (of what is not recorded), and George Alvis, one of Hickok's two black employees, was a blacksmith with $1,500 worth of real estate. Of the 1860 group, six owned real estate: three had expanded their 1860 holdings and three had acquired land for the first time. Two German-born employees, Christian Boyer and Henry Fink, especially prospered after leaving the Eagle Works. Forming a partnership, they had become brewers, and as individuals they held real estate valued at $15,000 and $13,000 respectively in 1870. Others continued to practice the same trade as at Hickok's but now worked for others. Two were molders, two were machinists, one was a cabinetmaker, and one who had been a patternmaker became an enginebuilder. Four others went into business for themselves: a grocer, a saloon keeper, a watch repairer, and a merchant. Of the remaining five, one had become a shoemaker, one a watchman. The others listed no occupation.

Unlike the cotton mill, the Eagle Works offered its young employees training in a skill, and some a future with the firm. A few had extremely long careers with Hickok. The 1880 payroll included ten who had been with the firm for at least ten years, five for fifteen years, and five for twenty-four years. Those who chose to leave often found their trade marketable with other employers or had enough self-assurance and means to set themselves up in small businesses.

The Harrisburg Car Manufacturing Company

The first detailed account of the Harrisburg Car Works appeared during its second year of operation. According to a reporter for the *Harrisburg Morning Herald* of May 19, 1855, the firm made passenger, mail, baggage, box, cattle, platform, coal, and hand cars. Its annual capacity was from 300 to 500 eight-wheel cars. Eighty-four or so employees performed all stages of work, from blocking out the raw materials to painting and decorating the finished cars. Ten men worked in the wood-turning machine shop, where they used the latest equipment to rough out lumber, sawing, planing, drilling, and mortising it. Thanks to their machines, they were able to do the work of 100 ordinary mechanics. Thirty workers in the wood-turning shop finished the rough timbers and shaped them into cars. Meanwhile, in the foundry, a dozen employees made a variety of hardware castings, including wheels. Another dozen in the iron machine shop finished and polished the castings. Twenty blacksmiths and a handful of patternmakers, painters, and the like completed the labor force.

Any discussion of the characteristics of car works employees is seriously handicapped by the rapid growth in the number of workers from year to year and a lack of means for identifying them in the census population schedules. From 84 employees in 1855, the payroll grew to 140 by 1860 and to 442 by 1870. The only census schedule occupational designations that identified car works employees were those that had "car shop" or "car works" in the title. Because fewer than an eighth of the 1860 employees were identified, no firm conclusions can be drawn for that year. Even the identification of 86 of 442 employees (nearly a fifth) a decade later, falls short of being representative. The two principal groups—car inspectors and car shop employees—probably included few if any of at least half the firm's employees who were factory artisans: blacksmiths, carpenters, machinists, molders, patternmakers, and painters.

With these limitations in mind, what can be said of the firm's work force as found in the census schedules? Before the Civil War, Harrisburg had few workers trained in the skills required for building railway cars. William T. Hildrup, general manager of the works, followed a strategy similar to that of Hickok at the Eagle Works: he hired a large number of youngsters and, assisted by a few older skilled employees, trained them to be factory artisans.[17] The scant data from the census is consistent with Hildrup's strategy (see Appendix I). In 1860, still early in the firm's history, well over a third were under twenty, and fewer than a fifth were more than thirty-nine. A decade later nearly half the employees were in their twenties, which probably included a good number who had been trained on the job. The percentages of workers under twenty and over thirty had declined. Accordingly, the percentage of those who were heads of household increased, while that of sons and tenants decreased. Ownership of real estate declined notably. In 1860 most of the older hands (the skilled workers who were training the youngsters) owned property. By 1870 these younger men became the dominant age-group in the shops. By then only in their mid to late twenties and married, most had not yet acquired real estate.

The percentage of blacks employed at the car works in 1870 was almost the same as the overall percentage of that group in Harrisburg, while the percentage of foreign-born employees exceeded their proportion in the community at large. All eight of the blacks were "carshop employees" (less skilled than "car inspectors"), all were born in Virginia and Maryland (marking them as part of the post–Civil War wave of black migrants to the community), all were between twenty-one and thirty years old, and none owned real estate. In other words they were cheap labor. By contrast, only four of the twelve foreign-born were under thirty (none was under twenty-three), more than 40 percent were car inspectors, and half owned real estate, suggesting that they were hired for their skills.

Whether the car works tended to hire longtime residents of Harrisburg or to retain or lose employees to better positions cannot be determined with certainty from the few identified employees in 1860. Of the sixteen found for that year, four had lived in Harrisburg in 1850. Two then were sons of a stonemason, one was married and a baker by occupation, and the fourth was an unmarried chairmaker. Of the nine employees in 1870 who had lived in the town a decade earlier, two had been patternmakers, one had been a carpenter, one a laborer, and five were, respectively, sons of a bookbinder, a clerk, a lime manufacturer, a printer, and a railroad employee. Four of the 1860 employees continued to live in Harrisburg in 1870. Three were carpenters who held real estate and quite possibly still worked at the car works. The fourth had become a railway employee and owned no property.

The Iron and Steel Industry

The production of iron and, later, to a lesser degree, steel became Harrisburg's leading industry. It embraced blast furnaces for smelting pig iron, and rolling mills, naileries, foundries, and forges where it was processed into wrought iron, nails, castings, and fabricated items. In 1850 Harrisburg boasted one blast furnace, one rolling mill, one or two foundries, and several forges. A nail works stood directly across the river in West Fairfield. By 1860 a second furnace and a second rolling mill had been erected. A decade later there were three furnaces in operation, three rolling mills, and a new nail works in the city. By 1875 a peak was reached with the addition of three new furnaces (for a total of six), and two of the three large rolling mills were new and improved. This did not include the massive Pennsylvania Steel Works at Steelton just outside the city or the enlarged nail works at West Fairfield.

Unfortunately, no significant runs of iron or steel company payrolls were discovered, and census schedules identify no specific employees of the McCormick, Bailey, or other iron companies. As a consequence, none of the army of iron and steel workers could be traced to a particular firm. The census reports, however, do indicate that large numbers of workers were engaged in iron and steel production. These groups, taken together, make it possible to make generalizations about the characteristics of iron and steel workers and the recruitment patterns of the industry even if not of individual firms.

Harrisburg's new iron facilities had to go further afield to secure workers than the other major industries (see Appendix I). In both 1860 and 1870 they attracted the lowest percentage of employees who had lived in Harrisburg at least ten years, the lowest percentage of native

Pennsylvanians, and the highest percentage of out-of-staters. In the latter year they also employed the highest percentage of foreign-born workers. At the same time, before the Civil War, iron and steel drew a larger portion of the sons of Harrisburg residents than other industries. That changed after the war, when the percentage of sons declined and the proportion of tenants, for the most part outsiders and foreign-born, increased.

The industry's reliance on strong backs meant that most of the employees had to be young but mature adults. Although a fifth or fewer ironworkers were under the age of twenty, those younger than sixteen were very rare. Most were between the ages of twenty and thirty-nine. Fewer than a fifth were age forty or older. Of the major firms and industries that employed adult rather than child labor, iron and steel workers were the least likely to acquire real estate. In spite of the relative maturity of its workers and the fact that most were married heads of families, only the children in the cotton mill and the young work force at the Eagle Works in 1860 had lower proportions of property owners.

About one in four ironworkers of 1860 reappeared in the 1870 census. Half were still in the industry. Two had become proprietors of foundries: James M. Bay, a molder who had lived as a tenant in 1860, and William Jennings, who had learned the trade of his father's foundry and taken over the firm after returning from the war. Seven other persisters, including one African American, were molders. Three, including the African American, acquired real estate between 1860 and 1870. Of the two rolling mill employees who remained, one had property in 1860 and had added to his holdings over the decade. Those who left the industry followed a variety of paths. Materially the most successful was Edwin Cursin. An English-born boiler-plate maker with real estate valued at $1,500 in 1860, he had become a grocer by 1870, holding land worth $23,000 and personal property valued at $5,000. Four other former ironworkers who acquired real estate by 1870 had become, respectively, a railway baggage master, a station engineer, a stonemason, and a blacksmith. A seventeen-year-old molder's apprentice in 1860 was listed by occupation as "at home" in 1870, and two others, age twenty-nine and thirty-seven in 1870, had no listed occupation. The iron industry offered its employees much hard work but few rewards.

Railway Workers

Of all Harrisburg major industries, only the railroads were controlled from outside. None of the several lines operating into and through the town was owned, managed, or operated by local residents for any length of time.[18] Neither did the various companies keep track of which employ-ees were residents of Harrisburg. And so, as with iron and steel workers, railway employees living in Harrisburg have been treated as a whole here, not according to the companies that employed them. Railway work-ers fell into three major groups: those whose duties were performed in the stations and yards, those who actually operated the trains, and those who, after the Pennsylvania built repair and maintenance shops in Har-risburg, serviced and repaired rolling stock. Cutting across all three were between half and two-thirds of the work force who listed themselves in the censuses simply as "railroad employees."

Large gangs of transient workers, many of whom were Irish immi-grants, actually constructed the lines. Those that appeared in the census of 1850, however, were listed simply as laborers, not as railroad work-ers. Once the lines were built, railway operations called for mature, dependable employees. The companies hired almost no one under six-teen and few under twenty-one (see Appendix I). The great majority were in their twenties or thirties. A fifth of the work force was over forty years old in both 1860 and 1870. The railroads also hired relatively few foreign-born workers and almost no African Americans. Reflecting their greater age, more than two-thirds of all identified railway workers headed households; the percentage who were sons or tenants was corre-spondingly lower. In 1860 more than a third owned real estate, though by 1870 that had declined to only a fifth.

About a fifth of the 1860 railroaders and slightly more than one in six of the 1870 employees had lived in Harrisburg at least ten years. Of the 1860 contingent, nearly a third (51) remained in town at least until 1870, and more than a fifth (35) of the total went on working for the railroads. Of those staying with railroading, six acquired real estate and four others added to property already owned, but a dozen had no real estate in either 1860 or 1870. Of those who abandoned railroading but remained in town, John DeHaven acquired the most property. He had become a stone-mason and somehow enhanced the value of his real estate from $2,500 to $50,000. Others who added to the value of land already held in 1860 included men who became a stationary engineer, a carpenter, a farmer, a hardware clerk, and a laborer. Two laborers, a blacksmith, and a watch-maker had no property in either 1860 or 1870, and one laborer's holding in 1860 remained unchanged after a decade. The value of property belong-ing to a railroader who by 1870 had become a shoemaker and to two others who had become laborers declined in value. Railroading was one of the more rewarding lines of work open to working-class males.

Combined census data on all identified industrial and railroad workers in Harrisburg in 1860 and 1870 provide reasonably accurate answers to several questions about the early impact of industrialization on this middle-size community. Although the process created a flood of new jobs, the overwhelming majority went to new residents—people who had lived in the community less than ten years (see "Total" column, Appendix I). Even so, four-fifths of the workers in 1860 and three-fourths in 1870 were natives of Pennsylvania. Had the census given county, town, or township of birth, most would probably be found to have come from the rural townships of Dauphin County bordering on Harrisburg or from nearby counties in south-central Pennsylvania.[19] The frequency with which family names of newcomers coincided with those of older residents suggests many were "country cousins" of those already there.

Industrialization added little to the percentage of foreign-born residents. In both 1860 and 1870 the proportion of immigrants working in the factories exceeded that for the town as a whole: 13 percent in industry as compared with 11 percent in the community in 1860, and 16 percent compared with 12 percent in 1870. This, however, was not an indigestible lump of strangers who would be difficult to work with or to assimilate. There were, after all, only 366 more foreign-born persons in Harrisburg in 1870 than would have been there had the proportion remained the same as in 1850. Moreover, except for a noticeable increase of Welsh and English, the new immigrants continued to come from Germany and Ireland as before.

At the same time, the proportion of African Americans in Harrisburg did not increase. The sizable number of former slaves who moved to the city after the Civil War merely offset a large-scale loss during the 1860s of long-established black families. In any event, so far as can be determined, relatively few blacks found factory work before the end of the century. The impact of industrialization on racial and ethnic groups in the city is discussed more fully in Chapter 10.

Another change that might have been anticipated from the rise of factories—the destruction of craft shop production—did not occur, as we shall see in Chapter 9. Consequently, large numbers of displaced local artisans, shorn of their skills by new methods of production, did not become the primary or even a major part of the industrial work force. Rather, Harrisburg's new factories blended a small corps of skilled artisans who continued at their accustomed trade in a factory setting with a large number of young men (and some women) going to their first or second job. In 1860, when no factory in Harrisburg had been in operation more than ten years, more than a third of the employees were under twenty-one, half were in their twenties and thirties, and nearly 60 percent were unmarried. Only 13 percent were forty or older, and more than a fifth owned real estate.

A decade later the work force was older and more settled. Fewer were then under twenty-one, more were adults in their twenties or thirties, more were forty or older, and a strong majority headed households. Land ownership, however, declined. Factory jobs may have contributed to the persistence of the working-class in the community. Of the 1860 industrial workers found in the census, 21 percent had lived in Harrisburg for ten or more years; a decade later, 28 percent of the 1860 workers were still in Harrisburg, and nearly half of them worked for the same firm or industry. Only 15 percent of a far larger number of factory hands in 1870 had lived there ten or more years, and fewer than a third remained with the same firm or industry.

That Harrisburg largely recruited its industrial work force from the outside might have been expected to provide tinder for social unrest or upheaval. In fact, it seems to have had the opposite effect. Instead of being bitter, alienated local craftsmen displaced by machines from their accustomed trades, the factory hands were primarily young people from the outside who presumably of their own volition came to Harrisburg to seek their fortunes. Their age, the fact that they were at their first job or had chosen rather than been forced to shift to a new occupation, and their decision to settle in a new community all worked in favor of their adjusting to the new situation rather than resisting it. Those who entered the factories were little different from those who had lived in Harrisburg for decades. Although many were strangers to the community, they were not aliens. Overwhelmingly they were American-born, even Pennsylvania-born, whites. Foreign-born and African American workers were too few in number to be a significant factor. As a consequence, unlike more notorious industrial centers, Harrisburg never had a work force that older residents could look down on because of a different ethnic or cultural background. Neither could employers easily use such differences to set workers against one another. Except for a small number of blacks and immigrants, Harrisburg never divided into "we" and "they."

Craft Workers: Fathers and Sons

Before the coming of the cotton mill and Porter's anthracite furnace, all commercial goods made at Harrisburg were produced at home or in craft shops. Although the community was long a satellite of nearby Lancaster and the roots of some of its craftsmen could be traced to that older and larger city, Harrisburg never developed into a comparable craft center. Lancaster's artisans fed on a rich German craft tradition dating from the mid-eighteenth century. There the celebrated Pennsylvania rifle and Conestoga wagon had been fashioned, and artisans continued to turn out prized silver and pewterware, fine furniture, organs, and clocks.[1] By contrast, Harrisburg could boast relatively few fine craftsmen. Customers desiring quality crafted goods often went or sent to Lancaster for them. Far fewer would have sought for such goods at Harrisburg, where artisans chiefly made and repaired tools or crafted plain, everyday goods that people could not or preferred not to make for themselves.[2]

By 1850, traditional craft production in America was in its closing days, done in by the combined forces of a cash-based market economy, the revolution in transportation, and the rise of factory production.[3] Formal long-term apprenticeships had largely given way to informal arrangements under which young persons received wages while learning a trade, many leaving for jobs without completing their training. The old craft terms "journeyman" and "master craftsman" continued to be used, but the more accurate "skilled worker" and "employer" increasingly supplanted them.[4] Further indications of the decline were the decrease in instances of families pursuing the same trade from grandfather to father to son and the ease with which skilled artisans shifted trades. Although several generations of Harrisburg Bernheizles, Colestocks, and Updegroves (who were carpenters), of Boyds (who were cabinetmakers), of Greenawalts (who were tanners), and of Zollingers (who were hatters) continued at the same or very closely related trades across most of the nineteenth century, they were exceptions rather than the rule. As common were those such as Charles Muench, the town's leading baker in 1850 (as measured by value of output and number of employees), who by 1860 had become its principal brickmaker. Others shifting at the time included two carpenters who became patternmakers and one who became a gasfitter; a blacksmith who turned to cigarmaking, and three shoemakers who respectively became a cabinetmaker, a plasterer, and a tinsmith.[5]

Customary accounts of how industrialization affected people who worked in craft production are grim. As usually depicted, factories with power-driven machinery replaced the craft shops and thereby deprived artisans of their livelihoods and ultimately destroyed a more humane way of life. With their skills rendered useless, these once-respected craftsmen had to choose between unemployment or working as machine tenders. Many if not most became factory hands. No longer owning the tools they used or possessing valuable skills, they had only their labor to sell. Gone was their economic independence, their pride in their work, and their status in the community. They had sunk into an enduring if not permanent working class.[6]

Certainly the industrial age ultimately resulted in factories replacing craft shops as the principal places where most manufacturing occurred. Also, over time, power-driven machinery did largely wipe out such crafts as cabinetmaking, cigarmaking, shoemaking, and weaving, to name but four. In most instances where industrialization eliminated crafts, however, the changeover was not immediate and was rarely complete. The process often took decades or even generations, which gave people time to adjust. Despite the rise of factory production, many established master craftsmen were able to continue at their old trade for years or even for the rest of their lives, albeit often less profitably. Young journeymen, alert to what was happening, watched for opportunities in other fields and shifted occupations. Masters in the dying crafts took on fewer, if any, apprentices, and not as many boys applied for such positions. Of those who went into industry, many became foremen or factory artisans. Large numbers moved to other crafts, became storekeepers, or otherwise avoided the factory. Nostalgia, preference for the old order, or dislike of the new should not obscure the fact that, although dislocations occurred, the changes in artisans' lives usually were neither sudden nor catastrophic.[7]

At the same time, industrialization, at least in the nineteenth century, had almost no impact on many crafts. Blacksmithing, carpentry, masonry, painting, plastering, plumbing, and the like continued to thrive and even expand.[8] To be sure, new technology, different business arrangements, and altered tastes would force changes on most of the persisting crafts. But for generations they held their own with factory production, many continuing strong even to the present. What happened at Harrisburg illustrates the process.

Of all Harrisburg workers in 1850, nearly half, some 853 persons, listed themselves as craft workers. It must not be assumed that all were independent artisans with their own shops or businesses. Only a few were master craftsmen who had gone through exacting apprenticeships and now operated small shops in the traditional manner, producing handcrafted goods to order for their customers. Others were merchant-craftsmen who operated combination craft shops and retail stores where they sold their own output as well as stock produced by other artisans or perhaps by factories. Still others, such as carpenters, for the most part hired out their services to a succession of customers. In fact, because of the vagaries of census responses, the majority were journeymen, or apprentices, or even casual workers employed by master craftsmen.

Harrisburg's "brewers" are an example. The 1850 census listed ten men with that occupation. Close examination reveals that only two, John C. Barnitz and Philip Garman, both American-born, were extensive property holders and probably master brewers. The other eight, five German-born, one Russian-born, and a single American-born son of each of the master brewers, probably worked for them. For instance, living in Barnitz's household were three other brewers: his son Jerome, age nineteen, and two apparently unmarried men, Barnhart Stroeh, age twenty-eight, and David Maley, age thirty-five. Garman's son Henry, age thirty-one, also a brewer, lived with his father. Three of the remaining four brewers were apparently recent immigrants: Joseph Slitz, who roomed at the Buehler Hotel, and Luke Koenig and John Sheader, who had German-born wives and American-born children age three or younger. Henry Frisch and his wife, whose oldest American-born child was eleven years old, had immigrated earlier. Similarly, of Harrisburg's twenty-four bakers, perhaps only eleven, including six who owned real estate, operated bakeries. The others probably worked for them. Of the town's fourteen confectioners, at least half were employees of the others.

None of the community's new industries directly displaced or competed with local craft shops. The Harrisburg Cotton Works manufactured coarse materials, such as bagging, not the types of fabric that craft weavers produced. Although the town's new anthracite furnaces replaced charcoal furnaces, none of the latter had existed in Harrisburg.

The new rolling mills made boiler plate and nails, but the manufacture of those products had long before been taken from the hands of blacksmiths. The Harrisburg Car Manufacturing Company introduced the building of railroad rolling stock, a product new to the area, and Hickok's Eagle Works developed types of machinery not previously made at Harrisburg.

Closest perhaps were the railroads, which undercut the jobs of boatbuilders and of canal boatmen (who were not artisans). There were seven boatbuilders in Harrisburg in 1850, and one each in 1860 and 1870. Although making railroad rolling stock at the car works would seem a natural alternative for those who were displaced, in neither 1860 nor 1870 did any former boatbuilders appear to be working there. The only 1850 boatbuilder still in Harrisburg by 1860 had become a coal merchant. The single boatbuilder of 1860 had shifted to retail merchandising by 1870.

Similarly, a shift to railroad employment could have been a logical adjustment for displaced boatmen. Their number declined from forty-five in 1850 to fifteen in 1860. (Because five of the fifteen were in jail when the census was taken, their usual residences could have been in other communities.) By 1870 the number of boatmen had shrunk to twelve. Of the 1850 contingent, six continued in that occupation through 1860, and one each became a railroad employee, a sand merchant, a general merchant, a brickmaker, a laborer, and a gentleman (which, since he owned no property, probably meant he no longer worked regularly). The remainder did not appear in the census. Of the 1860 boatmen, a decade later two continued in that occupation, one had become a coal dealer, one a hatter, one a laborer, and the others, including the five in jail, were no longer in the city.

An examination of the changing number of persons in the various trades in Harrisburg between 1850 and 1870 shows that the rise of factory production in the United States affected the crafts differently (see Table 9.1). During those two decades the population of Harrisburg increased nearly 200 percent. The number of those who were not craft workers (including factory workers and unspecified laborers) increased more than 230 percent, while the percentage of those listing themselves as craft workers increased less than 100 percent. Because many of the craft workers were in fact factory artisans employed in industry, these figures considerably undercount factory employees while overstating those working as traditional artisans.

The various crafts fell into three categories. The stronger ones, which included the building trades, the metal trades, the food trades, and such skills as barbering, all showed vigorous growth between 1850 and 1870.

By contrast, the weaker crafts—the needle trades, cigarmaking, shoemaking, coachmaking, and wagonmaking—grew only at minimal rates over the same twenty years. Those with the dimmest future—weaving, hatmaking, millinery, and cabinet- and chairmaking—all registered losses over the two decades.

Some of the strong crafts benefited directly from the rise of factories. The car works, the repair shops of the Pennsylvania Railroad, and Hickok's Eagle Works all employed numerous artisans. In 1860, for example, the car works hired approximately 9 blacksmiths, 42 carpenters, and 5 painters, and a decade later 30 blacksmiths, 130 carpenters, and 17 painters were hired. In 1870 the repair shops of the Pennsylvania Railroad hired 38 blacksmiths, 36 carpenters, and 6 painters. Hickok in 1870 employed at least 4 cabinetmakers, 4 carpenters, 2 blacksmiths, and 1 painter. Inasmuch as these three firms alone employed 49 percent of the number calling themselves blacksmiths in 1870, 43 percent of the carpenters, and 21 percent of the painters, it does not seem unreasonable to assume that perhaps half the blacksmiths and carpenters and a quarter of the painters actually worked as factory artisans.[9]

Interpreting what it meant for these persons to list themselves in the census by craft rather than as factory employees is difficult. It may mean that they had reluctantly entered into such work, saw it as only temporary, and, expecting soon to return to their regular trade, continued to think of themselves as craftsmen. But it is equally possible that they found sufficiently little difference between their work in the factory and what they had done on their own or in craft shops that they still regarded themselves as practicing artisans. In addition to these direct effects, all the strong trades benefited indirectly from industrialization. As floods of new workers migrated to the community, their need for housing and food created a demand for more carpenters, masons, painters, plasterers, bakers, and butchers, among others.

Although Harrisburg's new industries did not themselves displace local craft workers, general industrialization elsewhere, especially in the Northeast and the Midwest, did have an adverse impact on many of the city's artisans. The relatively slow growth of the needle trades, for instance, attested to inroads made by the sewing machine. As both private homes and factories making ready-made clothing acquired these machines, the need for skilled tailors and seamstresses fell. Individually produced hats and other headgear for both men and women similarly gave way to machine-made products. Although shoemaking continued as an individual craft, power-driven sewing machines in shoe factories (eventually even in Harrisburg) gradually eroded the market for hand-crafted shoes. Larger firms elsewhere also began producing such items as

Table 9.1. Numbers of Persons in Various Crafts, 1850, 1860, 1870

	1850 No.	1860 No.	1870 No.	1850–60 % Increase	1860–70 % Increase	1850–70 % Increase
Total population	7,834	13,400	23,105	71.0	72.4	194.9
Total workers	1,725	3,209	5,710	86.0	77.9	231.0
Total craft workers	853	1,239	1,684	45.3	35.9	97.4
Building trades						
Carpenters	156	232	394	48.7	69.8	152.6
Masons (brick, stone)	50	103	138	106.0	34.0	176.0
Painters	36	66	115	83.3	74.2	219.4
Plasterers	23	45	59	95.7	31.1	156.5
Total	265	446	706	68.3	58.3	166.4
Metal trades						
Blacksmiths	57	76	144	33.3	89.5	152.6
Tinsmiths	18	24	47	33.3	95.8	161.1
Total	75	100	191	33.3	91.0	154.7
Food trades						
Bakers/confectioners	38	61	86	60.5	41.0	126.3
Brewers	10	16	17	60.0	6.3	70.0
Butchers	27	50	73	85.2	46.0	170.4
Total	75	127	176	69.3	38.6	134.7

Table 9.1. *Continued*

Needle trades						
Dressmakers/tailoresses/seamstresses	56	130	113	132.1	−13.1	101.8
Milliners/mantuamakers	22	43	19	95.5	−55.8	−13.6
Female subtotal	78	173	132	121.8	−23.7	69.2
Hatters	13	13	5	0.0	−61.5	−61.5
Tailors	53	41	74	−22.6	80.5	39.6
Male subtotal	66	54	79	−18.2	46.3	19.7
Total	144	227	211	57.6	−7.0	46.5
Miscellaneous trades						
Barbers	19	25	48	31.6	92.0	152.6
Cabinet/chairmakers	48	36	22	−25.0	−38.9	−54.1
Cigarmakers & tobacconists	29	36	51	24.1	41.7	75.9
Coach & wagonmakers	24	22	40	−8.3	81.8	66.7
Shoemakers	90	127	124	41.1	−2.4	37.7
Weavers	10	12	7	20.0	−41.7	−30.0
Others	74	81	108	9.6	33.3	45.9
Total	294	339	400	15.3	17.9	36.1

SOURCE: Computerized population census schedules, 1850, 1860, 1870.

factory-made cigars and coaches and wagons in competition with local craft shops. Accordingly, cigarmaking, coachmaking, and wagonmaking slowed in Harrisburg.

Worse was the already widespread use of power-driven looms, and lathes and other new machines for making furniture. These inventions killed off the crafts of weaving and of cabinetmaking and chairmaking everywhere. Two measures of a dying craft were declining numbers of persons engaged in it and their increasing average age.[10] New England textile factories had by 1850 already rendered most hand-weaving obsolete. Harrisburg that year had ten weavers (only two of whom were under thirty) with an average age of forty-six. By contrast, the average age of Harrisburg bakers, carpenters, and blacksmiths that year ranged from 30 to 31 years. A decade later Harrisburg had eleven weavers, including two who remained from 1850 and two who were carpet weavers. Although three were younger than age thirty, the average age was forty-four. Of this group, only one remained a weaver in 1870, one worked at the cotton mill, and one was a laborer. The other six, including the two carpet weavers, were no longer in the community. By 1870 Harrisburg had seven weavers, five of whom were carpet weavers. Four of the group were immigrants from Germany. Except for two sons who were following in their fathers' trade, the group averaged fifty-five years of age. Cabinetmakers and chairmakers declined even more sharply in number, and their average age increased from the low of 28 in 1850 to 39 by 1870.

A smaller number of selected crafts makes possible a further analysis of changes in the lives of artisans (see Table 9.2). Carpentry, blacksmithing, baking, and barbering were all strong crafts. Although a few carpenters operated contracting or building businesses, the great majority either performed services for individual customers or took jobs at such industrial firms as the car works or the repair shops of the Pennsylvania Railroad. Many blacksmiths worked for the same firms; the rest operated small smithies, where they shoed horses, repaired tools, or made iron objects to order. By contrast, no bakers or barbers worked directly in industry. They tended to run small shops of their own with no more than one or two, if any, employees. The exceptions were a different baker in each census between 1850 and 1870 who hired from six to ten employees.

Shoemaking and cigarmaking were weakening crafts, and cabinetmaking was a dying trade.[11] For the most part, artisans in those crafts worked in small shops with few if any employees. However, as already noted, large firms located in other communities were stealing away their markets. No Harrisburg shoemaking firm had steam-powered machin-

Fig. 19.   Reuben Bowers at Work in His Smithie, 1898.   Independent craftsmen survived industrialization and increased in number. This scene suggests why the more efficient and productive factory system gradually replaced them.

ery before the 1880s, but shoe shops and boot shops had begun moving in that direction earlier. In 1850 one shoe shop employed twenty-five persons, a decade later two firms hired ten employees each and a third hired seventeen, and in 1870 Miller & Forney's boot- and shoemaking establishment had sixty-five employees. Quite possibly no more than half of the 124 persons listing themselves as shoemakers in 1870 were traditional, self-employed shoemakers; the remainder probably worked for Miller & Forney. Finally, in the mid-1880s, a full-fledged shoe factory with 800 employees began operations in the Allison Hill district, east of the industrial corridor.

Among other things, the average age of the craft workers in each of the selected crafts, except barbers and cabinetmakers, edged upward every

Table 9.2. Age and Ethnicity in Selected Crafts, 1850, 1860, 1870

Craft Workers	Year	Avg. Age	% Heads of Household	% White, American-Born	% Foreign-Born	% African American
Carpenters	1850	31.0	46.8	93.6	6.4	0.0
	1860	32.5	65.5	87.5	12.1	0.4
	1870	35.9	78.2	88.1	10.9	1.0
Blacksmiths	1850	31.1	57.9	89.5	10.5	0.0
	1860	33.3	69.7	84.2	14.5	1.3
	1870	38.4	76.4	77.1	21.5	1.4
Bakers & confectioners	1850	30.5	45.8	33.3	66.7	0.0
	1860	30.8	50.0	38.6	59.1	2.3
	1870	31.3	50.7	44.9	55.1	0.0
Barbers	1850	26.6	57.9	5.3	5.3	78.9
	1860	27.6	52.0	8.0	0.0	92.0
	1870	26.9	47.9	33.3	14.6	52.1
Shoemakers	1850	32.3	58.2	75.8	19.8	4.4
	1860	33.9	51.2	65.4	29.9	4.7
	1870	39.1	72.6	66.1	31.5	2.4
Cigarmakers & tobacconists	1850	28.8	51.7	96.6	3.4	0.0
	1860	33.4	61.1	75.0	5.6	0.0
	1870	30.8	58.8	82.4	17.6	0.0
Cabinet/chairmakers	1850	27.9	54.2	93.8	6.3	0.0
	1860	29.0	72.2	80.6	19.4	0.0
	1870	39.4	77.3	77.3	13.6	9.1

SOURCE: Computerized population census schedules, 1850, 1860, 1870.

decade. As might be expected, in most of the crafts so did the percentage who headed households. Whites born in the United States overwhelmingly dominated most trades in 1850, ranging from three-quarters of shoemakers to nearly all cigarmakers. However, minorities dominated baking and barbering. Two-thirds of the bakers were German-born, and more than three-quarters of the barbers were African Americans. Over the next twenty years, the percentage of foreign-born in each of the crafts except baking increased, the gains ranging from less than 5 percent among carpenters to more than 14 percent among cigarmakers. Similarly, blacks, who had been excluded from all the selected trades except barbering and shoemaking in 1850 made slight inroads into carpentry and blacksmithing and by 1870 constituted nearly a tenth of all cabinetmakers. On the other hand, they lost ground in their original areas of strength. Accordingly, the percentage of American-born whites declined in all the trades except baking and barbering. The largest drops, except for blacksmithing, came in the weak or dying crafts of shoemaking, cigarmaking, and cabinetmaking.

As a rule, higher proportions of the members of the selected crafts held real estate than among the population as a whole. Where in 1850 slightly more than one Harrisburg adult in ten owned land, carpenters, blacksmiths, bakers, and barbers (the stronger trades) all had higher rates; shoemakers, cigarmakers, and cabinetmakers had less than average rates. By the close of the prosperous 1860s at Harrisburg, a fifth of the city's adult population held real estate. However, rates were considerably higher for all the selected crafts. This remained true in 1870—except for barbers, where the dominant African Americans underwent considerable local upheaval during and immediately after the war, and for cigarmakers, who with the barbers had the lowest average age (compare "percent of all adults" 1850, 1860, 1870, Appendix F, with percent of real estate holders for the selected crafts, Table 9.3).

Although the average value of property held by the various craft workers at ten-year intervals cannot always be explained, certain factors are evident (see Table 9.3). The patterns of the stronger crafts paralleled the pattern of the general population. Large numbers of people acquired real estate by 1860, but the average value per holding was smaller. A decade later the percentage of holders again declined but the average value rose. By contrast, in the weaker trades (except for shoemaking) the percentage that owned real estate fluctuated much more. The average value of holdings increased steadily throughout. In the end, the high average value was due either to persistence in the trade and in the community or to ethnicity. All landowning cabinetmakers in 1870 had been in Harrisburg for at least a decade. Among the shoemakers, 44 percent

Table 9.3.   Changing Property Holdings in Selected Crafts, 1850, 1860, 1870

			Real Estate		Personal Property	
Craft Workers	Year	No. in Craft	Holders	Avg. Value	Holders	Avg. Value
Carpenters	1850	156	16.0%	$2,126	n/a	n/a
	1860	232	40.9	1,887	80.0%	$387
	1870	394	38.3	3,426	33.2	659
Blacksmiths	1850	57	12.3	3,625	n/a	n/a
	1860	76	36.8	1,354	63.2	767
	1870	144	25.0	2,747	32.6	398
Bakers & confectioners	1850	38	21.0	3,825	n/a	n/a
	1860	61	29.5	3,289	45.9	571
	1870	86	19.8	5,547	27.9	1,000
Barbers	1850	19	15.8	1,200	n/a	n/a
	1860	25	24.0	1,417	40.0	300
	1870	48	4.2	4,500	12.5	917
Shoemakers	1850	90	7.7	500	n/a	n/a
	1860	127	25.2	1,253	44.1	445
	1870	124	20.2	2,664	13.7	1,065
Cigarmakers & tobacconists	1850	29	6.9	1,100	n/a	n/a
	1860	36	25.0	2,567	44.4	825
	1870	51	9.8	7,400	33.3	706
Cabinet/chairmakers	1850	48	10.4	3,920	n/a	n/a
	1860	36	41.7	3,080	61.1	573
	1870	22	18.2	11,575	22.7	2,220

SOURCE: Computerized population census schedules, 1850, 1860, 1870.

n/a = not available.

of the landowners had persisted for at least ten years, 28 percent for twenty or more years. Although less than 14 percent of all shoemakers in 1870 were foreign-born, 40 percent of those owning real estate that year were German-born.

Personal property holdings were reported only in the censuses of 1860 and 1870. The most notable change between the two countings was the sharp drop in percentages of artisans listing personal property. Because this category included tools, it appears that as more craft workers became factory artisans, employers increasingly provided the tools with which they worked. At the same time, it is also evident that, except for blacksmiths and cigarmakers, craftsmen who continued to own tools had larger investments in them than before. More important than property ownership was the impact the coming of industrialization had on the lifetime careers of Harrisburg's mid-century artisans. Although by 1850 local craft workers had already suffered indirect consequences of factory production elsewhere, no local factories affected them yet. Three issues need to be addressed here: the extent to which the new factories displaced the artisans of 1850, the occupations to which the displaced artisans moved, and the occupations their sons adopted. The careers of both fathers and sons who remained in Harrisburg one or two decades after 1850 have been traced from the information provided in three successive censuses. However, as at other industrializing communities, Harrisburg residents were surprisingly mobile geographically. Each time the census was taken, half or more of the population was no longer in the city.[12] Some people, of course, had died; a few, though perhaps still in town, could not be identified. The others had moved elsewhere. The absence of detailed cross-indexing of census schedules for the whole nation makes any systematic attempt to trace the careers of those who moved impracticable.

Harrisburg artisans of 1850 were relatively stable (see Table 9.4). Among those in the selected crafts that year (embracing more than half the town's craft workers), nearly 42 percent were still in the community in 1860, and more than 27 percent in 1870. These proportions nearly matched those for physicians and lawyers, 47 percent of whom remained for ten years and 28 percent of whom stayed for twenty years. By contrast, only 14 percent of unspecified laborers remained until 1860, and fewer than 8 percent until 1870.[13] Persistence rates for both ten years and twenty years were fairly consistent among the various crafts. The exceptions were the cigarmakers of 1850, a much higher percentage of whom remained in both 1860 and 1870, and the barbers, 37 percent of whom remained until 1860 but only one of whom remained a decade later.

The most notable finding about craft workers in Harrisburg during the first two decades of industrialization was the extent to which they were not displaced or forced to become factory hands. Nearly a third of the original artisans of 1850 were still at the same trade in Harrisburg after ten years, 17 percent after two decades. If only those who remained are considered, more than three-quarters remained at the same trade in 1860, and 61 percent in 1870. Inasmuch as the selection included strong, weak, and dying crafts, the marked variations among them should be noted. Of persisting carpenters and blacksmiths, 82 percent were at the same trade after ten years, 74 percent after twenty years. Although baking and barbering were also strong trades, a high percentage of workers in each consisted of young transients. Some 59 percent of those who remained in Harrisburg continued in baking in 1860, and 36 percent by 1870. Patterns

Table 9.4. Persistence of Craft Workers of 1850 to 1860 and 1870

Craft Workers		In Harrisburg		Same Craft		Other Occupation	
		No.	%	No.	%	No.	%
Carpenters:	of 156 in 1850						
Remaining	by 1860	62	39.7	53	34.0	9	5.8
Remaining	by 1870	39	25.0	29	18.6	10	6.4
Blacksmiths:	of 57 in 1850						
Remaining	by 1860	25	43.9	18	31.6	7	12.3
Remaining	by 1870	18	31.6	13	22.8	5	8.8
Bakers & confectioners:	of 38 in 1850						
Remaining	by 1860	15	39.5	8	21.1	7	18.4
Remaining	by 1870	10	26.3	3	7.9	7	18.4
Barbers:	of 19 in 1850						
Remaining	by 1860	7	36.8	5	26.3	2	10.5
Remaining	by 1870	1	5.3	1	5.3	0	0.0
Shoemakers:	of 90 in 1850						
Remaining	by 1860	36	40.0	29	32.2	7	7.8
Remaining	by 1870	25	27.8	14	15.6	11	12.2
Cigarmakers & tobacconists:	of 29 in 1850						
Remaining	by 1860	19	65.5	17	58.6	2	7.0
Remaining	by 1870	13	44.8	11	37.9	2	7.0
Cabinet/chairmakers:	of 48 in 1850						
Remaining	by 1860	19	39.6	8	16.7	11	22.9
Remaining	by 1870	13	27.1	2	4.2	11	22.9
Total	in 1850: 437						
Remaining	by 1860	183	41.9	138	31.6	45	10.3
Remaining	by 1870	119	27.2	73	16.7	46	10.5

Source: Computerized population census schedules, 1850, 1860, 1870.

among the weak and dying crafts differed even more. A surprising 89 percent of remaining cigarmakers persisted at their craft in 1860; 85 percent persisted in 1870. Among shoemakers, more than 80 percent in 1860 continued at the same craft, then dropped to little more than half by 1870. The poorest showing was by cabinetmakers. Well under half of those remaining in Harrisburg in 1860 continued at their trade; a decade later only 15 percent continued in cabinetmaking.

To what occupations did those who left their crafts after 1850 go? Relatively few went into industry: fewer than 6 percent were working in factories in 1860, and less than 12 percent in 1870 (Table 9.5). More-

Table 9.5. Movement of Selected Craft Workers of 1850 into Other Occupations, 1860, 1870

Craft Workers		Other Craft	Business	Industry	Common Labor	Other
Carpenters:	of 156 in 1850					
	by 1860	0	4	2	1	2
	by 1870	2	1	4	0	3
Blacksmiths:	of 57 in 1850					
	by 1860	1	2	3	1	0
	by 1870	0	3	2	0	0
Bakers & confectioners:	of 38 in 1850					
	by 1860	1	3	1	0	2
	by 1870	1	3	0	0	3
Barbers:	of 19 in 1850					
	by 1860	0	0	0	2	0
	by 1870	0	0	0	0	0
Shoemakers:	of 90 in 1850					
	by 1860	2	0	2	1	2
	by 1870	1	1	4	1	4
Cigarmakers & tobacconists:	of 29 in 1850					
	by 1860	0	2	0	0	0
	by 1870	1	0	1	0	0
Cabinet/ chairmakers:	of 48 in 1850					
	by 1860	2	1	3	0	3
	by 1870	2	3	3	1	3
Total in 1850: 437						
Of 183 remaining by 1860		6	12	10	5	9
Percent of 1850 group		1.4	2.7	2.3	1.1	2.1
Percent of persisters		3.3	6.6	5.5	2.7	4.9
Of 119 remaining by 1870		6	11	14	2	13
Percent of 1850 group		1.4	2.5	3.2	0.5	3.0
Percent of persisters		5.0	9.2	11.8	1.7	10.9

SOURCE: Computerized population census schedules, 1850, 1860, 1870.

over, at least a third of those became supervisors or foremen, not mere factory hands. A few shifted into related crafts, such as cabinetmakers who became carpenters. A larger percentage went into business, opening grocery stores or other retail shops. Some shifted to clerical work or, as they grew older, went into politics or served in a variety of public offices. In other words, no significant number of once-proud, independent artisans became alienated machine-tenders driven by the rise of factories into "wage-slavery," to use a favorite term of that era's opponents of industrialization.

What of their sons? According to the census of 1850, the selected craft workers apparently had a total of 295 sons (Table 9.6). Of that number, 121 (slightly more than 40 percent) could be traced in the 1860 or 1870 census schedules for Harrisburg. Although a few of the sons were adults or near-adults in 1850 and had jobs, more were employed by 1860, and most by 1870. Because they were relatively young, several changed occupations from one census to the next. For this study the last occupation was used.

After twenty years of industrialization, more than half of the persisting sons of craft workers were also in a craft—a third of them in their fathers' craft, a fourth in other crafts. Fewer than a fifth had gone into industry, and an eighth each had become common laborers or gone into clerical work, business, or one of the professions. Some of the sons who began in a craft or as common laborers probably became factory workers, and some who worked in the factories as adolescents would later go on to other occupations. As was true of their fathers, relatively few craft workers' sons who remained in Harrisburg were forced into factory labor. They had other choices.

Throughout the period of industrialization, craft shop production remained viable at Harrisburg. The manuscript manufacturing census for 1880 was the last that allowed a measuring of craft shop production in the city. After that date, neither specific masters nor shops can be traced because census takers no longer gathered data by individual or firm name. Another limitation was that the manufacturing censuses between 1850 and 1880 listed only firms that produced at least $500 worth of goods during the previous year. How many shops there may have been with less output can only be guessed.

Charting the various craft shops in the manuscript manufacturing census across three decades reveals several trends (see Table 9.7). Overall, the number of shops producing $500 or more worth of goods steadily increased. It was also evident, however, that by 1880, with few exceptions, the average value of output per shop had declined. By then, factory production had eliminated or was badly undercutting such

Table 9.6. Occupations of Persisting Sons of 1850 Craft Workers

Sons of:	No.	%	Same Craft	Other Craft	Industry	Common Labor	Other
Carpenters in 1850	103						
with job 1860–70	43	41.7	14	9	13	3	4
Blacksmiths in 1850	42						
with job 1860–70	14	33.3	3	6	2	2	1
Bakers & confectioners in 1850	20						
with job 1860–70	9	45.0	1	4	0	2	2
Barbers in 1850	14						
with job 1860–70	7	50.0	6	0	0	1	0
Shoemakers in 1850	66						
with job 1860–70	26	39.4	9	6	1	6	4
Cigarmakers & tobacconists in 1850	22						
with job 1860–70	14	63.6	5	2	4	1	2
Cabinet/chairmakers in 1850	28						
with job 1860–70	8	28.5	2	2	2	0	2
Total sons in 1850	295						
Total with job 1860–70	121	41.0	40	29	22	15	15
Percent in job classification	—	—	33.1	24.0	18.2	12.4	12.4

Source: Computerized population census schedules, 1860, 1870.

Table 9.7. Craft Shops Producing at Least $500 Worth of Goods Annually

Type of Craft Shop	1850		1860		1870		1880	
	No.	Avg. Output	No.	Avg. Output	No.	Avg. Output	No.	Avg. Output
Baker/confectioner	7	$3,473	8	$4,260	16	$5,203	37	$3,389
Blacksmith	0	0	0	0	7	2,536	16	1,325
Builder/carpenter	0	0	0	0	13	4,596	21	4,249
Cabinetmaker	3	1,866	4	5,625	0	0	0	0
Chandler	3	3,658	2	3,975	3	2,092	0	0
Cigarmaker	4	1,662	0	0	4	8,500	24	3,325
Furnituremaker/upholsterer	0	0	0	0	1	2,000	9	1,178
Hatter/milliner	2	6,250	7	1,414	0	0	1	950
Painter/paperhanger	0	0	0	0	0	0	21	3,386
Plumber/gasfitter	0	0	3	3,333	0	0	12	4,416
Shoemaker/cobbler	7	3,143	22	1,910	21	2,734	38	854
Skin dresser	2	2,410	0	0	0	0	0	0
Tailor	4	3,125	11	4,358	7	3,415	1	950
Tanner	2	5,150	3	8,159	1	4,000	0	0
Tinner	2	5,900	4	1,338	7	8,176	14	2,000
Wagonmaker	0	0	0	0	5	1,820	4	1,275
Other	0	0	9	3,362	9	2,445	13	2,006
Total	36		73		94		211	

SOURCE: Computerized manufacturing census schedules, 1850, 1860, 1870, 1880.

trades as cabinetmaking, hatmaking, shoemaking, tailoring, skin dressing, and tanning. The value of output for other types of craft production increased, especially for the building trades: carpentering, plumbing, and home decorating (painting and paperhanging).

What happened in shoemaking, one of the declining crafts, provides insights into craft persistence at Harrisburg in spite of thirty years of industrialization. Well before the Civil War, shoe factories, especially in New England, had started cutting into the business of artisans who made shoes by hand. Even at Harrisburg, Miller & Forney by 1870 had forty male workers and twenty-five female workers. Although the firm did not have steam-driven machinery, it produced $72,000 worth of shoes in twelve months of operation and paid its employees an average of $1.33 a day (the differential for men and women was not given). Under the name "Forney & Brother" the firm in 1880 produced $75,000 worth of shoes in only nine months, with thirty-one male and twenty female employees. On average, the men that year received $2.00 a day, the women $1.00, for ten hours of work.

The appearance of the shoe factory by 1870 by no means eliminated the shops of artisan shoemakers. In 1870 Harrisburg had fifteen shops, each of which produced at least $500 worth of shoes (there had been twenty shops in 1860); by 1880 there were twenty-eight shops, with seven more combined under "all others" at the end of the list. The masters who ran these shops, resisting the inevitable, were still able to practice their trade in the traditional manner. With twenty-six male employees among them in 1870, they turned out $52,770 worth of shoes. This was two-thirds the output of the factory and a third more than the value of all shoes put out by the twenty shoemaker shops in 1860 ($39,921). A decade later, employing thirty males and one female, the shoeshop output by value ($32,456) was half as great as that of Forney & Brother and less than the output of 1860.[14]

Using the value of the shoes produced as the measure, George M. Groff was the most successful artisan shoemaker in 1880. He had come to Harrisburg to practice his trade sometime between 1850 and 1860 and was the only shoemaker who produced enough to be listed in all three subsequent manufacturing censuses. The best year for him had been 1860, when with the help of seven male employees, who received about 50 cents a day, he turned out $6,000 worth of shoes. In 1870, with only three employees, he made $2,400 worth of shoes. Although he now paid his workers approximately $1.00 a day, wages in his shop were considerably lower than at Miller & Forney. A decade later Groff had five employees—four male and one female. Together they produced $4,466 worth of shoes. He paid the men $1.75 a day and the woman $1.00, but employed them only nine months that year. At Forney &

Brother the men would have earned 25 cents more a day, and the women the same as at Groff's.

Eleven other artisan shoemakers produced at least $1,000 worth of shoes in 1880. Four of them appeared in earlier censuses. Samuel Urich in 1850 had been a seventeen-year-old employee in a shoemaker's shop. Ten years later he was on his own, holding $1,300 worth of real estate, which by 1870 was worth $1,500. Only in 1880 did he produce enough shoes ($1,000 worth) to be listed in the manufacturing census. That year Urich worked alone, twelve hours a day, for ten months. Wesley Groff (apparently a nephew of George) and William Meyers seem to have inherited shoe businesses from fathers who appeared as shoemakers in an earlier census. Also listed in the 1870 and 1880 manufacturing censuses was Samuel Barnhart, who in 1870 made $1,875 worth of shoes with the help of two men who received $1.25 a day for eleven months of work. Alone in 1880, when he was thirty-eight years old, he made $1,000 worth of shoes by working twelve hours a day for ten months.

Another sixteen shoemakers in 1880 produced between $500 and $900 worth of shoes each. Half had been in the business at Harrisburg for at least ten years. All worked alone except one, who was assisted by his son, and another, who paid out $60 in wages that year for occasional help. Two were German-born; the others were all natives of the United States. Two were sons of Benjamin Styers, the state librarian in 1870, who had come to Harrisburg as a shoemaker in 1850. The sons apparently learned the craft from him, or from their uncle William, before setting up their own shops. Samuel Tagg apparently inherited his shop from his father, Wilson, a Harrisburg shoemaker according to the 1870 census. The remaining shoe shop owners were probably newcomers who arrived after 1870.

All these artisan shoemakers, and probably others, avoided becoming factory hands, even at the cost of lower incomes. However, it must be remembered, these were masters. What of the thirty-one men and one woman who regularly worked for them, and the seven others who were hired for occasional work? The precariousness of being a journeyman, apprentice, or casual shoemaker by 1880 can be seen from the combined wages that year of the thirty-nine non-masters, who earned $7,390 (or $189 each) for irregular work periods ranging from eight to twelve hours a day for between five and twelve months. Unless they lived very marginally, they must have had additional income from other employment. It is even possible that they worked in craft shops to supplement their regular earnings as hands in the shoe factory.

Whatever the advantages of numerical or statistical calculations in measuring the impact of industrialization on the careers of craft workers and their sons, they impart little of the human side of such matters. The unpublished autobiographical memoir of Charles R. Boak, a New York editor who grew up in Harrisburg early in the twentieth century, relates anecdotally the complexities of the process. His account deals briefly with three generations of the Boak family: the author's grandfather, Abram Boak, a farmer who became an ironworker; his father, Charles L. Boak, who was a cigarmaker; and himself. Shortly after the Civil War, Abram Boak sold his small York County farm. "There were two stones to one dirt," he explained. He married, moved to Harrisburg, and was the person already noted above who walked each day to his job at an iron mill (probably McCormick's nailery) in West Fairview. Abram and his wife, Christiana, had eight children. Min and Bob, who never married, lived with their parents. Min worked in the old cotton mill (by then a silk factory) until it shut down, then remained at home. Bob became a mechanic. Daughter Kate married a master mechanic and foreman and had no children. The occupations of the other children and their spouses, except for son Charles L., are not known.[15]

Charles L. Boak resisted the trend toward industrial work and entered one of the crafts. With the permission of his father, he quit school in 1895 and apprenticed himself to a Harrisburg cigarmaker, Harry C. Knull. Five years later, at age twenty-one, he used his savings, a bank loan, and money borrowed from his father (the ironworker) to set up a cigar shop of his own. It was one of seventeen small cigarmaking establishments in Harrisburg at the time. When Knull retired, young Boak bought part of his stock and the trade name, "Ambrosia." Although "going into the cigar business was not uncommon then," Charles L.'s son later wrote, "successful small cigar factories were becoming fewer . . . because of the pressure of large corporations."[16]

By day, Boak's place of business on Broad Street was a small cigar factory and retail store; at night it took on the character of an informal neighborhood social club. Next to a large window in one of the back rooms, four cigarmakers sat at wooden benches. To each bench was attached a heavy slab of wood, something like a small chopping block, on which the cigarmakers first rolled their cigars, then clipped off the ragged ends with spring cutters. A female "stripper" worked with them, removing stems from the tobacco and flattening the leaves that would be used as binders or wrappers. The front of the room had display cases and served as the salesroom

At day's end the cigarmakers washed their hands under a cold-water spigot, dried them "on the long towel (changed every Monday)," and went home with a few "green" cigars in their pockets. After supper, men from the neighborhood and friends arrived to play "hasenpfeffer," Grandfather Abram Boak's favorite card game, or simply to talk. When

Fig. 20. Workers in Charles L. Boak's "Cigar Factory." Independent cigar-makers such as Boak survived well into the twentieth century despite the rise of cigar factories, but they gradually fell to changing tastes and cheaper production methods.

Grandfather left, the players switched to pinochle, the game they preferred.[17] The store remained open "until the topics of the day had been dismantled." Railroading was a common topic. Railway workers

> loved to talk knowingly of "old 38" or "number 16" and referred in railroad jargon to the various duties and phenomena characteristic of railroading in a way that silenced the other members who had to sit and listen—and complain later about those "dumb railroaders who don't know nuthin' but railroads." Other favorite subjects were the size of a member's coal pile, and federal and state politics. . . . What the political discussions may have lacked in cool analysis and perspicacity were brought to a fine balance by vigorous arm waving and warm vehemence.[18]

Master cigarmaker Boak put in long hours, starting early in the morning and rarely closing his store before 10:30 at night. Such workdays were bearable in part because he was free to leave the store several times a day to go out and cultivate the dahlias and peonies he grew in his yard. To his son Charles R. it seemed his father "scarcely knew the difference between business and pleasure." Moreover, time was not of great importance to him. He was usually ready to talk at length with customers or friends who came into the store or whom he met on the street. Some days he spent on his bicycle, riding around town selling cigars. Occasionally he delivered orders to customers in nearby communities by train. In his later years he provided assistance in settling estates and did other legal odd-jobs for people in the neighborhood. He also raised and caponized chickens.[19]

The aging cigarmaker fought "sturdily" against the "erosion of sales" that started in the 1890s and became permanent after 1930.[20] Still, Charles R. Boak recalled, his father

> continued to sell many cigars to the various Masonic Orders and to old friends, but even the Masonic sales declined in the 1940's and, of course, old friends died off. Young cigar smokers did not buy Dad's cigars. Many young smokers liked cigarettes. Hand-made cigars, of course, were more expensive than machine-made cigars even though Dad took a smaller profit in order to compete. He also used the best tobacco: Havana and Porto Rico fillers, Sumatra wrappers, Connecticut binders—always the best to-bacco he knew how to buy. But some people said such cigars were strong, stronger than modern cigars. They were like cigars which men smoked at the turn of the century when they drank beer with a strong taste, ate strong cheese, and wore heavy clothing even in the summer. Dad hadn't considered that tastes had changed much in forty years. His hadn't.[21]

In the 1920s Boak was offered a job running a local branch of a large corporation that produced machine-made cigars. Although the salary was "handsome," he turned it down, having decided to stick with the handmade cigar. He would cater to a "clientele who appreciated good cigars" even if that meant doing business on a small scale with a few older customers. "Our country was never so happy and prosperous," his son recalled him saying of the 1890s, "as when many people were in business." The subsequent rise of corporations and trusts would be the ruin of the country, he predicted. The benches of his cigarmakers began to go in the late 1920s, "never to be rebuilt." "For a few years there were four cigarmakers, then three, and so on until after World War II, when Dad was alone in the store." He was the last of the city's cigarmakers.[22]

This rosily nostalgic account by the cigarmaker's son offers a number of insights. Even some landowning farmers, such as Abram Boak, appar-ently saw more opportunities as unskilled factory hands than in agriculture. Once so employed, however, they did not become the nucleus of a permanent, oppressed, or downtrodden working class. Although some of Abram's children remained factory workers, others did not. And despite having a family of eight to support, he was still able to lend money to his son to set up in the cigarmaking business. That industrialization was already under way in cigarmaking did not preclude a newcomer from starting up in the traditional handcraft manner, making a respectable living for himself and his family, and remaining in the trade until his death midway through the twentieth century. Although cigarmaking was not an option for Charles L. Boak's son, the boy did not have to go into the factories when he came of age. His father had done well enough in the cigar business to educate him for an entirely different way of life.

The account of cigarmaker Boak's career points up another facet of the transition from craft shop to factory. Boak, it must be remembered, was the master cigarmaker and owner of the business. He was free to come and go as he pleased, to work in his garden, peddle his cigars around town and to nearby communities, talk with customers, and stay up late at night playing cards and discussing politics with his friends and neighbors. It must not be assumed that the apprentices and hired journeymen in craft shops enjoyed the same privileges as their employers. Although the employer-employee relationship may have been far more personal than later in a factory, and hours and other work conditions more relaxed and flexible, there is no indication that Boak's employees left to tend their gardens or to chat with friends and customers during the day. It appears that they remained at their benches until quitting time, when they left for their homes and families.[23] Nor were they the people who spent evenings playing cards and talking politics with their employer and his friends. Similarly, although Boak was able to keep afloat as cigarmaking declined and finally died, he had to sacrifice their jobs along the way.

Care must be taken not to romanticize the lives of those who worked in craft shops or to assume that what followed was always worse. The passing order had not necessarily been better for journeymen, for instance, than the one that took its place. The dying order provided Boak much satisfaction as a master-owner and even left a sweet aftertaste in the mouth of his son. Nothing is known, however, of the reactions of his employees, who were forced from the craft somewhat ahead of their employer. It is just possible that in the long run they were no more upset by the changeover to factory work (if that is where they went) than the new generation of customers were at the substitution of milder and cheaper machine-made stogies for the strong-flavored, hand-rolled cigars of the 1890s.[24]

Ethnic Minorities: Fathers and Sons

E very rapidly growing community in mid-nineteenth-century America had its "outsiders"—strangers or groups in the locality who for one reason or another did not belong and ended up at the bottom of the social and economic order. This was not simply a matter of being from another place. Given the geographical mobility of the era, newcomers were always coming and long-established residents were continually leaving. However, the majority of those entering, whether from a nearby farm or village or from a town or city in another state, found many familiar cultural patterns, social institutions, and expectations. Americanized English was the language. Protestant churches usually dominated religious life. There rarely was an aristocracy, but the "better families," headed by merchants, lawyers, physicians, and major property-owners, usually ran things and enjoyed social and political preferment. Most communities were officially governed by an elected body of representatives (whatever its precise designation), supported many of the same church denominations, provided public schools and meeting halls, and belonged to similar social organizations. Newcomers moving from one American community to another usually did not remain outsiders or strangers for long.

Lifetime residence in a community and familiarity with its institutions, however, did not always confer full membership. African Americans, for example, were the largest group who remained outsiders. Even though their forebears were among the earliest to come to America, race and status barred them from full acceptance. And people who came to the United States from foreign countries encountered different cultural patterns or institutions. They were outsiders long after they first arrived. Over time, however, they, but more often their children, could expect in varying degrees to become an integral part of the community.[1]

Harrisburg at mid-century had three major groups of outsiders: African Americans, Irish immigrants, and German immigrants. From the 1830s on, the capital city's black community was one of the larger in Pennsylvania; its immigrant population one of the smaller. Among the major cities in Pennsylvania, Harrisburg ranked sixth in total population from 1850 through 1880, then slipped to eighth by 1900. Its African American community stood third, except for 1860 when it was second and 1900 when it was fourth. Throughout those years, its proportion of blacks (approximately one-tenth) was greater than that of the other large cities in the state, including Philadelphia and Pittsburgh. By contrast, of Pennsylvania cities with populations in excess of 10,000 in 1870, only Reading and York had lower percentages of foreign-born.[2]

From 1850 through 1870, Harrisburg's three minorities, half blacks and half immigrants, made up approximately a fifth of the city's population (see Appendix A). After 1870 the proportion of blacks in Harrisburg declined even as their total number increased. Both the numbers and proportions of Irish and Germans fell. Although still the community's three principal minorities in 1900, by then they together made up only an eighth of the total. Blacks outnumbered the foreign-born three to two, while the combined Irish and Germans made up only slightly more than half of all immigrants.[3]

Those of African descent had been in Harrisburg from the beginning, brought there as slaves by the earliest settlers. Although free by 1850, they continued to occupy the lowest rung on the socioeconomic ladder and by the close of the nineteenth century had advanced the least. Their heritage as slaves and the racial attitudes of whites combined to keep them at the bottom of the heap. Next came the Irish, the larger of the two immigrant groups in 1850 but second in size thereafter. Their extreme poverty at the time of immigration and their Roman Catholicism worked against acceptance by the majority of native-born Americans. Most fortunate were the Germans, many of whom came to Harrisburg with some accumulated wealth and skills that made employment easier to find. Even their language was not an insuperable barrier in southeastern Pennsylvania. Unhindered by color, Irish and German immigrants could reasonably hope to advance economically. Previous immigrants from their homelands had already done so, as the ethnic origins of the established residents testified. Not so the blacks.

Despite a shared status as outsiders on the eve of industrialization, differences in backgrounds largely determined the occupations that members of the three groups held. Many of the community's essential but lowest-paid chores fell to the blacks and the Irish. Males from these groups did much of the hoisting, lifting, shoving, and shoveling. They also ran errands, hauled trash, carted goods, chopped wood, cleaned streets, and did the other undesirable and routine tasks that more fortunate residents, so far as possible, avoided. Blacks of both genders, and Irish and German women provided most of the servants who cooked, waited on tables, tended children, scrubbed floors, washed windows, laundered and ironed clothes, and did general housework. German males generally found work in the various crafts. How the advent of factory production affected these groups and their children provides yet another perspective on the overall impact of industrialization on Harrisburg's residents.

Minority Groups before Industrialization

African Americans

As late as 1850, Pennsylvania blacks still lived in the long shadow of slavery. Although the state had been the first to free its slaves by legislative act in 1780, the measure provided only for gradual emancipation. Slaveholders could keep their existing slaves by simply registering them with local officials. Although children born of slaves after 1780 technically were free, their masters held them as indentured servants to the age of twenty-eight.[4] As a consequence, a few persons were slaves as late as 1840; it is not known when the last indentured servants completed their terms. Only slowly did Harrisburg blacks win their freedom—first from slavery, then from having to live in the homes of their masters and employers—and finally gain control of their own institutions, such as churches and schools. The first federal census, for example, showed that ten years after the act of 1780 only one of Harrisburg's twenty-six blacks was free. By 1800, however, nearly three-quarters had acquired that status, and thereafter only one or two slaves were recorded in each census through 1830.[5]

All the town's blacks, whether slave, indentured servant, or free, lived in white households through 1810. That way, masters and employers could more fully control their time and labor, and if they were free their room and board constituted a part of their "pay." In any event, few freed slaves could have afforded housing of their own. By 1820 two-thirds lived in black households, and by 1830 three-quarters did so. By 1850 the 13 percent still with whites were chiefly single live-in servants.[6]

Emancipation, meanwhile, produced a flow of blacks into the urban centers of Pennsylvania. In the thirty years after 1820, the number in

Harrisburg grew from 177 to 886, an increase in proportion from under 6 percent to more than 11 percent of the population. Apparently most came from farms in Dauphin County and other nearby counties.[7] A portion came from the neighboring states of Maryland and Virginia— some manumitted slaves, some fugitives from bondage. Most newly freed blacks had no means to acquire farms of their own, and working for former masters had little appeal. Moreover, a place such as Harrisburg offered a greater concentration of jobs for unskilled workers, more social and educational opportunities, larger pools of potential marriage partners, and the solace of numbers in the face of white hostility. To runaway slaves, towns and cities offered the further advantage of greater anonymity.[8]

African American institutions began to appear in Harrisburg by 1817. That year, with financial support and direction from sympathetic whites, an "African church" and Sunday schools for both adults and children were established. Blacks, however, preferred to control their own institutions and did so as soon as means were found. For instance, one black man, Thomas Dorsey, that same year founded a school for "coloured children . . . both bound and free." In 1829, members of the African church broke away to found Wesley Union AME Zion Church and affiliated with the existing black African Methodist Episcopal conference in Philadelphia. Within a year it boasted 115 members (a quarter of Harrisburg's black community) and operated a school for black children that for a short time received a small subsidy from the commissioners of Dauphin County.[9]

The responses of whites to the growth of the African American community varied. One group founded a local antislavery society in 1836 and the next year sent delegates to a statewide convention in Harrisburg where the Pennsylvania Antislavery Society was formed. The local group seems also to have sponsored a public meeting at the courthouse in 1847 at which two nationally prominent abolitionists, Frederick Douglass and William Lloyd Garrison, spoke. The movement soon waned, however. Other residents were distressed as the number of blacks increased. Some formed a local auxiliary of the American Colonization Society to encourage the removal of blacks to Africa. Others, blaming a rash of suspicious fires in 1820 on blacks, pushed a measure through the town council requiring "all free persons of color" to register with the chief burgess and organized a "citizen's patrol" to "apprehend all suspicious and disorderly persons."[10] Throughout the pre–Civil War era, gangs of white boys repeatedly teased and harassed blacks on the streets and frequently disrupted their church services. Meanwhile, local newspapers either ignored them or alternately mimicked, ridiculed, patronized, and insulted blacks, making them butts of what passed for humor in their columns.

Far more troubling to blacks was the federal fugitive slave law, which provided that slaveholders had the right to pursue their "property" into free states and return them to bondage. Because there was no statute of limitations, successful runaways who had settled in the community, found work, and established families still faced return to slavery if discovered. Even free blacks, whether born in the north or manumitted by a slaveholder in the South, lacked complete security against seizure as a runaway slave. The laws and courts of Pennsylvania generally protected them, but whites administered those courts and could make mistakes.

The sight of fellow blacks, fugitives or not, being returned to bondage was the one thing that galvanized Harrisburg's African community into confronting whites. On two occasions, in 1825 and 1850, local blacks rose up in vain attempts to free runaways whose owners had captured them and were seeking permission from local magistrates to carry their property home. In both instances the "rioters," consisting chiefly of black family men and some property owners, gathered outside the courtroom and moved to free the fugitives as their masters led them out. Each time, white law officers thwarted the attempt (though one of three runaways in 1850 escaped) and arrested the black participants.[11]

Enactment of a harsher fugitive slave law as part of the Compromise of 1850 undercut the ability of Pennsylvania courts to protect blacks. It required any white called on to assist in capturing fugitive slaves and severely punished those who assisted runaways. To secure permission to carry an alleged fugitive home, a claimant had only to take a captive before a special U.S. commissioner, not a court of law, and swear ownership. The accused could not testify on their own behalf, and only if whites intervened or the commissioner had reason to doubt ownership were they released.

At Harrisburg a local lawyer, whose family had been the last to free its slaves, applied for and was appointed slave commissioner for the area.[12] The new commissioner zealously enforced the law. He also turned it into a racket by engaging local constables to run down fugitives in their off-hours, and he hired both white and black spies to ferret out former slaves who were concealing their status. Holding those captured on trumped-up charges until their alleged owners could be sent for, he remanded them south, collected his fees, and sometimes received rewards from grateful masters as well. In some instances his agents seized blacks known to be free and might well have carried them into slavery but for the intervention of excited crowds. A series of incidents pitting property rights of slaveholders against the rights of blacks and fair enforcement of the law frequently brought crowds of angry blacks and curious whites into the streets. In the course of two and a half years, the town turned against the commissioner, his agents, and the law. He resigned under pressure, no replacement was found, and relative calm returned to the black community.[13]

Economic gains came even more slowly for Harrisburg's African Americans. Whether free or bond, or residents of white households or black households, the occupations open to them were primarily as servers of whites. Males waited table or were coachmen, butlers, and general handymen. They also ran messages, chopped wood, hauled goods, and performed all types of casual labor. Women worked as cooks, cared for children, washed and ironed clothes, and cleaned houses—before returning home to perform the same tasks for their own families. The inclusion for the first time of occupational data in the census of 1850 reinforced

Fig. 21.   A Rag-Picker at the Turn of the Century.   Most Harrisburg blacks found employment as servants or casual manual laborers. Rag-pickers collected, sorted, and cleaned rags for use by various industries.

the point. That year preindustrial Harrisburg had 886 black residents.[14] More than three-quarters (681) lived in 175 family units in 152 separate dwellings. The remaining 205 roomed in black households or in the homes of white employers (see Table 10.1).[15] The great majority, 788 (89 percent) claimed birth in Pennsylvania; 98 came from other states, chiefly Maryland (67) and Virginia (13). A total of 195 black males worked at only sixteen different occupations. Because more than half (101) designated themselves as "laborer," the exact nature of their work is not known. A few (4 clergymen, 2 "doctors," and a schoolteacher) were the equivalents of white professionals and semi-professionals. In an era when white professionals were as likely to have acquired their status by apprenticeship as by formal education, it is improbable that the blacks had specialized schooling of any sort. The clergymen were probably charismatic preachers, the doctors practitioners who healed with folk remedies and herbs, and the teachers persons who were literate.[16]

Seventeen blacks apparently operated small businesses. Ten had barbershops (six working alone, the other each having from one to three

Table 10.1. Harrisburg's Black, Irish, and German Communities, 1850

	Blacks		Irish		Germans	
	No.	%	No.	%	No.	%
Total community	886	—	590	—	602	—
Born overseas	0	0.0	421	71.4	350	58.1
Born in U.S.	886	100.0	169	28.6	252	41.9
Singles not with families	205	23.1	221	37.5	109	18.1
No. in households	681	76.9	369	62.5	493	81.9
No. of households	175	—	93	—	118	—
Avg. per household	3.9	—	4.0	—	4.2	—
Households in which—						
Husband & wife both foreign	0	0.0	52	55.9	75	63.6
Husband foreign, wife American	0	0.0	19	20.4	33	28.0
No husband listed	23	13.1	16	17.2	4	3.4
No wife listed	2	1.1	6	6.5	4	3.4
Entire family foreign	0	0.0	30	33.3	24	20.5
Parents foreign, children mixed	0	0.0	9	10.0	9	7.7
Parents foreign, all children American	0	0.0	31	34.4	49	41.9
Husband foreign, wife & children American	0	0.0	20	22.2	35	29.9

Source: Computerized population census schedules, 1850.

employees); five ran oyster houses; and two were teamsters who each had a horse and cart and transported goods on demand. Seven blacks were skilled workers: four shoemakers, two coopers, and a butcher. Two worked as boatmen on the canal. The remaining fifty-four filled serving positions: thirty-four were waiters (in hotels and boardinghouses), sixteen were servants, three were hostlers, and one was a groom. Although the 1850 census made no provision for listing the occupations of females, nine black women were shown with jobs: five cooks, two servants, one washwoman, and one laborer. The seventy-seven other blacks (nine men and sixty-eight women) who lived in white households and had no listed occupations probably were servants. Blacks monopolized or dominated a few occupations: all of Harrisburg's waiters were black, and eighteen of its twenty barbers, seventeen of the twenty listed servants, and five of nine cooks. In sum, slightly more than half of the town's blacks were unspecified "laborers," over a quarter were servants, and not more than 16 percent held other occupations.

Ownership of land was another measure of the economic progress of African Americans. As late as 1825, local tax records listed only six black property holders.[17] By 1850 the number increased to twenty-eight (five of whom were women) with real estate and horses and carriages for hire valued at a total of $13,300. The federal census that same year showed thirty black property holders (including three women) with a total of $20,100 worth of real estate. The properties held by blacks were very low in value: eleven were worth between $100 and $400, twelve were worth $500 to $900, and seven were valued at between $1,000 and $1,800. As will be demonstrated, persistence in the community had much to do with the acquisition of real estate.

Irish- and German-Born Immigrants

Unlike the blacks, many of whose parents and grandparents were native Pennsylvanians, Harrisburg's Irish and German immigrants were, of course, born overseas. Most of the area's earliest settlers had migrated from the two countries, though most of those from Ireland had been Scots-Irish Protestants, not Irish Catholics. Once in America, their descendants tended not to remain outsiders, as did those who were black. Within a generation they had become an important part of the general American stock. Throughout the colonial and early national eras, a slow but steady flow of Scots-Irish, regular Irish, and Germans replenished and added to the original numbers. Then, in the late 1840s, massive waves of immigrants from the same countries suddenly poured into the United States—the Irish (overwhelmingly regular Irish) driven by the potato famine; the Germans, at least in part, motivated by the failure of democratic reform in Germany in 1848. Until 1850 there was no accurate count of the number of either group settling in Harrisburg.

Many Americans perceived this dual immigration at mid-century as a threat to the nation's cultural heritage and economic well-being. Halting that flow became the key doctrine of the strongly nativistic Know-Nothing movement that arose in the early 1850s. The Germans' love of beer and their patronage of beer gardens on Sundays antagonized the strong prohibitionist and sabbatarian sentiments of many nativists. At the same time, the German background of many older residents cushioned opposition to newcomers of that nationality. The Roman Catholicism of the Irish, and their extreme poverty, provided fuel for the Know-Nothings. That Harrisburg's hostility to the Irish was less harsh than in many communities may be attributed to two factors: famine Irish did not flock to the area in especially large numbers, nor did they stay long, and prefamine Irish already in the community paved the way for at least limited toleration of the impoverished newcomers. A similar though smaller wave had come to labor on the State Works two decades earlier. Over the years, those who remained blended successfully into the community. Foremost was entrepreneur Michael Burke, a donor of the land for St. Patrick's Church and a member of the borough council. Another prominent Irish-born resident, Father Pierce Maher, pastor of St. Patrick's from 1837 to 1868, had won the respect of most local residents.[18] As a consequence, neither the Irish, their church, nor their priest were totally new or strange to Harrisburg in 1850.

Even so, anti-Catholic incidents occurred periodically. Sermons attacking the Catholic faith were not unusual in Harrisburg, and the press often commented favorably on them. In October 1850 the death in a Protestant household of an Irish-Catholic servant who had a six-year-old daughter revealed the religious prejudices of both sides. According to the employer, the dying woman left the child to her. Father Maher sent for the child when the mother died, but was refused. A few days later, members of his flock seized the child on the streets and took her to his residence. He in turn placed her in the home of one of his parishioners. The Protestant woman thereupon secured a writ of habeas corpus and won temporary custody until a permanent guardian could be appointed. The guardian appointed was Catholic, as required by law.[19]

Irish and German immigrants, though spared the terrors faced by blacks, were constantly abused in the press. Newspapers of that era were quick to make sport of anyone except the socially prominent who ran afoul of the law for drunkenness, disorderly conduct, brawling or rioting, bastardy, or abuse of family members. As had long been true for blacks, the new Irish immigrants in particular came under considerable fire when Stephen Miller was editor of both the *Telegraph* and the *Morning Herald* between May 1854 and November 1855. A Methodist layreader, ardent prohibitionist, and strong advocate of Know-Nothingism from Philadelphia, Miller shamelessly exploited local fears and prejudices against the Irish and their religion. In reporting their scrapes with the law, he stressed their being Irish and Catholic and held them up to ridicule. He complained that Irish paupers were overrunning the country's jails and poorhouses and living off the taxpayer. He scanned news from all parts of the country for wrongdoing by priests and nuns. He denounced the pope for interfer- ing in American politics and "meddling" in the affairs of American Catholic churches. He welcomed the formation of the Guards of Liberty, a secret local Know-Nothing unit, and boasted that the American Party would save the nation by limiting immigration and extending the residency requirement for voting. Himself the grandson of German immigrants, he made far less of German newcomers, but their addiction to lager beer offended his prohibitionist views, and he demanded the closing of saloons and beer gardens on the Sabbath.[20]

Unlike the black community, all of whom were born in the United States, the Irish and German communities contained both foreign-born and American-born (see Table 10.1). In some families the immigrant parents had children who were born overseas, or in America, or some of each. There were also male immigrants who had taken American wives and whose children were all American-born.[21]

Germans in 1850 appeared to have been more strongly inclined to family living than the Irish. Only a handful of German households appear to have been headed by widows or widowers. Irish households had a higher percentage of women heads, and a much larger percentage of Irish lived as singles rather than in families. This difference may have been more apparent than real, however. For example, of the 160 single Irish males in Harrisburg in 1850, some 121 were laborers living in five hotels or boardinghouses. They probably were members of temporary railroad construction gangs working in the area that summer who moved on as soon as their work was completed. How many may have had wives and children living elsewhere while they worked in the Harrisburg area cannot be determined. Similarly, some of the women who appear to have been widows may have had husbands working in other communities and listed as singles there. In any event, by 1860 only 15 percent lived as singles, and the proportion of families apparently headed by widows fell to 11.5 percent.

Using the ages of children born in the United States to determine the approximate dates of immigration for families, no more and probably considerably fewer than 48 percent of Irish families (including couples with no children by which to date their immigration) could have been famine Irish. A slightly larger percentage of German families (53 percent) immigrated during the same five years, 1846–50. How many of the 221 Irish and 109 Germans not living in family units came in the great wave of migration cannot be determined.

Minorities and Industrialization, 1850–1870

Even as the fugitive slave problem and Miller's newspaper diatribes against immigrants attracted great attention, industrialization was beginning to change Harrisburg living patterns, even for the minorities. A comparison of the listed occupations for males in the three minority groups with those of all Harrisburg males in 1850 shows the relative status of the minorities as the process began (see Table 10.2). Although prejudice against the Irish ran high, their rural origins also worked against their moving to skilled urban jobs in America. The percentages in the various occupational groupings in 1850 show them as only marginally better off than blacks. Eighty percent of the latter were servants or laborers; 81 percent of the Irish were in those categories or did factory work. On the other hand, 20 percent of the Irish were in the higher-paying job categories, as compared with 17 percent of the blacks. However, if the 121 temporary Irish railroad construction workers are excluded, the Irish considerably outranked the blacks. Only 60 percent worked in industry or were servants or unspecified laborers, and 36 percent were in the higher-paying categories. German immigrants were not only best off of the outsider groups; they also had higher percentages who were in business and the crafts than was true for the community as a whole, and fewer industrial workers and servants.

As the overall population swelled between 1850 and 1860, the percentages of all males in the lower-paying occupational categories (servant, laborer, and industrial worker) grew. Of course, the combined higher-paying categories (professionals, businessmen, white-collar workers, and craft workers) declined proportionately. The largest percentage drops were for unspecified laborers and craft workers; industrial and railroad workers and servants made the greatest gains. For blacks and the Irish the reverse was true: the percentages of those who held higher-paying jobs increased slightly, while those with lower-paying jobs declined. The experience of German immigrants paralleled that of the general population, though shifts were smaller.

Table 10.2. Changing Occupational Patterns of Black, Irish, and German Males Compared with All Males, 1850, 1860, 1870

Category	No.	Professional	Business	Clerical	Craft	Industrial	Service	Laborer	Other
All males									
1850	2,336	5%	11%	10%	37%	9%	5%	21%	2%
1860	4,203	4	12	10	29	16	15	13	1
1870	7,443	3	12	11	22	20	15	19	0
Blacks									
1850	195	4	9	0	4	0	28	52	5
1860	263	3	11	0	5	2	35	35	10
1870	605	2	9	0	2	4	24	57	3
Irish									
1850	247	1	5	2	11	2	0	79	0
1860	239	2	7	1	13	16	6	55	1
1870	397	1	5	1	13	18	2	56	4
Germans									
1850	188	3	13	3	52	1	6	22	0
1860	438	2	18	2	45	9	3	21	1
1870	711	2	18	3	41	10	3	23	1

SOURCE: Computerized population census schedules, 1850, 1860, 1870.

For blacks the impact of industrialization was indirect. Apparently few found employment in factories or on the railroads, but they did enjoy new opportunities. For example, the number of different job categories they held increased from sixteen in 1850 to thirty-five by 1860 to forty by 1870. The construction of factories and mills, hundreds of new homes for the growing work force, and schools, stores, and other structures created a need for people to haul and handle materials, clear work sites, and clean up after construction. Many of those jobs went to blacks. Increasing prosperity of whites due to industrialization also provided more work for black females as cooks, domestics, and washwomen. The greatest gain among the higher-paying categories was for self-employed businessmen, probably because the growing African American community was able to support more of its own small-enterprisers. Among lower-paying jobs, the greatest shifts were from unspecified laborer to such jobs as waiter and porter. A handful did secure work in factories. If nothing else, between 1850 and 1860 the percentage of black males in occupations other than unspecified laborer or servant increased from one in six to one in five.

The Irish also enjoyed gains in the number and percentage who were engaged in business or the crafts, and the percentage of unspecified laborers declined sharply as more found jobs in industry or as servants. The number of German professionals, businessmen, craft workers, and white-collar workers all increased, even as the percentages in those fields (except for businessmen) declined. Percentages in lower-paying categories rose accordingly but shifted toward industrial and railroad work and away from the categories of laborer and servant.

Although the Civil War years enriched most of the city's industries and entrepreneurs, the war's overall impact on the occupations of males by 1870 was not nearly so beneficial. The percentages in the professions generally declined a little, and in the crafts they declined by seven percentage points. Clerical workers, industrial workers, and unspecified laborers all increased. For blacks the statistics were especially grim. All categories excepting industrial workers and unspecified laborers fell. Those with factory jobs increased a little, while laborers increased from slightly more than a third of the work force in 1860 to well over half, a condition worse than in 1850.

A closer examination of specific occupations shows more clearly what happened to blacks. They no longer dominated such relatively desirable jobs as barbering or running oyster bars and small restaurants, for example, nor were black women any longer the majority of hired cooks (see Table 10.3). They now shared these jobs with immigrants. Black men increased their holds on such occupations as

Table 10.3.   Changing Numbers and Percentages of Blacks in Specific
Occupations, 1850, 1860, 1870

	1850		1860		1870	
Occupation	No.	%	No.	%	No.	%
Barber	17	89	23	92	25	52
Carter	0	0	13	42	15	65
Coachman, driver	0	0	6	30	11	55
Cook	5	56	6	75	16	30
Domestic, male/female	0	0	3	6	109	24
Hostler	4	27	1	10	18	55
Laborer	102	21	94	17	346	25
Oysterman/restauranter	5	56	4	25	3	14
Porter	0	0	10	83	24	92
Servant, male/female	17	85	131	31	54	29
Teamster, trucker	2	67	1	6	18	35
Waiter	34	100	27	87	52	87
Washwoman/laundress	1	33	14	25	9	60

SOURCE: Computerized population census schedules, 1850, 1860, 1870.

carter, coachman, hostler, and porter, and black women were more
likely to be employed washing clothes and providing other domestic
services. Put simply, the percentage of blacks employed as laborers or
servants increased sharply by 1870, representing a substantial setback
that left them worse off than in 1850.

Only dismal sorts of work were open to black women. Because the
census listed the jobs of so few women in 1850, it is impossible to
measure with any precision what their status was. Inasmuch as 112 of
120 black women in 1860 were listed as cooks, domestics, servants, or
washwomen, however, precision would probably make little difference.
By 1870 matters worsened: 97 percent were servants or laborers.

Other than for African Americans, the war decade saw relatively
slight shifts in Harrisburg's occupational structure. Overall the largest
drop was in the percentage of craft workers, and the greatest gains were
made by industrial workers and unspecified laborers. For Irish males a
drop in the number of servants was the greatest change, with a lesser
decline of businessmen. Industrial workers and unspecified laborers both
had small increases. The number of German craft workers declined, with
many shifting into factory and railroad jobs. The proportions of German
laborers and clerks increased very little.

Ownership of real estate, as might be expected, generally mirrored changed occupational opportunities. The first decade of industrialization, the 1850s, witnessed a substantial increase in the number of landowners, even among the minorities (see Appendix F). At mid-century, African Americans had been little more than half as likely as the general population to own real estate, and their holdings were far less valuable. None exceeded $1,800 in value. More than three-fourths were worth less than $1,000, and more than a third were below $500 (see Table 10.4). Even the acquisition of small holdings required years of saving. Some 40 percent of blacks owning land in 1850 had lived in Harrisburg at least ten years, and nearly half of those had been in the census of 1830. During the 1850s the number of black property holders increased dramatically. So did the value of what they owned: by 1860 two-thirds were worth between $500 and $2,500, and the average value of holdings had increased more than 25 percent since 1850. Blacks were also acquiring property more quickly; 86 percent of those with real estate in 1860 had gained their land in less than ten years. Their average holding, however, was worth less than half that of landowners who had lived in Harrisburg more than a decade (see Appendix G).

An even lower percentage of Irish adults owned real estate in 1850. Those nineteen persons included Michael Burke and two Scots-Irish men, E. M. Pollock and George Beatty. The three were not included in the tables because their extreme wealth (as compared with the others) accounted for most of the property held by the group and distorted the pattern of holdings. Even with the exclusions, the average value of Irish-owned tracts was from two to three times more valuable than land owned by blacks. However, as was true for blacks, nearly half of all Irish landowners (including the three wealthiest) had resided in Harrisburg for a decade or longer (see Appendix G).

Again it appears that industrialization improved incomes. During the 1850s the Irish made more impressive gains in real estate holdings than either of the other minorities or the community at large (see Table 10.4). One adult in four, involving 60 percent of all Irish households, owned land by 1860. As with blacks that same decade, persistence had less impact on land acquisition than earlier. Although eleven of the nineteen landowners of 1850 still owned property in 1860, another twenty who had been landless acquired real estate during the decade. So did eighty-four newcomers who settled in the city after 1850. The value of both the average and the median holdings dropped, however, indicating that persons of limited means were acquiring that property. Despite jobs that were at or near the bottom of the occupational ladder (55 percent were

Table 10.4. Changing Patterns of Real Estate Holdings of Blacks, Irish, and Germans Compared with All Residents, 1850, 1860, 1870

Value of Real Estate	Persons (No.)	$100–$499	$500–$999	$1,000–$2,499	$2,500–$4,999	$5,000–$9,999	$10,000–$24,999	$25,000–$49,999	$50,000 & More
All residents									
1850	451	7%	16%	29%	19%	14%	10%	3%	2%
1860	1,460	9	23	35	12	10	8	2	1
1870	1,805	3	7	38	20	15	11	4	3
Blacks									
1850	30	37	40	23	0	0	0	0	0
1860	106	25	36	37	2	1	0	0	0
1870	69	12	16	57	12	4	0	0	0
Irish									
1850	16	6	19	50	13	13	0	0	0
1860	115	15	38	31	9	3	4	0	0
1870	150	3	12	55	15	9	6	0	0
German									
1850	28	25	14	36	18	7	0	0	0
1860	171	13	37	30	10	6	4	0	0
1870	287	1	6	22	9	32	18	10	2

SOURCE: Computerized population census schedules, 1850, 1860, 1870.

Fig. 22.   One of Harrisburg's "Invisible" African American Servants.   The label on the original of this photograph is "Dr. Rather and His Team, Third Street, August, 1896." The house in the background was the Eby mansion at Third and McClay streets.

laborers, factory and railroad workers, or servants), these once-landless tenant farmers from Ireland scraped together enough to buy small plots of land.

Even in 1850, nearly as large a percentage of German-born immigrants owned real estate as did the general population. A decade later both they and the Irish had greater proportions of landholders than the community at large. Unlike the Irish, however, the average value of German holdings increased during the decade. Moreover, concentrated in business and the crafts, the 31 percent of German-born males with wealth tied up in inventories and tools had more valuable holdings than the comparable 22 percent of Irish males with personal property.

As with occupational improvements, gains in property ownership during the 1850s reversed during the war decade. The percentage of Harrisburg adults owning real estate dropped nearly a third, to one adult in seven, but the average holding increased in value nearly 40 percent, offset in part by inflation. Once again African Americans suffered the greatest losses. In spite of a substantial gain in overall population, only two-thirds as many blacks held real estate as a decade earlier. The aggregate worth of that property had increased 8 percent (probably less than the rate of inflation); the average value of a holding advanced 65 percent. Among the Irish the number of property holders increased nearly a third though the percentage fell six points. The value of their holdings, on average increased more than half. The greatest gains were among the Germans. One eighth of German landowners had property valued at more than $25,000, nearly as great a percentage as had owned land as in 1860, and the average holding's value increased more than 135 percent.

Persistence in the community had much to do with whether a person owned land and what that land was worth. In 1860 many property owners who were landless a decade earlier had in the meantime acquired real estate (see Appendix G). Although the unsettling effects of the war and its aftermath resulted in considerably fewer landowners remaining between 1860 and 1870, those who stayed saw their property increase in value. The Irish were an exception. As other studies have concluded, the Irish went to extremes to acquire property. They often deprived their children of schooling and put them to work early to help the family acquire a piece of real estate.[22] This may account for the odd circumstance in 1870 that the average value of the holdings of Irish who persisted from 1860 was lower than that of Irish newcomers. Apparently those moving to Harrisburg between 1860 and 1870 came with enough money to buy tracts that were more expensive than those owned by longer-term famine Irish residents scrimping to pay for their first small plots.

A comparison of the city's occupational structures and real estate holding patterns over the two decades from 1850 to 1870 better measures the impact of early industrialization on the minorities (see Table 10.2). Although there were small shifts among professionals, businessmen, and clerical workers, white-collar positions overall in both 1850 and 1870 totaled little more than one-fourth of all males with occupations. The significant shifts were among the blue-collar workers, the major change being a decline of craft workers from more than a third of all males with occupations to one in five, and an increase of industrial and railroad workers from one in ten to one in five over the same two decades. Among the minorities this change had differing impacts. The proportion of blacks in crafts fell by half, and the percentage of industrial workers increased—but in both instances less than 4 percent of blacks were involved, a percentage about equal to that of black professionals. For blacks the principal shifts were a decline of servants and a nearly equal increase of unspecified laborers. Irish males had a large decline in the percentage who were unspecified laborers, a small increase in the percentage who were artisans, and a large increase in the proportion of industrial workers. Germans, more than half of whom had been crafts workers in 1850, sank to about 40 percent in 1870, with the majority of the change being to businessmen or industrial workers.

The major change in real estate ownership over the same two decades chiefly favored the Irish and German minorities (see Table 10.4). Where one in twenty of the former were landholders in 1850, nearly one in five held land by 1870, and the value of the average holding had increased nearly 50 percent. The percentage of German landowners increased from one in ten to one in four in the same period, and the value of their average holding increased 175 percent. Both groups did better than the population as a whole. Overall the gain had been from one in ten to one in seven, and the value of the average holding had gone up only 2.5 percent. Only blacks fell behind in terms of the proportion holding real estate. That dropped from one in seven to one in twenty between 1850 and 1870. For those fortunate enough to have land, however, the value of the average holding increased 127 percent in the same period.

The Restructuring of the Black Community, 1860–1870

How is the sharp deterioration of conditions for African Americans at Harrisburg during the decade that finally brought them full emancipation to be explained? If it had been only blacks who were pushed into lesser occupations, with fewer owning real estate, racial discrimination might plausibly have been the principal cause. However, with most of the major groups in the city experiencing notable gains in the 1850s followed by losses in the 1860s, more than race hatred was at work.

A likelier cause was the restructuring of the black community which occurred that decade. Between 1860 and 1870 a large number of the older, established, urbanized black families left Harrisburg and were replaced by recently freed slaves from the South. Why these simultaneous migrations took place is not clear. Some may have gone south to rejoin families or to use their educations and experiences to assist and uplift others of their race, but that presumably would have been a relatively small number.[23] Perhaps the abuse of blacks in the city during and immediately after the war, and the near approach of Confederate forces on two occasions, induced them to leave. But where would treatment have been any better, or economic prospects more promising? And were threats of enemy occupation sufficient to induce them to move instead of make temporary flights to safety? Because the answers to these questions are not known, the motives of those who left remain a mystery.

Similarly, why so many former slaves from Maryland and Virginia came to Harrisburg during or soon after the war is not known. Again, some may have come to be with relatives living in the community. Others may have been wartime refugees fleeing ahead of Confederate troops who escaped the purging of such persons from the city following the Battle of Gettysburg.[24] The policy of the Freedmen's Bureau after the war of providing free transportation to black refugees who had job offers in the North raises the possibility that they were drawn by the need of Harrisburg's new iron and steel mills for workers. But there is little evidence that these or any of Harrisburg's other industries hired more than a handful of black employees that early. Records of the Freedmen's Bureau, moreover, show only forty-eight persons transported to Harrisburg between April 1865 and the end of 1869. Most of those went to jobs as household servants, and few remained in the community long. Of the twenty-eight listed by name, only four subsequently appeared in Harrisburg's 1870 federal census schedules.[25]

Whatever the reasons for the dual migrations, by 1870 Harrisburg's African American community had two distinct segments. Half were persons used to freedom—natives of Pennsylvania or some other free state, or, if born in the South, either manumitted slaves or refugees who had come north before the Civil War. The other half were persons born in the South who had probably spent their lives as slaves in slaveholding communities and who migrated north during or soon after the Civil War.

Beyond lack of experience with freedom, these recent migrants differed from the others in several ways—for example, in literacy. More than 60 percent of the adults could not read or write, so their addition to the community raised illiteracy among black adults generally, to 43 percent in 1870 from 24 percent in 1850.[26] By contrast, fewer than 20 percent of black newcomers from free states were illiterate, and the rate of illiteracy among blacks born in the North who had lived in Harrisburg at least a decade was only 11 percent. Also, many of the southern newcomers came from rural backgrounds. As W.E.B. Du Bois pointed out in 1899, the lack of skills and experience with freedom handicapped the freed slaves when they moved to northern cities.[27] While familiar with agricultural tasks, they lacked skills that would have been useful in an urban center such as Harrisburg. This can be seen in occupational patterns; southern newcomers are almost totally missing in the higher-paying jobs and concentrated in those at the bottom of the scale. Although persistence in the community and occupational status were closely related for whites as well as for blacks, it is not clear whether staying resulted in higher-paying jobs or whether better jobs induced people to remain. Among black newcomers, being born in the North (or, if born in the South, having experience in the North) gave one a distinct job advantage over those who came directly from the South (see Table 10.5).

Turning to the ownership of real estate, persistence of ten or more years proved to be more important than whether one was born in the North or in the South. At the same time, the recent slaves, like the Irish, apparently made a greater effort to acquire land of their own, even if it was less valuable than the property held by their northern-born counterparts. Illiteracy too was directly related to whether one was southern-born and, if so, how recently a person had left the South. Northern-born blacks, whether persisters or newcomers, were far less likely to be illiterate than those born in the South, even those who had persisted at Harrisburg or lived in the North for ten or more years.[28]

Racism no doubt established the outer bounds for the advancement of blacks in the period, but it does not entirely explain the decline between 1860 and 1870 in jobs held or property owned. The large number of recently freed slaves simply were not prepared to assume even the lowly stations that Harrisburg's free blacks had struggled so long to achieve. However hardworking or ambitious, they could hardly be expected to adjust to life as free persons, move from a rural to an urban setting, fit into a new community, find good jobs, and become landowners—all in less than ten years. That perhaps two dozen did acquire real estate by 1870 was remarkable in itself and testimony to their efforts at self-improvement. The career of one illustrated how it was done. Turner Cooper, illiterate but newly freed, arrived in Harrisburg from Alexandria, Virginia, in 1868 with his wife and seven children. The census taker found him working as a brickyard employee in 1870. Assisted by his two elder sons, who lived at home and were listed as laborers, he had already acquired real estate valued at $1,500. Deeply religious, Cooper hated the "Bloody Eighth" Ward where he lived, so he sought a location several blocks east of the Capitol and beyond the transportation corridor. There he built himself a home, and with the aid of a white carpenter and five blacks, at least two of whom were also illiterate natives of Virginia, he began the Springdale neighborhood. By 1890 it was a "thriving, populous community of blacks and whites."[29]

What impact did early industrialization have on the children of the minorities—on their schooling, on the age they entered the work force,

Table 10.5. Distribution of Persisting, and Northern- and Southern-Born Black Newcomer Males in Occupational Categories, 1870

Occupation	Total n = 598	Persisting Harrisburgers n = 85	Northern Newcomers[a] n = 172	Southern Newcomers n = 341
Professionals	1.7%	5.9%	2.3%	0.3%
Businessmen	2.5	10.6	1.7	0.9
Craft workers	3.3	11.8	9.3	2.3
Industrial workers	4.0	3.5	1.2	5.6
Servants	28.3	24.7	35.5	25.5
Laborers	57.9	43.5	50.0	65.4

SOURCE: Computerized population census schedules, 1870.

[a]Includes 31 southern-born blacks who lived in the North 10 or more years.

on their occupational prospects? The relevant data in the census records from 1850 through 1870 has serious flaws. Although census takers asked whether children attended school during the past year, they did not ask for how long a period. With regard to jobs, census takers were instructed to list the occupations of males above the age of fifteen in 1850, of all persons above fifteen in 1860, and of all persons without regard to age in 1870. Nonetheless, in 1850 some boys not yet fifteen and a few girls were listed with jobs. In 1860 some boys and girls not yet fifteen were shown with jobs. Although these considerations complicate the picture, some patterns do emerge.

School attendance data, though scanty, were uniform throughout the three censuses. Although some children in 1850 started school when they were four or five, by 1870 six had become the customary age (see Table 10.6). The majority of boys attended through age fourteen at least until 1860; by 1870 they remained through age fifteen. Girls followed a different pattern. In 1850 the majority of those age five through fourteen were in school; in 1860 and 1870 it was age six through fourteen. Of the minority groups, far lower percentages of black children attended school in each age category and without regard to gender. Even of those between the ages of six and fourteen, barely half were in school. This compared unfavorably with the two-thirds and more of Irish and German boys who attended school, and the three-quarters of native-born

Table 10.6. Percentage of Children Attending School by Age-Group, Gender, and Ethnicity, 1850, 1860, 1870

Age	Group	Male			Female		
		1850	1860	1870	1850	1860	1870
0–5	American-born whites	12%	5%	2%	15%	6%	2%
	Blacks	5	2	1	6	3	1
	Irish	8	6	3	14	9	1
	Germans	8	3	1	6	7	3
6–14	American-born whites	77	77	75	78	74	72
	Blacks	51	41	50	49	52	53
	Irish	85	68	66	74	76	63
	Germans	83	67	70	69	71	62
15–19	American-born whites	24	26	19	20	23	18
	Blacks	9	7	15	2	10	9
	Irish	14	21	14	18	18	16
	Germans	10	13	11	3	11	10

SOURCE: Computerized population census schedules, 1850, 1860, 1870.

whites. A good number of youngsters, and especially males, attended school after the age of fourteen. Some of these were in high school, others simply were slow at completing their elementary schooling. Whereas between a fifth and a fourth of American-born white males in this group were in school, minorities lagged behind. Over the twenty-year span, American-born whites lost ground, possibly because the cotton mill hired chiefly young people of that background. Irish and Germans similarly fell behind a little by 1870, the Irish apparently allowing a higher percentage of their children to attend school than Germans or blacks.

As might be expected, first job patterns for boys ran counter to those for school attendance (see Table 10.7). While sons of immigrants remained off the job market until the age of sixteen, and only a few American-born sons worked before the age of seventeen, from a quarter to a third of young black males went to work two years earlier. Many of the younger working blacks were serving apprenticeships in barbershops or found jobs as domestics, servants, or waiters. By the time they reached

Table 10.7. Percentage of Male Children with Jobs, by Age and Ethnicity, 1850, 1860, 1870

Year	Age	All Males	American-born White	Black	Irish	German
1850	14	9%	8%	29%	0%	0%
	15	25	23	57	0	0
	16	39	30	50	60	80
	17	67	68	33	100	100
	18	74	67	85	100	100
	19	78	76	67	100	100
1860	14	19	17	26	29	15
	15	30	31	29	25	23
	16	37	33	44	50	50
	17	56	53	67	100	55
	18	69	73	50	80	60
	19	69	73	43	67	60
1870	14	12	11	13	18	9
	15	23	23	17	20	29
	16	38	37	29	44	39
	17	59	56	89	29	65
	18	77	75	75	77	95
	19	80	84	59	50	81

SOURCE: Computerized population census schedules, 1850, 1860, 1870.

the age of seventeen, most of those with jobs were common laborers or servants.

American-born whites were slowest to move into the job market, but by age nineteen they had the highest percentage at work. Irish and German boys followed a similar pattern. Even as the whites, both natives and foreign-born, moved into jobs when they reached the age of seventeen, eighteen, or nineteen, more blacks of those ages were not finding jobs at all.

A comparison of the occupations of specific fathers of 1850 with the last known occupations of their sons by 1870 shows how industrialization affected career choices for the sons (see Table 10.8). Even by 1870, perhaps in part because they were still relatively young, only one minority son had become a professional. That was Chester, the Harrisburg black who for most of his adult life lived and practiced law in the South.[30] About 10 percent of Irish and German sons followed their fathers into business, though that percentage was smaller than for their fathers. The gap between black sons and their fathers who were in business was even wider.

In spite of the decline of crafts, minority sons did better than their fathers in this category. Similarly, many Irish and some German sons became white-collar workers. Blacks, Irish, and Germans moved into industry, where few if any of their fathers had worked. Sons who were unspecified laborers were in each instance a smaller percentage than their fathers had been in 1850; about half as great a proportion in the instances of Irish and German sons. Probably because they were still young more of each minority were servants than had been true of their fathers.

Most significant were the relatively low percentages of sons of blacks and immigrants who persisted in the community and became industrial workers. Even if it is assumed that some of the unspecified laborers in fact worked as factory hands and that part of the craft workers were actually factory artisans, the percentages are surprisingly low. As in the instance of the sons of craft workers generally, young men who remained in Harrisburg found ways to avoid work as factory hands. Those jobs went to newcomers.

Minorities After 1870

Of the city's three minority groups, only the blacks were self-perpetuating. Few of them or their children ever passed the color barrier. By

Table 10.8. Comparison of Black, Irish, and German Fathers of 1850 and Their Sons, by Occupational Categories

					Occupational Category				
	No.	None	Professional	Business	Crafts	White-Collar	Industry	Servant	Labor
Blacks									
Fathers, 1850	84	15%	7%	11%	11%	0%	0%	11%	45%
Total sons	148	68							
Sons with occupation, 1850–70	48	—	2	2	23	0	4	27	42
Irish									
Fathers, 1850	40	8	3	15	10	3	0	0	63
Total sons	69	57							
Sons with occupation, 1850–70	30	—	0	10	23	17	13	7	30
Germans									
Fathers, 1850	68	4	4	16	50	0	0	4	21
Total sons	122	70							
Sons with occupation, 1850–70	30	—	0	10	73	7	10	7	10

SOURCE: Computerized population census schedules, 1850, 1860, 1870.

contrast, at least in small cities such as Harrisburg, the Irish and German communities gradually faded away as the older generation died off and their American-born children became part of the mainstream (see Appendix A). Only if new recruits from the old country replenished their number did the groups survive. Because relatively few new immigrants came to Harrisburg after 1870, the total foreign-born segment waned, and Irish and German immigrants became a smaller and smaller proportion of the total number of foreign-born. There were too few to support ghettoization, and the Irish and German neighborhoods that did appear were relatively short-lived. By 1900 the Irish-born and the German-born shrank to 1 and 2 percent of the city's population respectively and ceased to be numerous enough to warrant further discussion.

Between 1870 and 1900 the city's two black subgroups gradually became a single body. A sample study of 235 couples from the two principal black wards in 1880 provides insights into the process. That year, southern-born blacks still constituted about half the population. Well over half the heads of family and 45 percent of their spouses were born in Virginia or Maryland. Both partners were natives of Pennsylvania in only 27 percent of the instances, while 31 percent included one partner born out of state. The postwar newcomers and more recent migrants from the South were marrying into established local families, thereby speeding their entry into the community.[31]

Meanwhile, the occupational status of males improved. Although two-thirds were still unspecified laborers and servants, that was twelve percentage points lower than a decade before. For them, jobs were available at a tar works, at quarries and tanneries, and of course as haulers of goods, among other things. The number of blacks who were industrial workers had swollen to 19 percent, a major shift for a single decade. Factory jobs were chiefly in iron and steel, with some blacks traveling five miles to Steelton, south of the city, each day. Although it may safely be assumed that those were the lowest-skilled and lowest-paid jobs, even so, they provided steadier work and probably higher incomes than could be earned as casual laborers. Skilled artisans made up a tenth of the sample, as compared with only 2 percent of all black males in 1870. The percentage of professionals, after dropping to 2 percent in 1870, had climbed to more than 3 percent in 1880, the same as in 1860 and the same as for all residents of the city that year.[32]

Blacks also gradually regained the economic status lost between 1860 and 1870. Although they encountered both discrimination and segregation throughout the period, neither was absolute or rigid. The city's churches and lodges, for example, were segregated, and blacks had their own labor unions and cemetery, but blacks were admitted to the public library and the hospital, they rode the trolleys, they served on both petit and grand juries, and they sued and were sued in the courts with apparent fairness.[33]

The situation in housing and education was more complex. At first, housing for blacks followed the pattern in the South, being located along alleys at the rear of the homes of their wealthier, white employers. Eventually the principal enclave of blacks' homes became the district immediately east of the Capitol. In 1857 a merchant–real estate developer, William K. Verbeke, bought up a block in that district containing some twenty to thirty "huts" occupied by blacks. To provide housing for them, he purchased ten acres just over the northern boundary of the borough, an area that would be annexed to the city in 1860. There he sold lots to any displaced blacks who wanted to relocate, moved their houses for them, and allowed them to pay off their debt to him at the rate of one dollar a week. "Verbeketown," as the blacks called the area, became the nucleus of their second major location north of the Capitol. Still later they settled in East Harrisburg and in the southern part of the city near Steelton, where many found jobs.[34] Although blacks lived in all nine wards of the city in 1880, some 60 percent were concentrated in two of the central city wards and all but 8 percent resided in six wards. Few lived along the river or in the newer developments in the outer districts of the city. Two decades later, 45 percent lived in two wards, and only three of the ten wards had fewer than 5 percent.[35]

Despite state laws to the contrary, neighborhood grammar schools generally were segregated in practice. The black schools were staffed by black teachers and administered by principals of the same race. The city's two high schools, which were segregated by gender, remained completely white until 1879. The entry of blacks that year produced open hostility from the white community. Four years later two blacks graduated from Boys High School in a class of thirty-six. White resentment again flared when it was learned that one of the blacks ranked first in the class. Six or seven black girls failed to graduate that same year from Girls High School, allegedly because of "teacher prejudice." The next year, two blacks graduated from each of the schools. This limited integration of the high schools persisted for the balance of the century and after, but near-segregation remained the rule because only a few children from elite black families could afford the luxury of high school. Even so, recognition of the black community's stake in schooling was marked by the election of the first black to the citywide school board in 1878. Dr. William H. Day, a native of New York City with bachelor's and master's degrees from Oberlin and an honorary doctor of divinity degree from Livingston College, had come to Harrisburg as a public school teacher in 1872. Although he made his mark in education, he was also a respected religious and political leader in the black community. In all, Day spent fifteen years on the school board, the last two as its president. Being chosen by twenty-five white colleagues to head the board was an uncommon distinction for an African American in the United States at that time.[36]

These findings differ somewhat from existing research on the ethnic and racial experiences of northern industrial cities in the post–Civil War period. From Du Bois's 1899 study of Philadelphia to the present, scholars have pointed to significantly worsening economic, social, and economic conditions for blacks. At first the benefits of the industrial revolution passed them by because white owners refused to hire blacks for any but the most menial factory jobs. European immigrants encroached on the better-paying occupations traditionally held by blacks (barbering, carting, waiting table, catering food). Northern labor barred blacks from union membership and apprenticeships; and whites in the North, by refusing to engage black professionals or black artisans, in effect limited their clientele to members of their own race who paid poorly. As white business firms became larger and undersold them, the number of black enterprises and entrepreneurs declined. Meanwhile, de facto segregation in northern cities forced them into increasingly black ghettos, barred them from equal educational opportunities, sometimes restricted their right to vote, and usually kept them from holding any but the least important public offices. The common explanations concentrate on the growing racism of the era and increased competition from immigrants.[37]

The usual pattern of development in northern industrial centers was that factory production expanded rapidly during the late nineteenth and early twentieth centuries. This created a labor market that attracted a flood of immigrants from northern and western Europe until the late 1880s. Thereafter the flow shifted to masses of "New Immigrants" from southern and eastern Europe. As for blacks, relatively few moved to northern cities in those years.[38] These patterns all changed with the onset of World War I. Hostilities blocked the flow of immigrants from Europe, and at war's end restrictive legislation prevented a resumption at the old levels. Accordingly, a continuing need for cheap labor in factories and mills touched off the "Great Migration" of southern blacks northward after 1914. As they poured into northern cities in search of jobs and housing, resentment and hostility grew between blacks and the immigrants and their children.[39]

These patterns did not prevail at Harrisburg. Certainly the overall position of the black community declined between 1860 and 1870, but their condition had improved very slowly by the end of the century. Some moved into factory jobs on a limited scale by 1880. Although they were heavily concentrated in certain neighborhoods, Harrisburg had no great all-black ghetto. Rather, by the end of the century blacks were more scattered in pockets throughout the city than had been true in 1870. This is not to suggest that racism no longer existed or that Harrisburg had become an integrated community, but its blacks apparently suffered less than those in many northern cities.

Moreover, despite the growth of the African American community, its total number remained too small to confront the white majority directly. In the absence of intense provocation that might have produced outbreaks, Harrisburg's blacks adopted a strategy of accommodation. Their business and other leaders could not economically survive independent of the white community. They could hope to improve conditions for the race only by pushing steadily but gently for change.[40] As a consequence, race relations in Harrisburg, if not particularly good, were neither antagonistic nor marked by violence.

These differences may be traced to a number of factors, but especially to patterns of industrial development and immigration. Because industrialization at Harrisburg peaked between 1870 and 1880, then began a slow decline, the city's rate of population growth also slowed. The slowing continued until midway through the twentieth century, when it turned to positive loss. Similarly, Harrisburg's largest waves of both black and foreign newcomers came between 1860 and 1870, then slowed to a trickle. The city all but completely missed both the "New Immigration" from southeastern Europe and the Great Migration of southern blacks. Consequently, after 1870, there was much less competition between established blacks and immigrant newcomers for the lower-paying jobs. In addition, after 1914, settled immigrants at Harrisburg were not pressed for jobs or housing by an influx of new blacks from the South. Only on the eve of World War II did blacks move in again in large number. The flow of foreign-born never resumed.

Just as Harrisburg had class distinctions without class warfare after the rise of factories, it had racial tensions without racial clashes in the late nineteenth and early twentieth centuries. Because its minorities experienced no sudden bursts of growth late in the century, an increase in competition between blacks and immigrants was forestalled. The absence of strangers in the land similarly gave native-born whites little cause for alarm.

Labor Relations in
the Early Industrial Era

Payment in cash and the promise of steady work were among the considerations that drew them to Harrisburg's new factories—casual and floating workers in search of permanent jobs, farm laborers seeking a higher return on their labor, journeymen worried about the decline of prospects in their craft, and youngsters going to their first paid employment. There were unexpected benefits as well. Factory labor provided a kind of mass camaraderie rarely experienced by rural workers used to the relative isolation of the fields. Even in small urban centers such as Harrisburg, craft workers had found relatively few others engaged in the same line of work as themselves, and rarely were organized. Once employed in large groups in factories, they shared social as well as work activities with co-laborers. On civic occasions, for example, they trooped as a body in parades, carrying banners of the firms they served. One such occasion was the Fourth of July in 1854. Following the chief marshal, who was on horseback, the arrangements committee and governor of Pennsylvania riding in carriages, and various marching military units came groups of industrial workers. First, about 200 cotton mill operatives, "many of them intelligent, good-looking young ladies," passed in review, led by their plant superintendent. "Liberty and Independence," proclaimed one side of their banner, "Industry is the Strength of our Nation" declared the other. Sixty of Hickok's employees, marshaled by their foreman and his clerk, came next with a colorful banner featuring an eagle above the company's name on the front and bearing the slogan "By Industry We Prosper" on the back. "A respectable turn-out" of car works employees, their banner emblazoned with pictures of a railway car and the company's name on one side and a large car wheel on the other, completed the labor contingents that year. At the conclusion of the parade, between 6,000 and

8,000 people went to "Hanna's Woods," where a free picnic dinner was served (unfortunately not enough had been prepared and some went hungry), followed by a reading of the Declaration of Independence and patriotic orations. That evening at nine, crowds gathered on Capitol Hill to watch a fireworks display.[1]

Managers (abetted by a friendly press) worked hard at promoting good relations with their employees and the public. In January and February of both 1852 and 1853 a number of "gentlemen, citizens and strangers, interested in the successful development of American resources and American genius" sponsored a series of "industrial balls" for the "young ladies" who worked at the cotton mill. The men in attendance included "representatives of all classes of society, citizens, statesmen [members of the legislature], men of leisure, and the hardy sons of toil." They met on "a common level"—interest in the "successful development of American character and institutions." The mill girls were "not less interesting," bringing together "youth, beauty, talent, amiability, and accomplishment of manners, enriched with physical development and fortitude of bearing." Future generations would look back on such fine young women with pride not unlike that of the ancient Spartans for their mothers.[2]

Each summer, Harrisburg newspapers reported picnics and excursions for workers and their families, most sponsored by the companies, a few by the workers themselves. In 1867, for example, "the picnic of the season" was the Pennsylvania Railroad's second annual affair for its employees on July 13. At 7:30 in the morning, fifteen passenger cars and several baggage cars loaded with workers, their wives and children, selected guests, and supplies left the Harrisburg depot for the thirty-minute trip to a grove near Marysville, directly across the river. A second train at 1:30 in the afternoon brought latecomers. Two thousand attended the picnic and made use of the "tables, swings, 'copenhagen rings,' quoit grounds," and baseball diamonds. An orchestra from Lancaster provided music for the hundred or so who danced. At 7:00 that evening "the cheerful faces of the broad-shouldered sons of toil, the happy mothers and delighted little ones" reembarked for the ride home.[3] On August 8, Bailey's rolling mill and nailery workers went by train to Cold Springs for a picnic. August 10, notice was given of an iron molder's union picnic on Independence Island. The cotton mill employees' third annual picnic was held on August 17, with tickets costing 25 cents each (the equivalent of a half-day's salary for the army of youngsters who worked there). Last reported was "the grandest picnic of the season," to be held at Cameron's grove on September 24 and sponsored by the Lochiel Iron Works Library Association, "the only establishment of its kind."[4]

Mutuality of the interests of capital and labor apparently was accepted by both sides, especially in the early years. When, in its second year of business, the board of directors of the Harrisburg Car Works proposed giving General Manager Hildrup a silver tea set in appreciation of his efforts, he persuaded them instead to appropriate the money for relief of the families of workers who fell ill. "From that day to this," he wrote in 1884, "the Company has never allowed a sick employee or his family to suffer for want." Even in periods when workers were temporarily laid off for lack of work, funds were advanced to cover "their pressing needs."[5]

Hildrup took particular pride in the efforts of his firm to keep its employees working during the Panic of 1857. In the absence of orders for rolling stock, the car works took on railroad and canal construction projects and manufactured agricultural implements. Of course, more was at stake than concern for the welfare of the employees. President Calder, in another capacity, had construction contracts to fill and a ready force of car works employees at hand anxious for employment. Hildrup's principal concern was keeping intact the team of skilled workers he had so patiently directed by day and schooled by night.[6]

Workers occasionally responded generously to managers whom they particularly appreciated. Hildrup eventually got his silver tea set some twenty years later. On his fifty-first birthday in 1873, his employees presented him a set valued at $1,250, only $162 short of the daily payroll for all 796 employees that year. The nail works employees at West Fairview similarly honored their "gentlemanly" superintendent in January 1882 with a gold watch as a token of their esteem.[7]

At this early stage it was still possible for the managers to know intimately the capabilities as well as the needs of their individual workers. Just as they knew which of their laid-off regulars needed help in emergencies, so they carefully differentiated pay among their employees engaged in the same tasks. Their reports to the state usually gave a set daily rate for carpenters, blacksmiths, machinists, and others, but if Hickok was at all typical the actual pay scale for a given job ranged widely, with small gradations from worker to worker. For example, his two-week payroll for July 3, 1860, showed eight different machinists receiving respectively $11.66, $13.75, $14.00, $14.25, $14.35, $14.54, $14.57, $15.87, and $16.25.

Using the payrolls of Hickok's Eagle Works and a scattering of reports to the state by Harrisburg firms, it is possible to reconstruct the hierarchy of wages paid by some of the new factories (see Table 11.1). Not surprisingly, the gradations were usually based on skill; the cotton mill used gender and age determinants as well.

The cotton mill differed from the two other firms in its all but total reliance on female and child labor. Forty adult male employees in 1873 all earned $1.50 a day or more, nine adult females over age sixteen earned more than $1.00 a day, and 100 earned between 66 and 92 cents per day. Twenty-seven of the 131 children under sixteen who worked there earned 62 cents a day; the rest earned 50 cents. In fact, the wages received by adult males in all three industries were comparable and related to skill or managerial duties. Those at the top included three cabinetmakers at Hickok's and sixteen foremen at the car works. Those receiving between $2.50 and $2.99 a day included two engineers at the cotton mill and machinists, blacksmiths, molders, and woodturners at Hickok's. In the $2.00 to $2.49 range were foremen and machinists at the cotton mill; cabinetmakers, carpenters, machinists, molders, blacksmiths, and a painter at the Eagle Works; and patternmakers, blacksmiths, machinists, molders, carpenters, and painters at the car works. Those earning between $1.00 and $1.49 were almost entirely unspecified laborers at all three firms. The car works paid its apprentices $1.00 a day, except for thirteen apprentice painters who received 75 cents; the wages of Hickok's apprentices ranged from 50 cents to $1.43.

In general, factory jobs did provide steadier work and better cash incomes than farm labor or jobs in craft shops. These advantages were offset, however, by a greater dependency on regular paychecks and by periods of business sluggishness and depression. Casual workers in

Table 11.1. Wage Hierarchy of Industrial Workers, 1870–1873

Daily Wage Category	1873 Cotton Works		1870 Eagle Works		1873 Car Works		Total	
	No.	%	No.	%	No.	%	No.	%
$3.00 and up	0	—	3	3.8	16	2.0	19	0.2
$2.50–$2.99	2	0.7	7	9.0	0	0.0	9	0.1
$2.00–$2.49	18	6.4	25	32.1	369	46.4	412	35.7
$1.50–$1.99	20	7.1	14	17.9	375	47.1	409	35.4
$1.00–$1.49	9	3.2	6	7.7	23	2.9	38	3.3
$0.50–$0.99	231	82.5	23	29.5	13	1.6	267	23.1
Total	280		78		796		1,154	

Sources: Hickok's Payroll; Pennsylvania Secretary of Internal Affairs, *Annual Report, 1872– 1873*, pp. 369, 385.

small-town craft shops usually had time to garden and to hunt and fish. When work was slack they could at least fall back on those resources, however slim, until conditions improved. Farm laborers in such times might not be paid, but they could count on meals and a place to sleep. By contrast, the long and regular hours put in by factory workers in good times left them with little time or energy for gardening even if plots of land were available. When costs of production had to be reduced to meet the competition, employers lowered wages. When orders slowed or stopped, hours of employment (and wages, of course) fell or workers were laid off. This meant less, or nothing, for groceries, clothing, or rent.

Business during the last half of the nineteenth century was subject to considerable fluctuation. Periods with demand greater than the ability to produce resulted in soaring prices and profits. They were followed by times when demand dried up, prices slumped, and profits disappeared. The shifts often came suddenly and, whether good or bad, rarely lasted long. Worse still, however, business panics swept the economy at approximately twenty-year intervals. Before the Civil War the Panic of 1837 had been the worst, but Harrisburg and most of America was still agrarian then. The Panic of 1857, though severe, lasted only briefly. Following the war there were two major depressions—one from 1873 to 1878, the other from 1893 to 1897. Both were devastating in industrial centers. Many businesses failed, and those that did not slashed payrolls and reduced hourly or daily rates. The existence of reports on individual Harrisburg firms during the 1870s makes it possible to measure with some precision the impact of the Panic of 1873 on workers there.

After 1873 the reports of the Harrisburg Car Works to the state indicated the chaotic conditions that prevailed. In the report for 1876, General Manager Hildrup observed that the value of the firm's annual output ranged from $400,000 to $2,500,000 (both were exaggerations inasmuch as the range since 1868 had been between $565,000 and $1,586,000 and in 1876 was $306,000). Employees, he continued, numbered from 25 to 1,000, "varying as business in hand." Changes were so frequent and so variable that no reliable classification of employees for the year was possible. Boys earned from 5 to 10 cents an hour, men from 10 to 40 cents. "Time is only rated by us in hours, computing pay as per laws of State, making eight hours a day's labor, consequently none are employed by the day."[8]

Both the company's and the workers' difficulties are apparent in the firm's annual records for the entire ten-year cycle (see Table 11.2). The value of goods produced in the trough year, 1876, was less than a fifth that of the peak year, 1873. Reductions of both the work force and the wages of those kept on slashed the annual payroll by a like amount in the

Table 11.2. Harrisburg Car Works Output, Payroll, Employees, and Rate of Dividends, 1873–1882

Year	Output	Payroll	Employees	Dividend Rate
1873	$1,586,000	$305,000	796	12%
1874	645,000	102,000	n/a	6
1875	436,000	68,500	n/a	6
1876	306,500	61,000	25–1,000	0
1877	520,000	75,000	n/a	7
1878	323,000	72,000	10–400	8
1879	1,004,000	160,000	300–700	6
1880	1,760,000	206,000	647	8
1881	1,983,000	246,000	n/a	16
1882	2,301,000	286,000	n/a	33

SOURCES: Hildrup, *Harrisburg Car Co.*, pp. 17–25, 41–42; Pennsylvania Secretary of Internal Affairs, *Annual Reports, 1872–1873, 1875–1876, 1877–1878, 1878–1879, 1879–1880, 1880–1881.*

n/a = not available.

same period. Dividends fell to zero one year. President Calder wanted to close down until a profit could be made, but Hildrup persuaded the board of directors to keep the plant running so its trained work force would not be lost. In the end, however, little more than a skeleton force was in place much of the time. As soon as recovery set in, capital and management took care of themselves first. In 1882, with the value of output 145 percent what it had been at the previous peak in 1873, dividends were three times greater. By contrast, the payroll lagged, standing only at 94 percent of the 1873 figure.

The impact of the Panic of 1873 on the cotton mill and its employees was somewhat different (see Table 11.3). To meet the crisis there, the managers shifted to a higher percentage of female and child employees. Wages for all were cut , the pay of adult males by more than 20 percent and that for women and children by less than 10 percent. This amounted to a more than 16 percent payroll saving. As it turned out for labor, reduced wages were not temporary. By 1881 they had fallen even more sharply, and in 1884 they were still more than 10 percent behind the lowest depression average.

The annual average wages that Wister Brothers' furnace reported in the 1880 manufacturing census for the previous decade indicate the impact of the Panic of 1873 on the wages of Harrisburg furnace workers (Table 11.4).[9] Between the peak year, 1873, and the low point for each classification of worker (1878 for some, 1879 for others), wages

Table 11.3. Cotton Works Employees, Wages, Days of Work, 1873–1884

Item	1873	1877	1878	1881	1884
Employees	280	238	249	262	175
Days in operation	277	n/a	253	300	230
Annual wage payroll	$67,000	n/a	$48,000	$47,256	$25,560
Percent adult males	14.3%	10.9%	14.5%	n/a	n/a
Average daily wage	$1.95	$1.51	$1.52	n/a	n/a
Percent females/children	85.7%	89.1%	85.5%	n/a	n/a
Average daily wage	$0.68	$0.62	$0.69	n/a	n/a
Average daily wage	$0.86	$0.72	$0.77	$0.60	$0.64
Percent of change	—	−16.3%	+6.9%	−22.1%	+6.6%

SOURCES: Pennsylvania Secretary of Internal Affairs, *Annual Reports, 1872–1873, 1876–1877, 1877–1878, 1880–1881, 1884.*

n/a = not available.

Table 11.4.   Wages Paid Wister Anthracite Furnace Workers, 1873–1880

Classification	No.	1873	1874	1875	1876 & 1877	1878	1879	1880
Founder	1	$3.34	$3.22	$3.28	$3.11	—	$2.77	$3.66
Blast engineer	2	2.81	2.70	2.44	2.24	$2.09	2.24	1.50
Carpenter	1	2.46	2.21	2.00	1.83	1.25	1.25	1.65
Blacksmith	1	2.44	2.21	2.00	1.83	1.71	1.83	2.05
Keeper	1	2.08	1.87	1.68	1.54	1.44	1.54	1.70
Metal weighman	1	1.80	1.57	1.42	1.31	1.22	1.50	1.90
Keeper's helper	2	1.78	1.57	1.42	1.30	1.21	1.30	1.43
Filler/cinderman	14	1.67	1.47	1.33	1.22	1.14	1.22	1.35
Common laborer	7	1.48	1.33	1.20	1.10	1.02	1.00	1.10

SOURCES: *1880 Census, Statistics of Wages in Manufacturing*, p. 117. Number of anthracite furnace workers from Pennsylvania Secretary of Internal Affairs, *Annual Report, 1875–1876*, pp. 652–53.

dropped between 30 percent and 33 percent for all except the few at the top. The founder was cut 17 percent, the engineers 26 percent, and the carpenter 28 percent. Wages began to rebound almost as soon as the trough was reached. By 1880 the founder and weighman were both earning more than in 1873, and most others were getting between 80 and 84 percent of that year's wage. Only the carpenters and common laborers were as low as three-quarters of their 1873 wage. No explanation was given for the low wage paid the engineer that year.

In the same document, Wister Brothers also reported what they received per ton of iron and the labor costs involved in producing it for the same period. Not given were the other costs, the number of days worked, the number of workers (though that must have stayed somewhere near thirty), the amount produced, or the profits earned. Iron prices fluctuated wildly in the good years, from $33.25 in 1870 to $48.88 in 1872, or an average of $40.00 for the period 1870–73. By 1878, when it sold for $17.63 a ton, the drop amounted to 56 percent. By 1880 the price stood at 71 percent of the pre-Panic average. It appears that wages did not fall as far as prices or recover more slowly than prices, at least at Wister Brothers.

In part, iron companies restored profits by increasing efficiency—that is, by getting greater output from the same number or fewer workers. Calculations from another set of reports to the state (giving annual wages paid, total number of employees, and number of days in operation) show that the nailery in West Fairview successfully increased its output while reducing, or raising only slightly, the average daily wage paid its workers (see Table 11.5). As output between 1882 and 1885

Table 11.5.   Wages, Output, and Days of Operation, West Fairview Nailery, 1882–1885

	1882	1883	1884	1885
Annual wages	$159,491	$142,791	$139,665	$146,786
Days operated	271	291	281	274
Employees	412	403	397	406
Average wage/day	$1.43	$1.22	$1.25	$1.32
Kegs of nails	171,527	n/a	178,659	189,921

SOURCES: Pennsylvania Secretary of Internal Affairs, *Annual Report, 1881–1882*, p. 41; ibid., *1882–1883*, p. 6; ibid., *1884*, p. 95; ibid., *1885*, p. 61.

n/a = not available.

increased nearly 11 percent, the average daily pay per worker, after dropping for two years, was in 1885 more than 7.5 percent below 1882.

Nineteenth-century Harrisburg was neither a union town nor a town torn by labor dissension. This is not to say workers never organized or conducted demonstrations or strikes, but rather that such matters simply did not play a major role in the life of the city, or even in the lives of any substantial number of workers. Over the years, the types of organizations to which workers belonged and the issues that led to confrontations changed. In the 1850s only the nonindustrialized printers, who had a strong trade union, won a strike. Other worker organizations occasionally demonstrated in sympathy for strikers elsewhere or called for a halt to the influx of "pauper laborers." The unorganized focused in particular on hours of labor. Store clerks as individuals might complain through the newspapers, while the cotton mill operatives waged a strike on the issue.

At mid-century, labor disorders in Harrisburg usually meant carousing canal-boat workers, brawling Irish laborers, or demonstrations by canal and railway construction crews who had not been paid on schedule. None were long-lasting, and most if not all ended without notice or were handled quickly by local law officers.[10] The first successful strike of the period involved journeymen printers of the *Telegraph*, not workers in one of the new factories. The printers had organized in 1851 for the avowed purpose of raising wages. All the town's newspapers yielded and paid the new rates. But as editor Theophilus Fenn of the *Telegraph* saw it, wages were only part of what the union wanted. They also wanted to regulate hours of work, matters of ethics and morals, the qualifications of those to be employed, and Fenn's conduct toward them.

What actually happened was that in March 1852, Fenn, who had two large printing contracts from the state and was short of labor, hired Andrew Dunn, an inexperienced young Scot from Cincinnati, at the union rate. Dunn worked very hard, earning from $15 to $19 a week. The union journeymen complained that Dunn had presented no card from the printers' union in Cincinnati and gave him two weeks to produce one. Unwilling to submit to union dictation, Fenn refused to fire Dunn when he produced no card, so the journeymen went on strike on April 12. Fenn advertised for replacements and even recruited in New York City. His journeymen countered with a letter of their own to New York to discourage scabs. It was probably the cost of moving rather than labor solidarity that kept any from coming. Finally, on April 22, with no support from other local publishers, Fenn, as he put it, "yielded up possession of our office to the most unjust demand ever made upon proprietors."[11]

The following mid-March, Harrisburg mechanics held a series of well-attended meetings at the courthouse. Their objective was to raise "material aid" for journeymen mechanics of Baltimore who were on strike for higher wages. The sessions stirred considerable interest and sympathy, but some were alarmed. From the "tone of the speeches," the editor of the *Whig State Journal* suspected that a general uprising of mechanics in Harrisburg was being prepared for next year. Indeed, he warned, there were indications of a "general strike throughout the Union, and an open war between Labor and Capital."[12]

The matter was pursued in the columns of the *Telegraph*. A letter from "Oscar" to the editor spoke for the mechanic who was always sneered at by the wealthy young man and "smelt across the street" by the "aristocratic young lady." The day was coming when those who were excluded from the *"first society"* and seen as little more than serfs and tools of rich capitalists and employers would be appreciated. The mechanics of Baltimore, he continued, were striking not in a spirit of rebellion, as many claimed, but because wages there had failed to keep pace with the rising cost of rent and other necessities. Fenn responded to these republican sentiments with the observation that he had watched strikes for higher wages for twenty years and never knew any good to come to strikers from them. "Z" replied that Fenn might be right. A "peaceable application to the employer" for better wages was to be preferred. Still, experience taught that "man's selfish disposition" would not yield unless compelled. "Z" pointed to people who earlier had "petitioned and remonstrated" to no effect, then "struck from Lexington to Yorktown, until a glorious result followed."[13]

Hours of labor for children was another labor issue in Harrisburg newspapers at mid-century. An act of the Pennsylvania legislature dated

March 28, 1848, prohibited textile mills from hiring children younger than twelve or working them longer than ten hours a day. The act, however, permitted parents to enter into contracts by which children over fourteen could work longer. A year later the measure was repealed in favor of stronger legislation prohibiting children under thirteen from being hired, limiting work of thirteen- to sixteen-year-olds to ten hours a day and nine months a year, and requiring the completion of three consecutive months of schooling a year prior to employment. Firms violating the law were subject to a fine of $50 for each offense, and parents would receive the same penalty for consenting to, permitting, or conniving with employers to violate terms of the law.[14]

Young store clerks wanted the measure applied to them. "A Clerk," writing to the editor of the *Telegraph* on April 17, 1850, suggested that stores in the borough close at 8:00 in the evening. It was "absolutely necessary" that young men have leisure for the improvement of both their minds and their health. After toiling all day it was only fair that they have part of the evening to themselves. In a different context three years later, the *Telegraph* seemed to agree. It observed that mechanics worked ten hours a day, commencing at 6:00 in the morning and working until 6:00 in the evening with an hour off for breakfast and another for dinner. Why should not the ten-hour day apply to all other laborers, especially where toil was equal?[15] A few months later, however, the editor rebuked another clerk who wrote in pleading for reduced hours. The editor suggested that this and others in like position first set up facilities for self-improvement.[16] More than a decade later, conditions for clerks had improved little. "A clerk or salesman in a retail store, who labors twelve or fourteen hours a day," a new editor of the *Telegraph* wrote, "has a slim chance for personal improvement, a rather poor opportunity for recreation or pleasure. He becomes literally a slave."[17]

In the meantime the young operatives of the cotton mill struck for a ten-hour workday. Hours had been irregular, longer in summer, shorter in winter, but averaged "about 12 hours a day" and at the time of the strike "exceeded" twelve hours.[18] The operatives considered the matter for a while before finally presenting their demands. Refusing a counter-offer of an eleven-hour day, they walked out on October 19, 1853. Securing a drum and fife, they marched up and down the street in front of the mill, "much to the entertainment of the juveniles of the neighborhood." That evening the National Guard brass band marched from the armory on Capitol Hill to the mill, where the operatives joined them. Together they paraded to Market Square to the tunes of "national airs," some workers bearing transparencies. There, by the light of a large bonfire, "Colonel" George H. Morgan, Dr. Washington Barr, and others addressed them. Morgan, age twenty-four and a printer by trade, had been one of those involved in the strike against the *Telegraph* a year and a half earlier. Barr, about forty-two years old at the time, combined several occupations: physician, wagonmaker, and auctioneer. As quartermaster of the local National Guard unit, he may have arranged for the use of the band. The meeting adopted resolutions declaring determination of the operatives to remain on strike until their demands were met. At the same time, they disavowed any disrespect or ill-feeling toward the company or its managers. Their complaint was only that the required hours were too oppressive. After the meeting broke up, the operatives processed back to the mill, where they listened to a fiery address on women's rights by one W.C.A. Lawrence, gave themselves and Lawrence three cheers each, and broke up for the night.

The next morning, the band again appeared at the factory to provide music as the operatives marched to a vacant lot in the center of town. There they erected a stand and invited "Mr. Booth," one of their number, to address them. Afterward a committee presented resolutions calling for a compromise: a ten-and-a-half-hour workday. Following speeches by Morgan and someone named Carson, the procession formed again to march through the principal streets.[19]

The October 20 *Telegraph* posed as neutral: "If the burdens imposed are too grievous, and the profits of the mills large enough to warrant," the directors should grant the demand. No "claim of humanity," however, obliged the company to run at a loss, and to date not one cent had been paid in dividends. The alternative to accepting the eleven-hour day offered by the company might well be no jobs at all. By October 26 the *Telegraph* shifted fully behind the company. Outside agitators were behind the trouble, it charged. "A couple of emissaries from Lancaster" who had been instrumental in getting up a strike there had aroused "the dormant sensibilities of the 'down trodden' here." The speeches at the rallies had been "inflammatory and ridiculous," the company's counteroffer had been rejected by the workers "or by those who counseled and controlled them," and the better hands were leaving to seek employment elsewhere. Editor Fenn doubted that either coercion or laws regulating labor ever proved beneficial to workers. "Labor, like all other things, should be left to find its own best market." Instead of allowing themselves to be "trooped around the town at the tail of a band of music, under the lead of self-seeking politicians, seedy idlers," and the like, the workers should have presented their demands and, if not met, sought better jobs.

That same day the strike ended. According to the *Democratic Union*, no more than a third of the work force had struck. As soon as they were "left to themselves, without being molested or interfered with by men who have no interest in their welfare," all but three or four had "cheerfully" returned. The mill was "pleasantly situated" on the river, a "most airy and healthy" spot, the paper continued, and offered the "best wages paid in cotton mills." For eleven hours of work the operatives earned more than seamstresses, tailoresses, and most who did household work. By contrast, the company itself had yet to turn a profit, partly because of market conditions and partly because, in spite of spending much "in learning operatives," it still ran with few "*experienced hands.*" The stockholders, that "set of patriots" who brought the town its first enterprise and made the community prosper, needed some return on their investment.[20]

One other small labor incident before the Civil War involved workers at former Governor Porter's anthracite furnace. The *Morning Herald* of July 17, 1854, reported that Porter had three employees arrested for conspiracy in striking for higher wages and combining to prevent others from taking their places. In fact, six, apparently all Irish, appear in the docket books. Out on bail, they seem to have left town, because their bonds were forfeited.[21]

During the Civil War and for a few years after, Harrisburg workers apparently engaged in no major disputes with their employers. Trouble next appeared in the spring of 1871, when the management of the Lochiel Iron Works tried to step up production by its puddlers. Puddling dictated the pace of all other work in an ironworks and could be neither mechanized nor hurried. "Some iron takes longer than other iron," a puddler explained to a congressional investigating committee in 1883, with "heats" varying in time by as much as half an hour.[22] At Lochiel, moreover, puddlers had formed Labor Union No. 6, apparently affiliated with the National Labor Union, and adopted work rules. Ordinarily, from Monday through Friday they made six heats in ten hours; on Saturdays they made five.[23]

The *Patriot* announced on April 11 that Lochiel had suspended operations. Some 300 employees, including puddlers and other laborers, had struck because a puddler had been discharged for refusing to make an extra heat. Hildrup, acting president of the company, wrote to "correct" the article. The employee in question, after the third heat on Saturday, had been discharged for refusing to work longer. The workers demanded his reinstatement or they would go on strike. The threat, combined with a recent advance in the price of coal, resulted, Hildrup declared, in management's decision to close the works. In fact, as noted earlier, the firm had been financially shaky from the start. Plagued by delays in construction, cost overruns, poor management, increasing competition from steel rails, and finally a break in President William Calder's health, Hildrup of the car works had reluctantly taken over. Getting things moving quickly was his managerial forte.[24]

The president of Labor Union No. 6 presented a very different account of the affair. On both March 29 and March 30 the general superintendent, a man named Shaeffer, had asked the puddlers to make seven heats because Lochiel was "being pressed for rails." With union approval, the seventh heat was run. On March 31 they made the usual six heats, but because of an accident they were kept on the job as long as when they had seven. On Saturday April 1, in spite of the three previous long days, Shaeffer persuaded the men to do the usual five heats so as "to finish the order, to change rolls, and commence a new order on Monday." To the workers' surprise, all six heats on Monday were devoted to the order supposedly completed on Saturday. That shook the puddler's faith in Shaeffer. The next Saturday the men came in at 3:00 in the morning, but an accident prevented them from getting under way before 10:00. At the end of the third heat, when they had already spent ten and a half hours in the mill, Shaeffer asked them to remain and complete the customary five heats. Thinking they had worked enough, they decided to do no more. When the first puddler declined to work, he was discharged. The others, insisting they were equally guilty, threatened to strike unless he was reinstated.

Blaming the difficulties on the union, the company closed. The union president pointed out that the puddlers were actually helping the company by refusing to work. President Calder had recently told a group of workers that the firm lost $4 on every ton of rails rolled. The union leader concluded with the hope that "everything may be amicably settled between us ere long, and that this firm may think better of us when they plainly understand our object." Instead, the firm remained closed and completely reorganized its management and finances. When it eventually reopened, the union was gone.[25]

The onset of the Panic of 1873 not only led to wage cuts already described but also strengthened the hand of management in dealing with workers who resisted the reductions. On July 1, 1874, some 250 employees of the Pennsylvania Steel Company unanimously resolved to quit work rather than accept wage cuts ranging from 10 to 25 percent. Although this was not the entire work force, the operations they closed down in effect completely idled the facility. Three days later the Bessemer workers submitted, and the others submitted a week later—all on the company's terms.[26] A year later the puddlers at Bailey's Chesapeake Nail Works demanded a raise from $4.50 a ton to $5.50. According to the *Patriot,* their real goal was $5.00 a ton, the current rate at Pittsburgh. If the newspapers were correct, the Harrisburg puddlers already were earning more than their counterparts at Reading, Phoenixville, Birdsboro, Pottstown, Scranton, and Norristown. Puddlers at Elmira in New York were earning the same, while those at Conshohocken, the Lehigh region, and in Philadelphia earned more. The company conceded the $5.00 rate, and work resumed on June 3, 1875. Not long after, the rate returned to $4.50, and on November 29 the puddlers peaceably accepted a reduction to $4.00.[27]

Labor incidents at Harrisburg through the mid-1870s followed a pattern of their own. The only major strikes—at the cotton mill in 1853 and the Lochiel Rolling Mills in 1871—dealt with fair play, not with protecting wages or maintaining standards of living. The managers of the cotton mill insisted on longer working hours than the employees thought reasonable. They had gone along with long days when the mill began, but in its second year it seemed to them that the ten hours other workers were demanding and thought to be getting should apply to them. At Lochiel the problem was again honest dealing. The workers were willing to put in extra time to help the firm, but they did not want to be taken advantage of.[28]

From the company standpoint, prior to 1877, Harrisburg's industrial leaders never lost control during labor disorders, nor did they have reason to fear they might. Except for the printers at the *Telegraph*, few if any strikes involved the whole work force of a given plant. At no point did the workers do more than demonstrate peacefully and refuse to work. The strikers were either unorganized or they belonged to ineffective unions. They had no support system to assist them or their families when income stopped. The workers did not threaten to destroy property, nor did they turn to violence. Usually they were respectful of their employers, condemning the conditions under which they labored, not the persons who set the terms of employment. Under the circumstances, the managers did not need to react with force against the strikers or to hire strikebreakers. They had only to wait until the rebels' frustrations were spent. After a little symbolic language castigating would-be aristocrats—comparing themselves with the nation's founders, and their cause with that of the War for Independence—the strikers had only to check their pantries to realize the futility of continuing. Chastened, they returned to work. Lochiel's lockout and reorganization was not different, only a variation. During the process, the workers were without wages and their union died. Absence of evidence prevents us from knowing how many of the strikers reentered the mills when they reopened a few months later.

From the Riots of 1877
to Century's End

Harrisburg's most rousing labor disorders came in mid-July 1877 during the wave of railroad strikes sweeping the nation. In part it was a matter of contagion, with Harrisburg residents imitating their counterparts elsewhere. Curiosity played a part too, attracting many of those who were part of the large crowds. As elsewhere that summer, rowdy teenagers were noticed as especially prominent among the troublemakers. However, at the heart of the troubles in Pennsylvania's capital city (as in other communities across the nation) lay the bona fide complaints of workers and their families.[1] Since 1873 they had endured the wage cuts and layoffs already described. If not themselves activists, their sufferings and frustrations made them tolerant of, if not sympathetic to, attacks by others on the employing class. Certainly the city's leading industrialists saw the affair as rebellion against themselves and their property rights and organized to put it down. For the balance of the century only a small number of strikes occurred, most frequently among ironworkers. The usual cause was wages, either resistance to cuts by the companies or demands by workers for raises. Most strikes were short. The workers won only limited concessions on three of eight occasions.

The rioting that caught up even the usually quiet city of Harrisburg was part of the country's first widespread labor upheaval. The Great Railway Strikes of 1877, in turn, had grown out of the Panic of 1873, then in its fourth year. The trunk lines of the Northeast had taken early common action to forestall their losses by ending rate wars with one another and by agreeing to reduce the wages of their employees in unison. At the same time, most continued paying predepression dividends. Lower wages, reduced hours, increased workloads, and layoffs led to rebellion among the employees of the Baltimore & Ohio Railroad and the Pennsylvania Railroad. Before the affair ended, strikes and disorders spread to most of the nation's railroads except in New England and the South. The first outbreak occurred on July 16 at Martinsburg, West Virginia, following announcement of a wage cut on the B&O. The men refused to move trains through the yards or to allow others to do so. At the company's request, government officials first sent in the state militia, and when they proved ineffective called for the United States Army. By the time Martinsburg was unblocked, riots were raging against both the B&O in Baltimore and the Pennsylvania Railroad at Pittsburgh. Again both state and federal forces went into action. At Pittsburgh, hostile crowds of workers proved too much for local forces of law and order. When militia units arrived from the Philadelphia area, angry crowds drove them from the city amid considerable bloodshed. Mobs roamed the streets looting and pillaging on July 20 and 21. By the time federal troops arrived on July 23, Monday, the troubles there had burned themselves out.[2]

At Harrisburg many railroad employees belonged to one of two unions: the Brotherhood of Locomotive Engineers or the Trainmen's Union. The Brotherhood, organized nationally in 1863, limited its membership to the elite of the trainmen, the engineers. Because they "generally patched things up for themselves" and "didn't look after anything else," a few engineers joined with conductors, firemen, brakemen, switchmen, and others to form a new organization, the Trainmen's Union, in June 1877.[3] Relations between the two unions were not always harmonious. Trainmen at Harrisburg arrested during the troubles later complained that the Engineers started the affair, but once they got the Trainmen to support them they stepped back and let that group "stand the racket."[4]

Accounts of the strikes and rioting at Pittsburgh on July 19, 20, and 21 "kept the excitement at fever heat in this city," reported the July 23 *Telegraph*. "Excited crowds of men thronged the streets discussing the situation, and nowhere was anything else but the strike talked about." Railroaders, apparently forewarned of an imminent general strike on the lines, began to show uneasiness on July 21, a Saturday. That evening the first crowd, estimated at more than 3,000, formed at the Pennsylvania Railroad depot. "Demonstrative but not unruly," they attempted no violence. At 9:00 in the evening the Pittsburgh Express arrived from Philadelphia carrying 300 militiamen. The troopers disembarked and marched to an ammunition car on the siding, where each received forty rounds. Returning to the train, they discovered that the crew had deserted. An "extra engineer" was found to run the train and, in the absence of firemen willing to help, a "roundhouse cleaner" agreed to fire and the train left for Altoona. The crowd jeered and hooted and called "vile names" at the soldiers, who pointed guns out the windows to halt the throwing of stones. About 10:00 P.M. Mayor John Patterson, a onetime railroad clerk, learned of the disorders and dispatched a police officer and his men to the scene.[5]

Authorities wanting to load ammunition for shipment to Pittsburgh were concerned about the threat the crowd posed. To draw them from the depot, the police arranged that a young man named Finch (described by the *Telegraph* as "intoxicated and not a railroad man") began inciting "a first class riot." Two police officers arrested him and took him to the mayor's office. As hoped, 300 to 400 people accompanied the small party, throwing stones at the police (until the latter drew guns) and shouting "Bread or Blood!" "We Will Not Starve!" and the like. Those who did the yelling, declared the *Telegraph,* were not railroad employees but "turbulent spirits bent on raising a ruction." After the hearing, Mayor Patterson came to the door with Finch, who, he said, was not a railroad man. He had been arrested for being drunk and disorderly. Finch corroborated the mayor's remarks and was taken to jail as the crowd departed once more for the depot.[6]

While this was going on, a special train had taken on some 500 guns. Although "a large crowd" remaining at the depot had witnessed the loading, no one had interfered. About 10:50 P.M. a freight train halted in the yards to let the special pass. "Half-grown boys who wished to cause a disturbance" removed its pins. After several false starts, it too cleared the yards. When the Cincinnati Express arrived at 12:55 A.M., members of the crowd immediately boarded it to search for black troopers rumored to be en route to Pittsburgh. Only a few white militiamen were found, and apparently the train was allowed to proceed. About 3:00 A.M., except for a few "watchers," the group broke up for the night.[7]

Fearing imminent conflict, authorities were anxious to keep guns from the crowd. Officers of the Harrisburg City Grays, a local militia unit, ordered all armaments and ammunition removed from their armory to the state arsenal just outside the city limits. The militiamen carried out their instructions during the night, then took turns standing guard at the facility. Meanwhile, reports of disorders at the capital reached state officials. Governor John F. Hartranft, traveling in the Far West, headed back to Pennsylvania. His adjutant general, James W. Latta, ordered General J. K. Seigfried of the National Guard to put his regiment, the Eighth, on duty at the state arsenal, where all the state's ammunition was stored. So as not to incite the crowd in the city, Seigfried, with troopers from the Pottsville area, disembarked from their train two miles above Harrisburg and marched to the arsenal, arriving about dawn on Sunday,

July 22. As additional units arrived from other areas, they too skirted the city by one route or another and made their way to the arsenal. To prevent strikers from seizing and using Mexican War cannons that stood on the grounds, officers ordered them spiked.[8]

Some 300 railroaders assembled at 9:00 A.M. on Sunday morning and voted to strike. Apparently following instructions that accompanied the warning of the coming general strike, they refused to move any freight trains or local passenger trains through the yards or to allow others to do so. On the other hand, they were not to block through passenger trains, especially those carrying the mail. For the most part, that policy characterized the difficulties at Harrisburg. Also seeking to avoid trouble, "the better class" of local railroaders called on at least one Harrisburg attorney and urged him to use his influence to keep authorities from calling in the military. If troops came in, they feared violence would follow.[9]

The first incident involved a train attempting to leave for Altoona. Strikers mounted the locomotive, persuaded the engineer and fireman to quit, and then drove the engine to the roundhouse. In the confusion another freight loaded with coal tried to leave for Philadelphia. Strikers halted it too, again talked the crew into quitting, and moved both that engine and a "shifter engine" to the roundhouse. At 1:00 P.M. the crowd moved to block a passenger train, but "cooler heads" prevailed and it was allowed to proceed to Philadelphia. When a later mail train arrived with none of its usual New York or Philadelphia newspapers, rumors spread that tumultuous riots were in progress in those cities.

Meanwhile, Mayor Patterson, though uncertain of his powers, took steps to keep matters in check. He asked the city's editors to print no "extras" because they would only inflame the already excited citizens. He also closed all saloons until the following Wednesday. Concerned that weapons might fall into the hands of dangerous elements, he directed gun and hardware dealers discreetly to take all armaments and ammunition from their stores. That evening he issued a proclamation urging all citizens to return to their homes and remain there until the excitement ended and differences between the railroads and their employees could be "amicably and satisfactorily adjusted."[10]

Railroaders, fearing that outside elements might wrest control of their strike from them, sought a moderating influence. They pressured H. L. Tolbert, "a local speaker of note prominently identified with the greenback movement," to address the crowd that evening. Tolbert was reluctant to speak out, in spite of his lack of sympathy for rioting then in progress at various rail centers. The excitement, he told some 4,000 people assembled at Front and North streets, grew out of widespread unemployment. Joblessness induced some to adopt socialistic or communistic ideas and to disturb society. The government, he suggested, should put surplus workers on farms in the West and provide them with rations until the first crop was in. Inasmuch as federal authorities gave rations and farm implements to the Indians, he said, there was no reason to do less for workers. The use of militia units instead of sheriff's and citizens' posses to quell disorders was a mistake, he argued. It only excited crowds and brought the country to the edge of revolution.

Tolbert recognized the right of workers to quit when they were not able to support families on the wages they earned. They also had the right to ask others not to take their jobs. But, he warned, the use of force and intimidation to back up that request would be wrong. Disavowing both communism and socialism, he declared that he stood with the workers and hoped they would win. The general public was with them too—it was always on the side of the weak. "Act like men," he urged, help preserve law and order, and respect the property of the railroads, "for they are indeed in a very bad way." If the lines surrendered, the railroaders must be sure that all of their number were returned to work. The military, he said, should be respected as long as it did not attack peaceable strikers. If attacked, the law of nature allowed them to defend themselves just as the people of Pittsburgh, when provoked, had protected themselves.[11]

Whatever the impact of this strange jumble of ideas on the crowd, it did not end the troubles. When an express train arrived at the Pennsylvania depot from the west at 8:00 P.M., a group allegedly containing no railroad men surrounded it and pulled the pins connecting the cars to prevent it from leaving. One of several troopers from Philadelphia who were on the train said his group had been sent home because they were unwilling to fight fellow workers. Although the crowd "kindly relieved" them of their weapons, the militiamen were otherwise not treated with any hostility.

Standing by as these events transpired was "a great gathering" of "nearly all of our own citizens—good, bad, and indifferent," Mayor Patterson later testified. Addressing the crowd from the top of a car, he asked "if they were ready to assist the police." If so, they were to step forward and with their help he would disperse the crowd. At this early stage, "as there was no violence done as yet," no one stepped forward. "I presume the great majority of the people were in sympathy with the strikers—looking upon it as a strike or dispute between the employees and officers of the road—and their sympathies were with the employees." Only violence and destruction would change their attitude. Patterson signaled the train to move out, but the engineer declined because of alleged obstructions on the track ahead. Finally, about 3:00 A.M., after

Patterson left the scene, the train departed and the crowds drifted home.[12]

Meanwhile, Reading Railroad employees, who had not received their wages for May or June, struck at midnight. On July 19, with trouble brewing, the company had announced that the workers would be paid (how much was not clear) on July 27. Unappeased, the workers now joined the other strikers. Throwing down their tools, they began cutting locomotives from trains and running them into the roundhouse. They also blocked all freight traffic.[13]

Monday, July 23, marked the turning point at Harrisburg. A crowd, somewhat smaller than the previous day's, again gathered that morning to watch the action. With no one attempting to run trains, all remained quiet. In fact, the strikers were moving to adjust their dispute with the companies. During the night, they organized to protect railroad property. They were also careful to avoid the morning crowd at the Pennsylvania Railroad depot. At a general meeting of their own at the Broad Street market they drafted resolutions: wages must be restored to the scale of 1873, the classification of employees into different grades must end, the companies must reinstate all workers to their prestrike positions, and a local railroad official must be removed because of his conduct toward them. The strikers wanted trains on the Middle Division defined as forty-five empty cars westward and forty-five loaded cars eastward. Every train was to have an engineer, a fireman, and a full crew. They called for the removal of troops and promised there would be no violence or destruction of property "providing we can prevent it." Until these demands were met, they would allow no freight or troop trains to move through the yards. Harrisburg newspapers praised the responsibility of the trainmen, contrasting it favorably to that of the "tramps" and "strangers" who were at the forefront of the disorders and "would delight to see a general insurrection."[14]

As General Seigfried, in his somewhat exaggerated report, later analyzed the situation at Harrisburg, the difficulties persisted because of the failure of the citizenry to stand against the rioters. The railroads ran no trains because they were in the hands of the mob, business and manufacturing were at a standstill either by voluntary or enforced idleness, citizens either sympathized with the strikers or were terrorized by them, the local sheriff was vacationing at the New Jersey shore, and other officials were "powerless and inactive."

Two events, however, soon galvanized a number of citizens into protecting their property and restoring law and order.[15] The first occurred when a number of militiamen from Philadelphia straggled into the Harrisburg area. Some were near Marysville, across the Susquehanna and about eight miles north. Others were at Rockville and Progress, villages on the Harrisburg side of the river north and east of the city. The troopers had been en route by train to Pittsburgh when a threatening crowd at Altoona blocked their way. Now the men were trying to get home. Fearing that they would be attacked if they entered Harrisburg, those at Marysville apparently sent word that they wanted an escort into the city. Accordingly, on Monday afternoon, a crowd of between 150 and 300 people, ignoring the protests of the toll collector, stormed across the camelback bridge to "interview" them. Thinking they were under attack, many of the militiamen fled to the mountains behind Marysville. Two hours later the crowd recrossed the bridge, "fetching" with fife and drum some twenty or thirty disarmed soldiers. "Half grown boys and negroes" bearing the bedraggled militiamen's weapons and forming a hollow square around them, paraded the dispirited band into Harrisburg, up Market Street to the depot.[16]

Several militiamen on the eastern shore also reached Harrisburg. Most were picked up singly or in groups of two to five. They too surrendered their arms without protest. Military officers learning of these incidents were dismayed. According to General Seigfried, "the humiliating spectacle of the troops, prisoners to a motley crew of boys and roughs, paraded through the principal street of the town, guarded by their own arms, which had been taken from them, aroused the indignation of the better class of citizens, who naturally became thoughtful and asked if that strange guard, which surrounded the troops of the State, were the masters of their lives, honor and property."[17] On the other hand, according to the *Patriot,* the captors took the troopers from Marysville to the Boyer Railroad Hotel, there supplied their wants, and promised them aid in returning home. Soon after, troopers from Rockville and Progress straggled into the city and were promptly surrounded by railroaders. With assurances from Mayor Patterson that their arms would not be used against them but would be boxed and shipped to Philadelphia, that group too turned over their weapons and were taken as guests to the homes of railroaders.[18]

The other turning point Seigfried noted was an attack on Harrisburg gun shops that night. Apparently concerned that the militia at the arsenal had to be stopped there or "we'll have a second Pittsburgh," a crowd with their "regular leaders at their head" proceeded to a gun shop on South Second Street. There men and boys "yelling and cursing like fiends" took guns. Mayor Patterson and some police arrived and persuaded them not to rob the "worthy mechanic" who owned the store. Except for a few pistols "stolen, it is supposed, by boys who are thieves by nature," the guns were surrendered. Later the troublemakers went to a second gun shop (its stock had already been removed) and then to a pawnshop, where they took the guns on display.[19]

By then, citizens concerned about restoring law and order were organized to halt the disorders. Early Monday morning, Mayor Patterson had called on "many of the prominent citizens," warned them of probable violence that night, and told them that whatever was to be done must be done then. Later he sent police officers to notify "the better class of citizens" in person that their services would probably be required soon and that they should be ready to report to the mayor's office when two taps were sounded on the courthouse bell "at any time, day or night." That evening at 7:00, Dauphin County Sheriff William Jennings (the independent ironmaster who had been a colonel in the Union Army) arrived on the scene. The strikes had delayed train movements from Atlantic City, where he was vacationing. He promptly took charge. After consulting his attorney regarding a sheriff's powers and authority, he met with a number of prominent citizens and prepared a proclamation for Tuesday morning. Meanwhile, the mayor reported the preparations that had already been made, graciously surrendered authority to him, and with his small police force worked in close cooperation with the sheriff.[20]

While dealing with the scattered assaults on the gun shops, Mayor Patterson received word of a greater threat. A mob was organizing to burn the offices of the *Telegraph* because of the position it had taken against the disorders. The alarm was sounded and a large force turned out. Patterson estimated the "law and order posse" at between 300 and 500; the sheriff thought there were between 100 and 150.[21] Most were armed with clubs (Jennings did not want them carrying rifles) but also had pistols concealed in their pockets. Jennings, accompanied by Patterson and one police officer, approached the crowd and attempted to speak. Unheeding, some began to break into stores. The sheriff and mayor then returned to the posse and led it against the sizable mob.[22]

The confrontation took place at the foot of Market Street near the depot. Jennings approached a line of men standing there and told them what he proposed to do. One man in the line had a loaded gun in his hand. Jennings went up to him, took the gun, and when the man responded with "some impudence," seized him by the neck and tossed him into the crowd. "We cowed them," Jennings later said, "and the parties who replied and gave us impudence, we arrested them at once." The crowd broke, many fleeing over a bridge that crossed the canal. Although clubs were used on those who resisted, not a shot was fired and no one was killed. Jennings then organized his force into companies and sent out squads to patrol the streets throughout the night. Because it was known that the posse would arrest any groups that formed, the streets soon cleared.[23]

The next day, Jennings organized his companies into a regiment numbering between 900 and 1,000 and marched them around the principal streets. "We ran the town on military principles for about one week," he later testified. "We had an officer of the day detailed to patrol the town at night, and we had the fire department under command, and everything in readiness if there should be any further trouble." Rounding up troublemakers began as soon as the streets were cleared Monday night. Jennings "sent squads out to arrest and take these men out of bed who had been prominent and active as rioters" and put them in jail. Eight or ten were arrested that night and forty-five or forty-seven more during the days immediately following. Of those, forty were indicted, tried in the county courts, and either fined between $20 and $500 or sent to jail for three to eight months. Others were held for a while and eventually released.[24] Although the tramp of Jennings's patrols could be heard through the next several nights, the troubles were over after July 24. Once more, trains ran without interruption, factories reopened as vital raw materials resumed their flow, and business returned to normal.

So detailed an account of a single labor disorder in the city's history may seem unjustified, but no other event better revealed the class cleavages that existed in nineteenth-century Harrisburg. The community had very few labor disputes and none that involved anywhere near as many of its residents, none was so fully covered by the press, and no other Harrisburg strike became the subject of a legislative inquiry.

Both newspaper accounts and testimony at the inquiry afterward agreed as to who was responsible for the disorders at Harrisburg. Just as at other trouble spots across the nation, Harrisburg railroad workers for the most part only tied up trains and were not involved in any violence. The looting, pillaging, and destroying of property allegedly was the fault chiefly of strangers and tramps, half-grown boys, and locals who never were known to work for a living.[25] According to Mayor Patterson, about a third of those arrested at Harrisburg were railroaders whose crimes were limited to interference with the running of trains. Only one of those arrested at the confrontation on Monday night worked for a railroad. The mayor attributed the violence that did occur to "outsiders and strangers" and to "boys fourteen to twenty-one years of age—bootblacks and all classes."[26] Sheriff Jennings agreed. Mill workers, furnace workers, and other factory hands of Harrisburg did not engage in rioting—at least "not the men that worked." In his opinion, "the parties who were most active and violent were those who did not work at any time." When asked if that element belonged to the city, he replied that

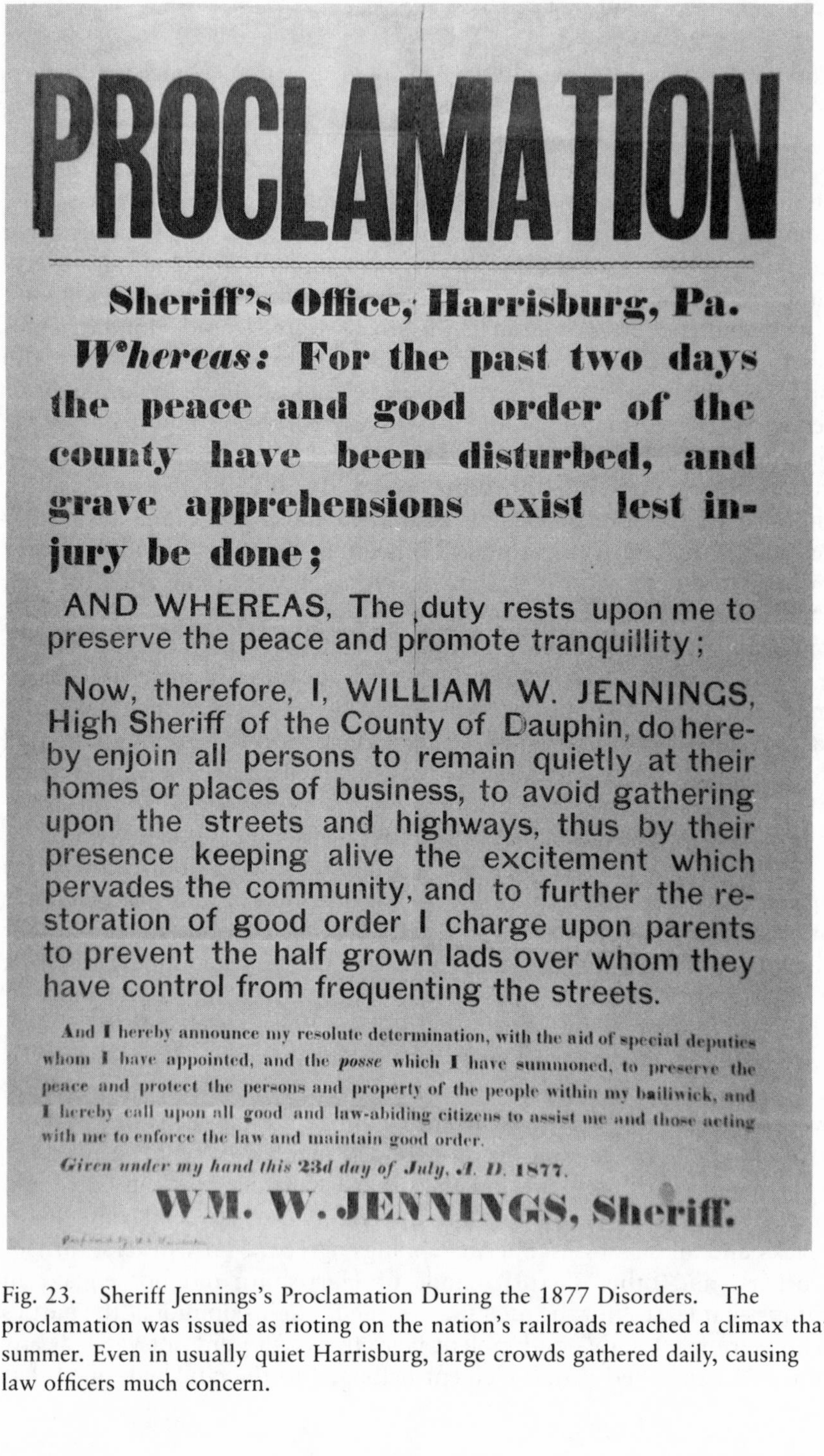

PROCLAMATION

Sheriff's Office, Harrisburg, Pa.

Whereas: For the past two days the peace and good order of the county have been disturbed, and grave apprehensions exist lest injury be done;

AND WHEREAS, The duty rests upon me to preserve the peace and promote tranquillity;

Now, therefore, I, WILLIAM W. JENNINGS, High Sheriff of the County of Dauphin, do hereby enjoin all persons to remain quietly at their homes or places of business, to avoid gathering upon the streets and highways, thus by their presence keeping alive the excitement which pervades the community, and to further the restoration of good order I charge upon parents to prevent the half grown lads over whom they have control from frequenting the streets.

And I hereby announce my resolute determination, with the aid of special deputies whom I have appointed, and the *posse* which I have summoned, to preserve the peace and protect the persons and property of the people within my bailiwick, and I hereby call upon all good and law-abiding citizens to assist me and those acting with me to enforce the law and maintain good order.

Given under my hand this 23d day of July, A. D. 1877.

WM. W. JENNINGS, Sheriff.

Fig. 23. Sheriff Jennings's Proclamation During the 1877 Disorders. The proclamation was issued as rioting on the nation's railroads reached a climax that summer. Even in usually quiet Harrisburg, large crowds gathered daily, causing law officers much concern.

many did: "it brought our worst characters to the surface, of course."[27] Attorney David Mumma said that of those who took guns from the militiamen he knew only one: "a loafer who does not do anything and never did a day's work when he had it." The others were a "lawless class of men, mostly strangers. I did not know them, though I know a great many of our citizens."[28]

It is not difficult to believe that a large number of teenage boys joined the crowds or played a significant role in the strikes that summer. Many were idle, the disorders offered the best show with the most excitement in town, and the crowds provided them with both an audience and personal anonymity. The alleged number of strangers and outsiders, however, is less convincing. If strangers constituted a large if not major part of the crowds at the nation's many rail centers during the strikes, where did they all come from and how did they get to those centers when trains were not running? It seems unlikely that very many would have been country people who rushed to the cities to witness or join in the excitement. Neither does it seem likely that they were unemployed persons or loafers who rushed from one rail center to the next as the strikes spread across the country.

Instead, they were probably residents of Harrisburg (and of the other centers where the troubles occurred) who simply were not known to the news reporters, lawyers, public officials, or the "better class of citizen." Given the high rate of mobility in American society at the time, rarely more than half and often far less of the population of any city lived in a single community as long as ten years.[29] A good part of most urban populations consisted of relative newcomers who would be regarded as strangers by established residents. Moreover, the failure of reporters, lawyers, public officials, or other prominent citizens to recognize someone did not mean that person was an outsider. Because of the nature of their jobs, Harrisburg factory workers had few occasions to go to the center of the city, to visit the courthouse or Market Square, or to promenade along the principal streets. After working ten to twelve hours a day, six days a week, relatively few spent their evenings or Sundays in areas frequented by "the better classes." During the strike, however, many were unemployed or their places of employment were shut down, so they were free to gather and watch the excitement. Some, no doubt, participated in the disorders. The fact that after the confrontation on Monday night the authorities knew where the supposed unknowns lived and sent police to arrest them in their beds undercuts the contention that they were total strangers to the city.

Harrisburg's crowds of 3,000 to 4,000 constituted more than a third of all the male residents of the city between the ages of thirteen and sixty-nine. Though a very large percentage, they probably were not strictly representative of the male population as a whole. Those who assembled probably included more of the idle, the unemployed, and the adventuresome young than of the hardworking, prudent, property-owning older men of the community.[30] The large gatherings, in spite of their presumed bias against wealth and power and their sympathy for the cause of the strikers, avoided or opposed violence and the destruction of property. Whenever events threatened to get out of hand, the great majority, despite the strong provocations of unemployment, reduced wages, and the arrogance of the railroad companies, stood and watched, left the scene, or willingly supported public officials when called on to join posses to restore law and order.

Little in events at Harrisburg supports the contention that the railway strikes brought the nation close to revolution. Crowds there did not appear to have been part of a downtrodden and alienated laboring class who, seething with revolt against their wealthy oppressors, joined with strikers to paralyze the transportation system until suppressed by the military. At most the crowds seem to have sympathized with the railroad workers in their quarrel with the railroad companies. A majority, including those on strike against the railroads, wanted to resolve the difficulties without resorting to violence, destroying property, or showing disrespect to public officials, law officers, or militiamen. Although they disarmed the few militiamen who fell into their hands, they were careful to prevent the arms from falling into irresponsible hands. They wanted only to prevent the arms from being turned against themselves. This they accomplished by the force of numbers, not arms. Revolutions are not born of such attitudes and behavior.

On the other side, Harrisburg's leading newspapers, Mayor Patterson, Sheriff Jennings, and others in authority acted responsibly to maintain law and order while showing remarkable restraint. Although these groups spoke out against violence and the destruction of property and harbored sharp biases against the "lawless classes," they were careful not to inflame the situation. Militia officers had troops march around rather than into the city. When the disorders were at their worst, local officials did not call for state militiamen or federal troops to restore order, but relied on sheriff's posses and local police. Although the mayor and the sheriff took steps to keep arms from falling into the hands of the mob, they also saw to it that the "law and order posses" carried clubs rather than rifles. Even business leaders and prominent citizens who strongly supported the mayor and the sheriff did so quietly behind the scenes so as not to stir up the hostility of the crowds.[31]

There can be little doubt that industrialization had sharpened class distinctions at Harrisburg and that the depression of 1877 exacerbated them. Working people apparently wanted redress of grievances, however, not social upheaval, so they halted far short of insurrection against the system. Gone even from their rhetoric was the prewar talk of "the rights of free men" and "liberty from tyrants." In 1877 and after, their voiced demands chiefly concerned bread-and-butter issues: wages, decent hours and workloads, and job security. For their part, those representing commerce and industry, especially those who had been officers in the Civil War, talked rhetorically of "rebellion" and "insurrection" and of quelling the troubles with a little grapeshot. But even the majority of them shared the consensus of Harrisburg's residents that differences should be resolved without armed conflict.

Following the tumultuous summer of 1877, Harrisburg workers remained quiet for the remainder of the depression. With returning prosperity after 1878, however, wages once more became an issue. During the early 1880s workers fought to improve their incomes; later in the decade they resisted wage cuts. As already noted, although the economy improved and profits swelled, employers resisted restoring wages to predepression levels, and some even introduced further cuts. Common laborers, because they failed to organize, could not resist reductions, far less force wage increases. Not so the iron puddlers of the region, whose skills gave them some bargaining power. They formed the rudiments of area-wide trade unions and attempted to act in unison.

The first failure occurred in March 1880. Puddlers at Harrisburg and nearby Fairview and Duncannon who were earning $5.00 a ton demanded $6.00, refused a compromise of $5.50, and walked off their jobs. Between five and six weeks later they returned to work at the original pay scale, the demand for iron having in the meantime fallen. Two years later, on January 2, according to state reports, seventy-five puddlers at two Harrisburg firms again struck for higher wages. After tying up the work of 530 other employees for two weeks, they again failed and returned to work at their old wages.[32]

Newspaper accounts related a fuller and more complex situation in early 1882. All employees of the Lochiel Rolling Mill had reportedly walked out on the last day of 1881, demanding an advance of wages. Lochiel officials claimed no knowledge of a strike at their firm. The mills, they said, were idle because of the holidays and a failure of the Pennsylvania Steel Company to furnish them with blooms to work. Because they had received no strike notice or demands, they expected work to resume that morning, January 2. The absence of further information in the newspapers suggests that the "strike" at Lochiel did not materialize.

At the same time, during negotiations at the Chesapeake Nail Works, puddlers threatened to strike. If that plant closed down, it was for no longer than one day. On December 17 the puddlers had given notice that unless wages were advanced from $4.50 to $5.00 a ton they would strike on December 31. When that deadline arrived they extended it until January 14 so that puddlers at West Fairview, Duncannon, and other places "in the middle district" could act in conjunction with them. A reporter talked with representatives from both sides and learned that negotiations were proceeding "in a business-like manner"; the puddlers' demand was based on a rise in the selling price of iron. "Do the men take much interest in the iron market?" he asked. A company spokesman replied that it was "the same to them that the stock market is to a banker and his clerks." They watched the rise and fall of prices and knew whether fluctuations were healthy. If so, they prepared for an increase. "Do they consider it healthy now?" the reporter queried. The official replied in the affirmative, saying he had never seen conditions better and that they were improving every day. Demand for iron so exceeded supply that not a mill in the country could keep pace with orders. The reporter asked whether only the puddlers were keeping track of the market and demanding a raise. "Bless you, no," came the reply. "Nailers, feeders—all are more or less interested." If the puddlers got their raise, "which it is fair to presume they will," the others would soon follow, "although the distinctive associations do not act in any way with each other." Did proprietors generally accede to wage increases when the market was healthy? the reporter wanted to know. "It is as much to their interest as to the men," the spokesman replied, adding that some firms gave raises in advance of demands. He anticipated no trouble whatever at Chesapeake.[33]

Matters turned out differently from what the spokesman expected. When a delegation called on the management at Chesapeake for an answer on January 14, they were told that the firm had always been the first to raise wages, "and it was about time that somebody else should take the initiatory step." The men met that afternoon and discussed the reply. For several years they had made the first demand and as a consequence frequently were idle for two weeks or longer before getting an advance. By contrast, puddlers in other firms got raises as soon as Chesapeake yielded without striking or giving aid to those who did. The Chesapeake puddlers decided unanimously that they would do just as well to continue at the old rate of $4.50 unless all puddlers went out together. That evening they "lit their fires" and resumed work. According to the *Patriot*, the workers spoke "in the highest terms" of Mr. Bailey and the management, agreeing that his treatment had "been of the best."[34]

Given the dearth of information about labor relations in Harrisburg at this time, the *Patriot* articles were especially revealing. Skilled ironworkers in the area were organized in unions according to skill. The various ironworkers' unions, and even different units of the puddlers' union in the area, apparently did not always coordinate their activities. They did, however, closely follow market conditions in the industry, and their wage demands reflected changes in the selling price of iron. When the market thrived, the companies usually met and sometimes anticipated wage demands. However, they also often allowed a brief strike before giving in. The puddlers at Chesapeake appear to have been the most aggressive in pressing wage demands, but they got little assistance from puddlers in other firms who benefited from their actions. At the same time, the tone of the company spokesman's remarks, and the puddlers comments on the managers, reflected a mutual respect.

Four years later, the ironworkers were better organized and affiliated with the Knights of Labor, then approaching its zenith. Near the end of July 1886, thirty puddlers' helpers at Central Iron's "Hot-Pot" ceased work. Their "kick" was that the iron they were working was too hard. They elected a committee of five to take demands to the plant superintendent, who in turn referred them to Gilbert McCauley, general manager and treasurer of the company. The helpers wanted "easier iron" to work, an extra worker to clean "boshes," and an increase of 2 cents per heat. With a total of sixty to seventy men idled by the strike, McCauley met with the committee and explained that the company had no other iron on hand to use. He was willing to hire extra help to haul sand and push buggies, however, and offered a one-cent-per-heat increase in wages. Meeting afterward with a news reporter from the *Harrisburg Morning Call*, he complained that by quitting without notice the workers had shut down the mill—"an unheard of movement, and as the men now admit, a very unreasonable act." Central Iron was always willing "to deal honorably and justly" with its employees and had "always given them steady employment." This, he claimed, was the firm's first strike.

The helpers, in turn, complained to the same reporter that the press had credited the puddlers for the walkout and had linked the helpers with the Knights of Labor, which had nothing to do with the strike. Meanwhile, one puddler claimed that the puddlers did not support the helpers' demands and declared that McCauley had offered a "very fair adjustment of the matter." Members of the Knights of Labor, he continued, agreed with this assessment and regarded both Central Iron and Chesapeake Nail as the "best of their kind" in the state to their employees.[35]

The *Patriot* offered additional information. Ten employees of Mc-Cormick's works claimed that the iron worked at Central was particularly hard and that consequently a heat took more time there. Moreover, the trouble was not between the puddlers and their helpers, as some papers reported, although to save hiring a bookkeeper the pay of helpers passed through the hands of the puddlers. But it was the company that set the rates and paid the helpers' wages, with nothing coming from the pockets of the puddlers. On July 29 the workers' committee accepted McCauley's offer. With their pay now the same as at Bailey's other firm, the Chesapeake Nail Works, the helpers returned to work.[36] Although they believed they were entitled to the full amount of their demand, there was no ill-feeling. McCauley, they said, had "acted the part of a man and gentleman and treated them as such."[37]

A month later, all Harrisburg area puddlers belonging to the Knights of Labor demanded an increase in pay from $3.50 to $4.00 a ton. Clearly wages for puddling had been falling: from $5.00 a ton in 1880, to $4.50 in 1882–83, to $3.50 in 1886. As the *Patriot* explained, puddlers did not earn $3.50 a day. Four employees—two puddlers and two helpers—worked on each double furnace. The team received $3.50 for every ton of iron (calculated at 2,240 pounds) they puddled. Because a team made an average of 4,800 pounds a day, they earned about $7.50 as a group. Puddlers got $2.35 each, helpers got $1.40. On average, a puddler took home $50 a month and a helper took $30.

Only puddlers from Bailey's Central Iron and Chesapeake Nail indicated an intention to strike, an action that would tie up some 600 ironworkers in the two plants. A union committee called on Bailey several times but got no satisfaction. He would meet the demand only if laborers in all the other iron mills got similar raises from their employers. When a city-wide Knights of Labor committee called on Bailey, he refused to meet with them because only one of their number was in his employ. After rejecting a compromise offer of $3.75 a ton, seventy-five puddlers at the two facilities went on strike August 30. On September 4 perhaps 400 plate-mill men joined in sympathy. A city-wide meeting of the Knights attended by some 800 members sanctioned the strike. But many ironworkers were already seeking work elsewhere, some finding jobs at McCormick's mills and other places that paid $4.00 a ton. After sixty-one days the puddlers accepted a compromise of $3.85.[38]

Three other reported strikes rounded out the century. One involved silk mill operatives, another rolling mill employees, and the third consisted of a few skilled shoe factory workers spontaneously walking off their jobs. Early in August 1887 about 100 girls in the weaving department of the silk factory (which succeeded the cotton mill in the mid-1880s) went on strike. The company had announced a thinly disguised speed-up. Whereas the girls had each operated a single loom and received 6 cents for every yard produced, beginning August 8 they would receive 5 cents a yard and each tend two looms. According to the manager, with a single loom a girl had made only twenty yards a day, but with two they could make between thirty and thirty-five yards. Instead of earning $1.20 a day, an employee would be able to earn between $1.50 and $1.85. Not only would they earn more than before, they would earn more than many comparable workers elsewhere. He did not add that the change would enable the firm either in increase its total output between 50 and 75 percent or to reduce its work force by nearly half. Neither did he point out that the rate cut of 1 cent a yard would reduce company costs an additional 16.7 percent. By the end of the first day of the strike, half the strikers had returned to their looms. The thirty most determined held out a week before surrendering in total defeat.[39]

What little is known of the rolling mill strike comes from the limited information published in a state report. Apparently a rollers' union in one of Harrisburg's mills called a strike on August 30, 1887, with the purpose of blocking a wage cut. Although the strike involved thirty-five rollers and idled a total of 165 workers, it did not succeed in closing the plant. Even so, the company appears to have compromised on the wage cut.[40]

Another ineffective strike, by eighteen cutters at "the big factory of the Harrisburg boot and shoe company on Vernon Street," began on October 7, 1895. The firm, founded some seven or eight years before, employed approximately 600 workers, nearly all of whom lived in the Allison Hill district of East Harrisburg. In spite of "dull times" beginning in 1894, the firm had not cut wages. According to the company, the cutters, who were earning better wages "than anybody else in Central Pennsylvania," were asking for an increase during a "slack season." Paid 2.3 cents a pair, the majority of cutters, according to the company's books, made between $12 and $14 a week. However, the pay envelopes the workers showed to reporters indicated that they earned only $6.00 a week. In any event, the cutters demanded 3 cents a pair—a 30 percent increase. After a warning that they would be replaced, two or three returned. Although the firm secured a few scab cutters, the strikers persuaded them to quit. A general meeting of Harrisburg shoe cutters offered sympathy and paid the fares of strikebreakers to leave town. Cutters elsewhere helped by discouraging would-be scabs from seeking jobs at the struck facility. After only a week the strike ended. Most of the old cutters returned at their original wages, but two or three were still holding out.[41]

A half-century of industrialization produced remarkably few strikes and other labor disturbances in Harrisburg. Those that did take place were usually short and resulted in few victories for the strikers. Although able to cow the workers with controlled force during the 1877 disorders, Harrisburg employers rarely found such tactics necessary. Their usual policy was to avoid confrontation. They dealt with workers in a "manly" fashion, treating them as "men," which in the terminology of the day meant showing them respect. When disputes arose they met their employees (though not their unions) face-to-face, explained their position, and either offered a compromise or held firm. If the workers persisted and went on strike, the entrepreneurs simply closed down until necessity brought the workers to terms. By closing down and not attempting to run their plants with strikebreakers, they did not go to war with their workers. As a consequence, picket lines to dissuade scabs from taking jobs were not necessary, and police did not have to be called in to protect strikebreakers. Defeat of the workers, while disappointing to them, left far less bitterness than violent repression would have. Frustrated in their attempts to bring about improvement through direct action, it remained to be seen whether laborers would instead turn to that other major form of organization to effect change—politics.

Industrialization
and Harrisburg Politics

T he interconnections of local politics and industrialization at Harrisburg involved a number of considerations. These included the structure and perceived functions of local government and whether the coming of industry altered them; the extent to which officeholders were representative of the social structure of the community as a whole; who (without regard to official position) in fact governed the community; how the community responded politically to the problems produced by industrialization; and how the handful of industrial entrepreneurs advanced their interests without stoking the fires of class warfare that appeared in many manufacturing centers, particularly in Pennsylvania.

Because no serious political history of Harrisburg has been written, these and related matters cannot always be ascertained easily. Texts of municipal charters and state laws related to local government are readily available, but records of the pre–Civil War borough government have been lost or misplaced. Minutes books of the city council, indicating the subjects but not the substance of debate, and ordinance books that record the texts of adopted measures, have been preserved but have not been indexed. Persons elected to office, and their political affiliations, are available in newspaper reports of annual local government election returns.

Information on the motives and reasons behind actions or omissions of local government is extremely limited. Community leaders left few letters, diaries, or memoirs explaining their conduct in office, and the charges and countercharges of partisan newspapers on the eves of elections provide largely suspect or biased information. Further complicating local politics were the city's relationship to the state government and to the Republican machine that ran it through most of the period under discussion. In sum, although the forms and personnel of local government can be reconstructed with relative ease, the substance of the political process can at best only be pieced together from spotty and often biased evidence.

Well into the nineteenth century, governments of American towns were usually simple in structure, limited in power, and run by amateurs elected for short terms.[1] In drawing up the charter for the borough of Harrisburg in 1791, Pennsylvania's General Assembly simply duplicated the structure it provided for nearby Reading in an earlier charter. After seventeen years Harrisburgers petitioned for a change, complaining that experience had shown the circumstances and needs of the two communities to be quite different. However satisfactory for Reading, the structure was "not well calculated for the good government of the borough of Harrisburg."[2] Accordingly, the Commonwealth granted a new charter that remained in force from 1808 until 1860, when Harrisburg became a city.

The charter of 1808 provided for annual elections of a chief and assistant burgess and nine council members. In 1837 the terms of council members, but not of the burgesses, were extended to three years, with three standing for election annually.[3] During the two weeks or so preceding borough elections, meetings of interested citizens, volunteer fire companies, and other informal groups nominated candidates for office. With the election only days away, ward or borough party caucuses determined official slates, and the respective party newspapers announced and endorsed them. So far as can be determined from the local press, nominees announced no programs, and local parties neither adopted platforms nor held public meetings to discuss issues or build support. Apparently, interested residents discussed politics informally during the week before the voting. On the afternoon of the third Friday of each March, the qualified voters assembled in the courthouse to cast their ballots. Election officials tabulated the results and announced the victors. One week later the newly elected officials assumed their duties.[4]

During the 1850s the onset of industrialization and swelling population touched off a movement to enlarge the community and to have it designated a city. Changes in the structure of government provided by the charter of 1860 were not great. The charter established six wards (by 1900 there would be ten) and prescribed the number of representatives from each that would sit on the common council. Initially there were thirteen councilmen (with staggered terms) and a mayor, with veto powers over council measures. All were elected for three-year terms. As the population grew, the number of council members changed.[5] In 1867 the General Assembly mandated that Harrisburg have a select council that would share power with the common council—then a year later abolished it.[6] Following adoption of a new state constitution in 1873, the legislature the next year passed a law (which was still in effect in 1900) requiring all Pennsylvania cities to have, among other things, both common and select councils. Under the law's terms, each ward in Harrisburg had one member on the select council and two on the common council. Terms of mayors were reduced to two years.[7]

Municipal charters in Pennsylvania granted local governments only minimal powers. Moreover, the General Assembly could arbitrarily alter charters at any time. Harrisburg's Charter of 1808, for example, conferred on the burgesses and the council only those functions absolutely essential for maintaining community life. In general, they were "to promote the peace, good order, benefit and advantages" of the borough. More specifically, they were charged with "regulation of the markets, improving, repairing, and keeping in order the streets, lanes, alleys and highways"; determining depths for "vaults, sinks, [and] pits, for necessary-houses"; making rules for the foundations of buildings, party-walls and fences"; and providing for the "safe-keeping and preservation from injury" of the property within the borough belonging to the Commonwealth. The state took over this last function in 1824. The council and burgesses had the power to assess, apportion, and appropriate taxes on real estate as necessary, but (as prescribed by law in 1839) rates were not to exceed 1 cent per dollar of valuation. The chief burgess, assisted by elected constables, was to enforce the laws of the borough and maintain law and order in the community. He also sat as a magistrate in cases involving persons accused of disturbing the peace or committing other crimes against the borough.[8]

In somewhat vaguer but more felicitous language, the city charter of 1860 conferred essentially the same functions. The council received "full power and authority" to make the laws and ordinances it deemed necessary "to preserve the peace and promote the good order, government and welfare of the said city, and the prosperity and happiness of the inhabitants thereof." Specific mention was made of providing and maintaining roads, streets, and alleys. The mayor, in addition to his veto power over legislation (which the council or councils could override), was responsible for executing and enforcing those ordinances. He was further charged to "preserve the peace and promote good order" and to "secure the comfort and happiness of the citizens and of all strangers and sojourners in the city." As had been true of the chief burgesses, mayors were to sit as magistrates in cases involving violations of local law.[9]

To whatever extent industrialization brought problems to the community, they rarely found their way into Harrisburg's politics. Most of the time of the municipal council was taken up with administering the public markets, paving streets and putting in sidewalks; regulating inns, taverns, and other public houses; controlling livestock within the borough; keeping the streets clean and unobstructed; halting mischief and noise-making by teenage boys; and dealing with health problems caused by accumulated waste and puddles of stagnant water. Only occasionally did providing such services as police and fire protection and clean drinking water intrude.[10] By state law, public schools were independent of the city government. Initially each school district in the city elected the school board that governed it. A single board of education for the city was set up in 1869.[11]

Local newspapers, the principal reporters of the community's political life, had little interest in local matters. Their chief concerns were the state and national politics of their financial backers. Because Harrisburg was the capital of the Commonwealth, each political faction in pre–Civil War Pennsylvania sought to maintain a newspaper in the city to trumpet its agenda for members of the legislature. New journals regularly appeared on the eve of national political campaigns, only to disappear after the elections. When a party or faction won control of the state or national administration, that group's local editor was rewarded with printing contracts at the state's disposal and sometimes with political plums, such as the postmastership.

Out of the pre–Civil War period came the town's two principal newspapers of the next century, the *Telegraph* and the *Patriot*. Established in 1831 to promote the Whig cause in a Democratic stronghold, the *Telegraph* was the town's leading newspaper well into the twentieth century. Although Theophilus Fenn, its founder and for more than twenty years its editor, won few political converts, his spritely style won the widest readership. It was when Simon Cameron took over behind the scenes in the 1850s that the *Telegraph* moved through its Know-Nothing phase into the lists as a Republican organ. George Bergner, its editor from 1855 until his death in 1874, enjoyed two stints as postmaster of Harrisburg: from 1861 to 1866 and from 1869 to 1874. Not surprisingly, the *Telegraph* consistently supported the state Republican machine. It continued publication until 1948. The politically dominant but badly splintered Democrats had several papers with numerous publishers and editors before the Civil War: the *Oracle,* the *Democratic State Journal,* the *Keystone,* the *Democratic Union,* and the *Patriot,* to name only the more important. Those in business in 1858 merged to form the *Patriot and*

Union. Eventually called simply the *Patriot*, it became Harrisburg's Democratic organ.[12] Today it is the city's only major newspaper.

Well into the second half of the nineteenth century, local news received short shrift from most editors. Advertisements, serialized fiction, legal notices, and items on health, manners, and religion took up at least half of the available space. The portion alloted to news went overwhelmingly to short summaries of foreign and national affairs and to detailed reports on state politics. Some newspapers ignored local events altogether, perhaps on the assumption that in so small a town everyone already knew what was going on there. The *Telegraph* (and later the *Patriot and Union*) introduced a regular column of local items and occasionally devoted an editorial to local issues. Fires, floods, freakish weather, and strange happenings; meetings of church groups, lodges, volunteer fire companies, and military organizations; coming entertainment; crime (from minor disturbances of the peace to theft and murder) and the perennial problems of boys with too little to do; and derogatory squibs about drunks, the Irish, and African Americans—these were the topics of most of the columns. Local political items were filtered through a partisan lens dedicated to winning the next state or national election for the papers' respective political sponsors.

The industrialization of the community, reported on and warmly welcomed by the local press, stirred no political controversy—nor did the consequences of industrialization: the armies of newcomers seeking jobs in the community, the occasional labor dispute, the extensive child labor at the cotton mill, the periods of unemployment, the pollution of waterways and the atmosphere. The rising factories with their hundreds of workers were hailed as progress. Labor disputes were the result of outside influences or ignorance of the laws of supply and demand. Child labor warded off child delinquency and what otherwise might be suffering by the poor. Periods of unemployment, like spells of bad weather, were to be expected and endured. If anyone was to blame, it was usually the pauper immigrants. As for polluting streams, where had everyone always disposed of their waste? Smoke indicated a busy and prosperous community.[13]

Local happenings became important only when they touched somehow on state or national political issues. Thus, before the Civil War, when local blacks tried to block enforcement of the fugitive slave law in the area, and local Know-Nothings organized and sought to stir up the town against Irish and German immigrants, Harrisburg's newspapers devoted columns of space to these events.[14] They also thoroughly covered the one major local incident growing directly out of industrialization: participation of Harrisburg residents in the nationwide railway strikes of 1877.

Officeholding by municipal officials was divided between those who served only a term or two and those who served for many terms. Of the principal officials of the borough between 1830 and 1860 (the chief burgesses, assistant burgesses, and council members), most served a single term and then either declined renomination or were not reelected. Among those serving multiple terms was Charles F. Muench, who established the record, sitting intermittently on the council for nine years and as chief burgess for four. William Kline was assistant burgess for one year and chief burgess for six. Harrisburgers elected Christian Seiler as chief burgess on five occasions, while George Beatty and Henry Critzman each served two terms. Eleven others in the same period had only single one-year terms.

The same pattern prevailed under the city charter between 1860 and the end of the century. The majority of mayors and council members served a term or two and then passed back into the obscurity from which they arose. In the forty years before 1901, thirteen men served as mayor. Although only two held multiple terms—John D. Patterson and John A. Fritchey each served three consecutive terms, followed, after an interim, by a fourth—their combined tenure amounted to eighteen years.[15] Most of the burgesses and mayors, including the multi-termed Kline and Patterson, were not sufficiently important or remembered well enough to have so much as a short biographical sketch in any of the histories of Harrisburg. Even contemporary reporters covering elections were so unfamiliar with some council members that they put down wrong names or spelled them differently from one account to the next or even from one page to another.

Democrats were the dominant party in Harrisburg politics, holding the office of chief burgess or mayor for fifty-one of seventy-five years between 1830 and 1905. Before the Civil War that dominance usually paralleled the politics of both the state and Dauphin County. Except for 1835–39, 1848–52, and 1855–58, for example, all Pennsylvania governors were Democrats. Although both Pennsylvania and Dauphin County swung to the Republican column in 1861, Harrisburg did not. Republican governors presided over the Commonwealth except for the 1883–87 and 1891–95 terms. Until 1873, Harrisburg's mayors were all Democrats except for eighteen months in 1868–70. Republicans did rule as mayor for the next fourteen years, but thereafter they were out of office except between 1896 and 1899. Dominance by one party rarely meant exclusion of the other. Members of the opposition party almost always held seats on the city council and in a few instances were in the majority.

From at least 1830 until 1900, persons elected to the borough and city councils represented a wide range of socioeconomic groups within the community, not just the elite. To be sure, in the early years, one or two members of the leading families almost always sat on these bodies, as if, by agreement, monitoring their common interests. By the end of the century, however, direct participation by entrepreneurs was exceptional. The occupations of the remaining council members reflected economic changes occurring in the community in those decades. During the preindustrial years, membership consisted chiefly of physicians, lawyers, merchants, storekeepers, an occasional banker, and artisans. With the rise of industry by 1860, the first few factory owners and industrial workers appeared. As both industry and government grew larger and more bureaucratic by 1870, managers, foremen, and clerks from industry, railroads, and the government were elected to the councils.[16]

Although no entrepreneurs happened to be sitting on the council in 1870, its membership otherwise nicely illustrated the transition taking place in the city's economy. Of twenty-five members, six represented the older order: a grocer, a hatmaker-merchant, a stoneware dealer, a painter, a cabinetmaker–furniture dealer–undertaker, and a brickmason. Another five were transitional businessmen: a coachmaker-manufacturer, the co-owner of a small shoe factory, two owners of brickyards, and a commission agent who handled railway freight. From the new industrial community came five managers: a rolling mill superintendent, the manager of an ironworks, two railway express agents, and the superintendent of the Pennsylvania Railroad's local stockyard. Representing lower levels of industry were two railway clerks, two railroad engineers, and a locomotive machinist. There were also three government clerks: one who worked for the county commissioners and two who worked for state officials. The remaining member was an unspecified "agent."

An analysis of the listed occupations of all persons sitting as members of the borough council in 1840 and 1850, and the city common and select councils each tenth year from 1860 through 1900, reveals something of the degree to which that body represented the various classes in the city (see Table 13.1). Professionals, who had been the third largest category in 1840 and 1850, became proportionately fewer after the Civil War. Businessmen, usually the largest group, also declined, from between two-fifths and three-fifths of the total before the war to approximately one-third between 1860 and the turn of the century. Perhaps most surprising were the relatively high percentages of workers (craft and factory workers and unspecified laborers) elected throughout the period. This group usually ranged from a fifth to a third of both the common council and the select council (and once reached half the com-

Table 13.1.   Occupational Categories of Harrisburg Council Members, 1840–1900

Category	1840	1850	1860	1870	1880	1890	1900
			Common Council				
	n = 11	n = 12	n = 13	n = 25	n = 36	n = 18	n = 20
Professional	18%	25%	0%	0%	3%	6%	0%
Business	45	42	62	36	33	11	35
Managerial	0	0	0	16	8	28	10
Clerical	0	0	8	28	11	6	15
Worker	27	33	31	20	33	50	25
Other	9	0	0	0	11	0	15
			Select Council				
					n = 9	n = 9	n = 10
Professional	—	—	—	—	22%	11%	10%
Business	—	—	—	—	44	22	30
Managerial	—	—	—	—	11	22	10
Clerical	—	—	—	—	0	11	20
Worker	—	—	—	—	22	22	30
Other	—	—	—	—	0	11	0

Sources: Harrisburg city directories for 1840, 1880, 1890, 1900; computerized population census schedules, 1850, 1860, 1870.

Business = manufacturers, merchants, shopkeepers and storekeepers; Managerial = foremen, managers, superintendents; Worker = craft workers, industrial workers, laborers; Other = "gentlemen," others, unknown.

mon council). However, using census reports of landholdings as the measure, worker council members were among the more successful members of their class and not typical of the working class as a whole.

The same measure can also be used to determine the degree to which the council members failed to represent certain large constituencies in the city. Between 1840 and 1860 landholding by council members increased. Then, by 1870, both the number of council members holding real estate and the average value of their holdings dropped. Because the census of 1840 did not include real estate holdings, what they held in 1850 was used. Of eleven council members, only four reported owning real estate ten years after leaving office, four showed no holdings, and three did not appear. Those reporting no property had been landless while in office, or failed to list their holdings, or in the course of a decade had lost what they owned. Three of the four landed council members had holdings valued in excess of $10,000.

Data for the period from 1850 through 1870 is both direct and more complete. All eleven council members of 1850 who appeared in the census (one did not)—and a decade later all thirteen—held real estate. Three council members did not appear in the census of 1870. Of the remaining twenty-two, five were landless and seventeen were landowners. As the council grew in size over the same three decades, the proportion holding land valued at $10,000 or more rose from five of eleven in 1850 to six of thirteen by 1860, then fell to six of twenty-five in 1870. The average value of holdings followed the same course, rising from $8,909 in 1850 to $11,985 in 1860, then sinking to $6,159 in 1870.[17] Unfortunately, the property holdings of common council members were not available after 1870. By law, members of the select council had to own real estate in the city.

As for ethnic representation on Harrisburg councils, Scots-Irishmen born in Ireland and German-born citizens were members at least as early as the 1830s. In 1839 Michael Burke became the first known Irish Catholic elected to that body. From the 1860s through 1900, one or two foreign-born members at a time regularly sat on council. Apparently no African American was so honored.[18] Although councils were more broadly representative of the population than might have been expected, at least two major groups of Harrisburgers—the 10 percent who were African American and the slightly more than half who were female— were not represented. Greatly underrepresented were the three-quarters of the population who owned no real estate.

Being represented, however important symbolically, did not ensure protection or advancement of the interests of any group, especially of the lower classes. This was because, in politics, holders of public office were not necessarily the ones who governed. Even if those who already dominated a community economically and socially did not themselves hold office, they were widely suspected of manipulating those who held office from behind the scenes. Also, whether those elites and their families continued to live in the community and take an active interest in its affairs or were absentee employer-owners had much to do with who governed and how.[19]

The relationship of Harrisburg's entrepreneurs to its municipal government was not simple. As a group they and their descendants did remain residents of the city and took pride in their contributions to its well-being. At the same time, they appear generally to have left the day-to-day administration of local government and many of the major policy decisions as well in the hands of elected officeholders. This was in part because they were too busy with their own affairs and in part because municipal government did not touch their interests enough to make involvement worthwhile. Nor were the officeholders merely tools or stooges of the elite. With few exceptions, Harrisburg's entrepreneurs were Republicans who had little influence or control over the nominating process of the Democratic party, which, except between 1874 and 1887, almost continuously dominated the city government.

But the entrepreneurs certainly had no difficulty making their views known or influencing matters that concerned them. One or two of their number usually sat on the city council, especially from the 1830s to the 1870s. Their control of the leading industries, banks, and businesses; their financing and sometimes outright ownership of the local press; the roles they played in the leading churches, charities, and clubs; and their close ties to the medical and legal professions assured them a full and careful hearing. They also swayed decisions with their donations of money, property, and services to causes they favored.

On particularly critical matters the entrepreneurs sometimes found it easier to turn to the state legislature for protection than to attempt to manipulate the municipal government. The state's powers were greater, the outcomes were more certain, and one of their number (Simon Cameron, from 1857 to 1877, and J. Donald Cameron, from 1877 until 1897) controlled the all-but-omnipotent Republican machine.[20] Tracing the operations of the machine is difficult because of the style set by the elder Cameron: he "never wrote when he could speak. He never spoke when he could nod. He never nodded when he could wink an eye. And he never winked an eye when he had the proper lieutenants who could anticipate his wishes."[21] The device most frequently used was a "special" or "local" act that applied to a specific corporation, community, or group. Thanks to legislative log-rolling or pressure from the machine, these measures passed with little study or debate.[22]

The entrepreneurs' reliance on the state began early. When they were erecting Harrisburg's infrastructure in the preindustrial era, what they wanted and needed lay more often within the province of the state than that of local government. It was the state that chartered their banks, turnpikes, canals, railroads and later their corporations. It was also the state that provided lucrative contracts for building the State Works, and loans, grants, and subsidies for bridges, turnpikes, and canals. Once, when the entrepreneurs were divided, those hoping to establish a private, profitmaking waterworks were thwarted by others, who secured permission from the legislature for Harrisburg to construct a publicly owned facility.

Legislative help was forthcoming even more once the Cameron machine began to take shape in 1857. Determining which or how many of the Harrisburg entrepreneurs made use of this form of assistance is difficult. Not all of them, of course, were equally concerned with the issues taken to Capitol Hill. For example, tax rates on factory properties affected most of them, but whether large industrial properties just outside Harrisburg should be annexed touched only those who owned them, and securing the voters of Harrisburg for the Republican machine mattered principally to the Camerons.

One of the early recourses to the legislature involved altering the proposed boundaries of Harrisburg when it became a city in 1860. The borough council that year, after vigorously opposing previous efforts, led the move to incorporate.[23] It engaged one of the town's prominent older lawyers, John A. Fisher, to draft a charter and define the new city's limits. Fisher's draft proposed adding extensive areas north and south of the existing limits as well as the river and its islands as far as the western shoreline. This was to enable the city to protect its water supply by draining swamps and stagnant ponds and prohibiting privies and the dumping of waste within the watershed. However, some of the extensions to the south and the east included holdings of the McCormicks, the Camerons, the Calders, and other entrepreneurial families. To placate them, lands these men were believed to want in the city were included, and four of them were included on a proposed commission to lay out streets and alleys. The drafters thought it only "right and proper" to include the major landowners to ensure the least harm to their property.[24]

The bill ran into trouble in the legislature. The state senator from Swatara Township (where Cameron's country estate was located) introduced a substitute bill excluding "only a couple of hundred acres of farm and other undivided property" from the proposal.[25] In fact, the excluded properties at the south included Cameron's estate, Paxton Furnace, and an adjoining ten acres belonging to the McCormicks, two large distilleries with accompanying farms where by-products were fed to a thousand or so hogs, and farmland belonging to two unidentified bank directors. At the east, land belonging to the "State Printer" (George Bergner) and to a partner of a recently formed bank (William Calder of Cameron, Calder & Eby) were also excluded. The reason, it was said, was that if these properties were included they would be taxed to help pay off debts contracted by the city prior to incorporation and not for any benefit to the owners. That argument applied with equal force to any of the proposed annexations and, if adhered to, would have prevented Harrisburg from ever extending its boundaries.[26] Apparently the real objection was that if the city marked out streets and alleys the value of those tracts as sites for large manufacturing plants would be destroyed. A unanimous resolution by the borough council endorsing the original bill did not prevent the legislature from adopting the substitute measure.[27]

The original purpose behind the next intervention is not clear. For some reason one of the city's former Democratic mayors drafted a bill to amend the charter. A seven-member select council (one member from each ward and one at large) was to be co-equal with the common council. Its members differed in that they had to own real estate in the city. Further, instead of the common council electing police constables, the mayor and president judge of the county were to appoint them. The Democratic newspaper, the *Patriot*, strongly objected. Cameron's paper, the *Telegraph*, however, saw merit in the measure, possibly because it might strengthen the Republicans, who had elected their first mayor and were making inroads on the common council.[28] Certainly Cameron, who controlled the Commonwealth, chafed at the idea of the government of the city where he lived being in the hands of the enemy. In any event, with little advance notice or discussion, the Pennsylvania legislature adopted this measure just before the municipal elections of March 1867.

Whatever was expected from the changes, the results disappointed the Republicans. Democrats not only elected the next mayor and all seven common council members up for election, but six of the seven new select council members as well. A year later they won every open seat on both councils.[29] After the second debacle, the *Telegraph* called for redistricting the city. Further, the outcome persuaded "all upright and honorable men of both parties" that elections should be held at the "regular time," in the fall, when "everybody turns his attention to politics." The *Telegraph* was "astonished" that the Democrats, who benefited from the present arrangement, did not agree.[30]

In April the legislature dutifully repealed the previous year's work. At the same time, it amended the charter further to correct certain matters of concern to Cameron and other entrepreneurs. The new measure divided the city into nine "fair and convenient wards, giving all parties an equal chance to be represented in the council." It allowed the city to borrow $50,000 to pay off existing debts but forbade new expenditures in excess of revenues. Taxes hereafter were to be based on services provided, with three classes of real estate rates. Areas with full services paid the full rate; areas such as the Fifth and Sixth wards, where the city provided neither gas nor water, paid only two-thirds of the full rate, and manufacturing areas paid half-rates. Property assessments no longer were to be made by elected officials, but by persons appointed by the county court. With these matters settled, the measure annexed the properties that had been excluded at the time the city was incorporated: the iron mill district to the south belonging to the Camerons, the Mc-

Cormicks, and the Baileys, and the land to the east held by Simon Cameron, William Calder, and William T. Hildrup, among others.[31]

Redistricting helped the Republicans a bit, giving them three new seats on the common council. With the election held in October in conjunction with state races, the press gave even less attention than usual to local issues.[32] But a wave of reform was soon sweeping the nation at large. Revulsion at the abuses of "Grantism" in Washington and widespread corruption at the state level produced the short-lived Liberal Republican reform movement of 1872. In Pennsylvania it resulted in a call for a new state constitutional convention. The Cameron machine had no choice but to run with the tide.[33] Among other things, the Constitution of 1873 provided that legislation should be general in application, all but eliminated special and local legislation, and prohibited the legislature from incorporating cities, towns, or villages or changing their charters. It again separated state and national elections from those at the local level, with the former in the autumn, the latter in the spring.[34] The next year, the legislature enacted a general law governing third-class cities (those ranging in population from 10,000 to 100,000), prescribing for them both select and common councils, and mayors with veto powers. It fixed the terms of mayors and the select council members at two years, members of common council at one year. Seven and a half pages of the law detailed expanded powers of city councils, placed limitations on real estate taxes (10 mills per dollar of valuation) and new debts (no more than 2 percent of total property valuations per annum), and spelled out the process for annexing land to cities.[35]

Despite the improvements resulting from the new state constitution, reform was not permanent in Pennsylvania. As one history of the Commonwealth put it, "Pennsylvania balanced the scales by voting both to reform the machinery of government and to keep Cameron and his lieutenants operating that machinery."[36] So far as Harrisburg was concerned, the killing of special legislation came too late. Most of the favors the industrialists wanted had already been granted.

Meanwhile, the special concern of the Camerons—increasing the strength of the Republican party—was being dealt with by the national government. Both Cameron and his paper, the Telegraph, championed suffrage for African Americans, but the issue was not popular among Pennsylvanians. Accordingly, Democrats vigorously opposed it, and the Republican machine tried to ignore it.[37] Nonetheless, when the Fifteenth Amendment was ratified in 1870, blacks began to vote. Whatever Cameron's personal views on racial equality, an expected partisan advantage would be increased Republican voters both in the Democratic south and in northern cities with growing black populations.

In Harrisburg, black votes made a difference, helping to bring the Republicans to power at last. The effects were most notable in the Eighth Ward. Although a significant white foreign-born working element lived there, it was the city's largest black enclave. In 1869, the last election in which blacks were excluded, the Republican candidate for council received only 37 percent of the ward's vote. With blacks voting in 1870, the Republican candidate for mayor, though losing the city, carried the ward with 53 percent of the ballots, as did the Republican candidate for council who won. Thereafter, the Eighth Ward regularly elected Republicans to the city council with majorities ranging from 52 to 70 percent and contributed 81, 53, 11, and 34 percent of the margins of victory for Republican mayors elected in 1872, 1874, 1876, and 1879.[38]

Although encouraged to vote, blacks were not invited to run for public office. When William R. Dorsey won election as a constable of the Eighth Ward by seven votes in 1871, he became the first of his race to hold an elected office in Harrisburg. There would not be another for seven years, until William Day was elected to the school board.[39] Increasingly unwilling to simply supply votes for the machine and its candidates, Harrisburg blacks in 1882 demanded that the Republican mayor, whom they helped to elect, name a few of their race to the police force. The Telegraph supported them, noting that they were "an integral part of our city, pay taxes, support public institutions and by their votes keep in power in this city the party which gave Mayor [John C.] Herman his office." The Patriot pointed out hopefully that when Philadelphia's Republican mayor similarly refused such appointments blacks there helped replace him with a Democrat.[40] Meeting a few days before the election, aggrieved blacks adopted resolutions threatening that if their demands were not met they would "pursue our own respect and protection." A week later, the Sixth and the Eighth wards (both heavily black and usually solidly Republican) each elected one white Democratic council member with the help of black voters.[41] Although it appears that no blacks were added to the police force, two won offices in the Eighth Ward in 1884. Major Simpson, a Republican, was elected an alderman, and Nicholas L. Butler, a Democrat, became a judge of elections.[42] Even if others not noted were elected to public office, the loyalty and votes of Harrisburg's African Americans to the Republican party went largely unrewarded.

Harrisburg's entrepreneurs and other privileged groups apparently got much of what they wanted from municipal government—or, just as important, blocked much that they did not want. Moreover, the Cameron machine on Capitol Hill stood ready to provide (at least until 1874) special legislation for matters beyond the scope of local government and to overturn harmful decisions made at the lower level. Still, Harrisburg rarely experienced anything resembling open class politics. At least three factors played a part. The lack of one dominating employer or industry deprived laborers of a single, unifying common enemy. Iron manufacturing involved the most capital and greatest number of workers, but Harrisburg had other industries and significant other economic activities: merchandising, transportation, craft production, and government administration.[43] The failure of the working elements to develop strong labor unions deprived them not only of the economic strengths and loyalties fostered by such bodies but also of the organizing skills that might have carried over into politics. Finally, the policy of Harrisburg industrialists of simply closing their plants during labor disputes was of major importance. Elsewhere, and especially in nineteenth-century Pennsylvania coal and iron communities, employers regularly brought in strikebreakers, workers responded with picket lines to protect their jobs, violence ensued, and employers called in local and state police to "restore law and order." With their own governments turned against them, workers in such places often organized politically in an attempt to wrest control of local government at the ballot box.

Scranton, Pennsylvania, for example, experienced both class warfare and class politics during the 1870s and after. Anthracite miners in the city and the area had a relatively strong union. To break it, the companies closed the mines. Violence followed, which the companies turned to advantage by tarring the union with complicity in the Molly Maguire murders. Two years later, when the railway strikes erupted, the mayor of Scranton tried to prevent trouble. With only ten police officers against a mob numbering between 1,500 and 2,000, he deputized "Special Police" armed and trained by the Lackawanna Coal & Iron Company. As might be anticipated, coal operator W. W. Scranton soon took command of those he had selected and, in the melee that followed, six people were killed and fifty-four were wounded. Both state and federal troopers arrived and policed the city for weeks. In municipal elections the next February, Terence V. Powderly, candidate of the Greenback-Labor party and soon to be Grand Master Workman of that era's most important labor union, the Knights of Labor, became mayor for the first of three terms.[44]

At least in part because employers did not use local police to defeat strikers, Harrisburg workers felt no need to organize politically. Accordingly there were no local labor-reform parties, no significant support for national third parties, and no attempts to take over one of the regular parties. Neither did Harrisburg's workers run candidates for public office or succeed in electing them, except in the first two elections after the demonstrations in 1877. At the election of state and county officers that November, Greenback candidates appeared on Harrisburg ballots for the first time. They received between 436 and 520 votes for the various offices, or about 12 percent of the 4,242 total. The Greenback candidate for district attorney, who had no Democratic opponent, did best, securing just over 37 percent of the vote.[45]

At elections for city council members the next February, Greenbackers ran independent slates in each ward except the First and Second, where there were combination Republican-Greenbackers. One fusion candidate in each of the two wards won, as did two straight Greenback candidates from the Sixth Ward. The strongest Greenback candidate in each of the other wards drew from 2.5 percent of the vote, in the Ninth Ward, to 40 percent in the Sixth Ward. In the First, Fifth, and Seventh wards they got between a fifth and a fourth of the total vote; in remaining wards they garnered between 11 and 15 percent. In this the third party's best showing, Greenback council candidates overall received 13.4 percent of the vote, and fusion candidates got 3.9 percent.[46] The presumably prolabor council members were, of course, too few to alter the course of city government. None was reelected at the end of the one-year term.

Republican Sheriff Jennings, who had led the charge against the demonstrators, did not seek reelection when his term expired in November 1878. The Democrat who replaced him won a three-way race in which the Greenback candidate received only 462 of nearly 5,600 votes. In that same fall election, the Greenback candidate for governor drew 8.8 percent of Harrisburg's vote. By contrast, other Pennsylvania cities that had an overbearing single industry, or where local police had been used as strikebreakers, gave him much greater support: Reading (where eleven were killed and many were wounded during the railroad strikes), 16.8 percent; Altoona (whose single industry, the Pennsylvania Railroad, had been the antagonist of the strikers), 18.4 percent; Wilkes-Barre (the scene of recent coal-strike clashes), 21.9 percent; and Scranton (which suffered both coal and railroad disorders in 1877), 35.8 percent.[47]

Following the victory in February 1878, support for the third party waned, and Harrisburg politics gradually returned to normal. In February 1879 the Greenback party offered no candidate for any local office. Republican John D. Patterson, mayor during the disorders, won reelection to a third term with 55 percent of the vote (he had received 52 percent in 1874 and 64 percent in 1876). Democrats took control of both the select council and the common council. In 1880 Greenbackers ran council candidates only in the Fifth, Sixth, Seventh, and Eighth wards and drew 2.5, 13.2, 10.5, and 1.0 percent of the votes respectively. The last gasp came in 1882, when the Sixth Ward's slate of four candidates for council attracted only 5 percent of the vote.[48]

Another way in which politics at Harrisburg differed from politics in many other industrial centers was that the city was never run by a boss who depended on an ethnic-based machine. A familiar pattern of community development was for local governments to pass through two stages during the nineteenth century. Initially a patrician elite of landowners, merchants, and professional people ruled, followed in the post–Civil War industrial era by bosses who controlled the votes of masses of foreign-born workers.[49] If Harrisburg's wealthier founding families constituted a patrician elite that dominated the early government of the borough, their overt rule ended sometime before 1830. By that date, if not earlier, they were sharing offices—and presumably power—with persons of humbler backgrounds. When the rise of industry attracted the usual Irish and German immigrants to the city, particularly between 1850 and 1870, the Irish were too few to do much more than add flavor to the Democratic party while the Germans tended to be inactive politically. After 1870, when industrialization slowed in the city, the flow of "old immigrants" fell off. With little of the "new immigration" from southern and eastern Europe coming to Harrisburg, the foreign-born as a percentage of the total population declined.

But could not the immigrants, combined with the significant African American presence already in the city, have made boss rule based on ethnic votes possible? Certainly the Democratic and Republican newspapers charged each other with operating such "machines"—the Democrats allegedly catering to Irish voters, the Republicans to African Americans. However, the newspapers were remarkably vague about such details as who the "bosses" were, how they operated, and what the consequences were. Rather than a boss-directed, self-perpetuating ethnic machine controlling Harrisburg politics, those who administered the local government might with greater precision be seen as temporary, constantly shifting cliques or "rings" elected by the great majority of voters who were American-born shopkeepers, craft workers, factory hands, and laborers.

Two major factors worked against rule by a typical city machine. For one, the municipal budget provided too few city jobs and too little graft to support anything so grand as a "machine." Council minutes show that only a dozen employees worked for the city in 1860—perhaps fifty by 1900. Annual budgets in that period grew from less than $13,000 to a little more than $250,000. Of that, 25 percent to 75 percent (depending on the date) went to interest and taxes on the city debt, largely owed for construction of the waterworks and later for enlarging them. Police and

Fig. 24.   A Street in Harrisburg's "Bloody Eighth" Ward.   Located "behind the
Capitol," the Eighth Ward was home to many of Harrisburg's immigrant and
African American families. Its black and Irish voters were allegedly pawns of
"political machines."

fire protection, streets, sewers, and gas lighting used up most of what
remained. Holders of the city's bonds probably did well, and paving,
sidewalk, and sewer projects may have added to the incomes of a few
contractors. If the bondholders were local, most would have been the
banks and members of the entrepreneurial families, not politicians, and
few local contractors came from the immigrant community. Whatever its
share of the city budget, graft could hardly lubricate, much less fuel, a
machine.[50]

The city's relatively small and declining number of minorities also
made putting together a political machine based on ethnic groups difficult. Together, the foreign-born and the African Americans made up only
22 percent of the population between 1850 and 1870, then sank to 13
percent by 1900. While numerous enough to contribute to electoral

victories, those segments of the population dominated neither party. Also much to the point is the fact that the ethnic votes did not go to a single "machine" but were divided: the Irish supporting Democrats, blacks supporting the Republicans.

Harrisburg's politics, as so much else about the community, tended to be steady, bland, and relatively unexciting. Industrialization and its problems had little impact on the city's political processes beyond protecting the interests of the entrepreneurs. Addressing the interests of the working classes would have required leadership to articulate those concerns, the organization of labor unions to fight for better wages and working conditions, and labor involvement in reform politics. But the factory system at Harrisburg was never so all-embracing, never so harsh, and never so disruptive of the lives and labors of working families as to arouse them to such efforts.

14

The Impact by Century's End

By 1900 Harrisburg bore little resemblance to the sleepy, pre-industrial river town of 1850. It had changed considerably, even from the industrializing city of a quarter-century before. Its population, now 50,000, had more than doubled since 1870, but the rate of growth had slowed by nearly two-thirds. On the other hand, the area of the city was nearly the same. Only the Tenth Ward, a residential district to the north between Maclay and Division streets, had been added. Housing was more compact throughout the city, with acres of homes now occupying the large, previously vacant tracts owned chiefly by the entrepreneurial families. The city acquired its first telephones in 1878, and electricity began replacing gas lights in 1885, when one of the nation's earliest successful commercial power companies opened there. Although Strawberry from Third to Market Square became Harrisburg's first paved street in 1886, most byways remained unimproved in 1900.[1]

Most of the area west and southwest of the capitol building had become commercial. Impressive new multiple-story buildings were beginning to dwarf the church steeples and the courthouse cupola that once stood out in the skyline. Gone for the moment was the Capitol, however. It had burned to the ground in February 1897, but plans for constructing a massively domed new building were under way. The Harrisburg Hospital had arisen on South Front Street a block north of the old Harris Mansion. The garden of the Haldeman home at Walnut and Front was torn up to make way for a city library soon to be built. The Harrisburg Club, meetingplace of the male elite, stood at Market and Front streets. A new post office building occupied the block at Third and Walnut, across from the south Capitol grounds.[2]

Margaretta Cameron Haldeman, daughter of Simon Cameron and

widow of Richard Haldeman, lived with her children at the Harris Mansion on lower Front Street near the Settler's Grave. Although the large and fashionable residences of the wealthy stretched a bit farther north along Front Street than thirty years before, construction there had peaked, and several properties had passed into probate for distribution to children and grandchildren. Because development of the riverfront into an attractive park still lay in the future, exposed sewerpipes continued to spill untreated and sometimes noisome waste into the Susquehanna. The stagnant waters of the all-but-unused Pennsylvania Canal remained at the heart of the industrial corridor to the east. As before, sluggish Paxton Creek, now straightened and confined, served the area as an open sewer.

Two new bridges—one for regular traffic, the other belonging to the Pennsylvania Railroad—spanned the Susquehanna. The Walnut Street toll bridge, erected by the People's Bridge Company, offered competition and an alternative route to "Old Camel-Back," the tolls from which still went to the first families. In addition to horses and carriages, it carried a streetcar line connecting Market Square with the western shore of the river. Somewhat south of the old Cumberland Valley Railway Bridge, the Pennsylvania Railroad Company had erected a new bridge to accommodate larger rolling stock and to allow through freight to bypass the city. Branch streetcar lines now ran to East Harrisburg and Steelton.

If the rail traveler who passed through Harrisburg in 1850 and 1871 rode the length of the industrial corridor once more in 1900, many changes would have been evident. The area was even more heavily industrialized than in 1871, but additions to existing firms, rather than new enterprises, accounted for most of the differences. Just below the city's southern boundary stood the now gigantic mills of the Pennsylvania Steel Company. Since 1871 that firm had added a blooming mill, replaced its old Bessemer plant with three eight-ton converters, erected four blast furnaces, and put fourteen open-hearth furnaces into operation. As it passed into the city, the train swept past the Lochiel furnace and rolling mill, once managed by the McCormicks but now held by the Steelton firm. To the right of the track stood a new enterprise, the Elliott & Hatch Typewriter Company. Next came the Central Iron & Steel Company, combining what previously had been the separate firms of the Baileys and the McCormicks in South Harrisburg. Both had built extensively during the 1870s—the McCormicks erecting two furnaces, Paxton No. 2 and Lochiel in 1872 and 1873, and the Baileys completely rebuilding their rolling mill in 1878. Indicative of the stagnation mark-

Fig. 25.   The Harris Mansion on Front Street.   Built by city founder John Harris in 1766, the mansion was later the residence of banker Thomas Elder and Senator Simon Cameron. Today it houses the Historical Society of Dauphin County.

ing both firms since that period was the only recent construction, Bailey's Universal Mill, built in 1892.

The Pennsylvania Railroad too had replaced its Italianate station at Market Street with a new structure. After pausing there, the train moved northward to the city limits, passing most of the 1871 factories and other facilities: the Philadelphia & Reading Railroad station; Hickok's Eagle Works; and the Pennsylvania's own enlarged repair shops, freight yards, storage sheds, and roundhouses. As before, intermixed with the factories were many of the same brickyards, planing mills, coal docks, and warehouses. Only a burial-case manufacturing firm, a flour mill, a meat-packing plant, a match company, and the Harrisburg Heat, Light & Power Company were new. One old giant, the bankrupt Harrisburg Car Works, now housed the Harrisburg Pipe & Pipe Bending Company.

During the community's first half-century of industrialization, the number of people with occupations listed in the census increased nearly tenfold, from 2,300 to 20,500. The ratio engaged in white-collar work as opposed to blue-collar work remained remarkably unchanged: one to three. There were changes within those broad categories, however. Executives and managers, manufacturers, professionals, and merchants, taken together, had filled half the white-collar positions in 1850. The other half were sales and clerical workers. By 1900 nearly two-thirds were salespersons and clerks, the latter reflecting the growth of both industrial and government bureaucracies. Changes were more pronounced among blue-collar workers. In 1850 half had worked in the crafts, little more than a tenth worked in the town's first factory or on the railroads; the remainder were unspecified laborers and servants. By 1900 half were industrial workers, 15 percent remained in the crafts, and approximately one-third were unspecified laborers and servants. Despite their lower proportion, the number of craft workers had increased nearly three times (from 853 to 2,420), offering opportunities for newcomers to enter as well as for established artisans to continue in many trades.[3]

The slowing of industrialization in the city by 1880 largely accounted for the limited changes in Harrisburg's ethnic makeup by 1900. Unlike communities where industrialization continued apace, the capital city's factories no longer beckoned additional labor. As a result, the percentages of both foreign-born and African American populations declined steadily. By 1900, immigrants accounted for only 5 percent of the city's population (see Appendix A). Germans were the largest contingent (40 percent), with the Irish, British (English, Welsh, and Scots), and Russians trailing, with percentages of from 14 to 16 percent each. African Americans had fallen to 8 percent of the population.

Despite the economic losses and widespread unemployment of the Panics of 1873 and 1893, homeownership appears overall to have increased in Harrisburg between 1870 and the end of the century. Although data in the censuses are not wholly comparable, 1,805 Harrisburgers (or one in seven adults) owned real estate in 1870. By 1890, of 8,311 dwellings in the city, one-third were owned and two-thirds were rented. Of the owned homes, only a quarter were encumbered by mortgages. The depression of the 1890s caused some slippage. By 1900 little more than a quarter of dwellings in the city were owned, and more than 40 percent of those were mortgaged.[4]

Industrialization continuously reshaped the outer appearances of Harrisburg during the last half of the nineteenth century. It also changed the composition of the population, occupational patterns among the residents, and the degree to which they owned homes. Beyond these aspects, however, did the coming of factories change the city in other ways, and what does Harrisburg's experience indicate about the impact of second-stage industrialization on communities across the nation?

The Destruction of Work Worlds That Never Were

When deindustrialization struck the smokestack industries of Pennsylvania and other rust-belt areas in the 1980s, it touched off lamentations at the passing of a newly discovered near-golden age for factory workers. Displaced mill hands and assembly line operatives, as well as those who provided services to them and to the firms they worked for, began to look back wistfully on the "good times" of the previous quarter-century. Then, workers had steady employment, high wages, job security, paid vacations, good health and other fringe benefits, and the promise of solid pensions for their old age. One could buy a home, equip it with the latest appliances, own a car, and even perhaps a cottage at the lake, or a recreational vehicle or boat. If they wanted to, the children could go to college. In retrospect, life had been comfortable and predictable.[5] Or so it was remembered. To those familiar with generations of ideological criticism of industrialization, such reactions may have come as something of a surprise. The closing of what once had been called "dark satanic mills" where workers were exploited, deprived of their humanity, and reduced to mere appendages of the machines they tended would seem rather an occasion for rejoicing than regret.

Major transformations in American economic life, such as industrialization and deindustrialization, have always caused enormous dislocations, considerable suffering, and great uncertainty for those living through them. But they also have created new opportunities. With the passing of time the worst features of such changes softened. For people able to adjust to the new order—and many could not—what started as a curse became beneficial, what was new and frightening became familiar, what had been disruptive became more predictable. Once past the initial disruptions, relatively few people, however nostalgic about life when they were younger, would have returned to the old order even if that were possible. Today, for example, except for the minority whose lives have been severely disrupted by deindustrialization, there appears to be no greater desire to restore the industrial order of the mid-twentieth century than industrial workers at that time longed to return to the life of the preindustrial artisan. At the same time, it should not be surprising that in the midst of such changes those experiencing the dislocation fashion worlds that never were from the combination of their wrecked plans, fears of what lay ahead, and selective memories.[6] Unfortunately, scholars may someday use these unreal portrayals of industrial America as the ideal against which to criticize post industrial society, much as some have latched on to an idealized world of independent artisans and yeoman farmers to measure the shortcomings of the industrial order.

According to critics of industrialization, the world of the factory hand compared most unfavorably with that of the craft worker it replaced. Gone were such features as control of one's time and the pace at which one worked when employer-imposed rules and time clocks took over. The harsher relationships of manager, foreman, and machine tender replaced the almost familial relationships of master, journeyman, and apprentice. Skilled masters no longer served as mentors and role models, nurturing youngsters through the difficult years of adolescence and carefully preparing them for productive lives. Skilled journeymen forced into factory jobs lost any particular advantage over unskilled laborers. As machines divided and subdivided skills, and workers no longer contributed more than a small part of any ultimate product, the dignity of labor and pride in one's work disappeared. When ownership of the tools of one's trade and mastery of a marketable skill faded, so did the independence of the artisan.[7]

The problem with this view lies less in its depiction of factory work, which all too often was harsh and grim, than with its idealized portrayal of artisan workers before the coming of factories.[8] The craft system in the United States had deteriorated long before the advent of industrialization. Indeed, like many European institutions, it had not weathered the passage to America well. Here a thin and widely scattered population, and shortage rather than abundance of artisans, forestalled the development of guilds and legal machinery for upholding craft standards. The "entrenched though anemic" institution soon withered. The Revolution and the rise of republicanism undercut the authority not only of king and bishops but also of lesser figures, including master craftsmen. The steady inflow of immigrants, many with skills, and the constant moving of Americans, made policing the system impracticable. A rash of printed manuals early in the nineteenth century often revealed to apprentices more of the mysteries of a trade than their masters could teach them. The introduction of cash wages gradually replaced the customary exchanges of services, mutual obligations, and goods between masters and apprentices. And, increasingly, half-trained apprentices threatened the jobs of journeymen who could not advance to masterships because of the high cost of setting up shops of their own.[9]

In many if not most communities, the half-mythical, self-reliant crafts-men never constituted a majority of those who worked with their hands. At least as many were servants or unskilled casual laborers who drifted from job to job. Only a minority of workers could possibly have enjoyed the benefits of the "world of the artisan." At Harrisburg on the eve of industrialization, for example, two-fifths of all male workers were un-specified laborers or servants, an eighth worked at the town's first indus-trial plant (the anthracite furnace), and slightly less than half were en-gaged in craft production. By 1860 a little more than two-fifths were "laborers" and servants, one-fifth were in industry, and not quite two-fifths were in craft production (see Appendix E).

Even those figures considerably exaggerate the number of traditional craftsmen. Those calling themselves "carpenters," for example, were self-selecting—that is they told the census taker they were carpenters. How many had gone through apprenticeship, formal or otherwise, as opposed to picking up the trade at their father's or an employer's side can only be guessed. In 1860, census takers at Harrisburg for the first time distinguished masters from journeymen and apprentices. More than three-fifths continued to carry the simple undifferentiated designation "blacksmith," "carpenter," "painter," or "tailor." Many of those may have learned the trade piecemeal rather than gone through an apprentice-ship. In other instances, the person answering the census taker's ques-tions may have been a member of the family who did not know the artisan's level of skill. In any event, of those who were differentiated, only one in five were masters, fewer than a fifth were apprentices, and slightly more than three-fifths were journeymen. Moreover, a fourth of the journeymen were at least thirty-five years old and unlikely ever to become masters.[10]

In the age of craft production, it was master artisans, not the other craft workers, who enjoyed self-sufficiency and independence. The scale of their operations certainly meant there would be more direct personal contact with their masters than would be possible in the later factories, but it cannot be assumed that the system necessarily or characteristically produced relationships that were more harmonious. A master's treat-ment of an apprentice, depending on a variety of factors including the temperaments of both, might be anywhere from abused and overworked servant to favored son. Journeymen, unless their wanderings from job to job are romanticized into "independence," were hired skilled employees who usually furnished their own tools and were paid piecework wages at the local market rate. Most probably never achieved the status of mas-ter.[11] In sum, only a small part of the preindustrial working class held the status of independent artisan-craftsman. So it was not a "world of the artisan" that industry finally undermined, but only the vested interests of the remaining masters.

What Industrialization Did and Did Not Do at Harrisburg

What industrialization did not do at Harrisburg was in many ways as important as what it did do. Certainly it did not eliminate the skilled trades or turn all craft workers into factory hands or machine tenders. A few crafts did fade away, and others did decline, but the number of artisans in most trades grew. Overall, artisans more than any other workers persisted in the community decade after decade, and more of them owned real estate. Although a significant portion of the journeymen took jobs in industry as factory artisans or foremen, most of the established masters were able to remain at their old stands and practice their old trade, even if profits declined. Some craft workers shifted to related trades or became storekeepers, while a handful of artisan entrepreneurs expanded their operations to quasi-factory scale. More artisans' sons than fathers found their way into industry. Approximately half, however, followed their father's trade or went into one of the other crafts.[12] Until thorough studies are made of artisans and their sons in many communities—both those who moved into industry and those who remained at their trade—the full story of the factory system's impact on craft workers will not be known.

Several other undesirable consequences often associated with industrialization did not develop at Harrisburg. Studies of industrial centers elsewhere have frequently found that large influxes of immigrant workers touched off interethnic-group rivalries for housing and jobs and produced hostility between established residents and newcomers.[13] Inasmuch as casual laborers, youngsters in the community, and migrants from nearby villages and farms easily supplied the demand for workers at Harrisburg, hordes of alien workers were not attracted there. Furthermore, there is little evidence that industrialization created a sullen, alienated, semi-permanent class of wage slaves in the community. Many of the first generation of industrial workers perceived the factories as offering opportunity, not exploitation. However unattractive they appear to middle-class observers, jobs at the cotton mill offered working-class youngsters a short-run chance to earn a cash wage. Few sought long-term careers there, instead moving after a brief stint to other jobs. And Hickok's Eagle Works and the Harrisburg Car Works offered training to unskilled young people that frequently led to better jobs in those plants or elsewhere. At the time and for many years after, workers perceived the railroads as good employers that paid well and offered secure jobs.

Of local industries that hired adults, iron and steel mills apparently offered the least attractive jobs, but they paid higher wages, offered steadier work than other local industries, and drove their workers less than the major iron and steel plants at such places as Pittsburgh.[14] Nonetheless, Harrisburg iron companies worked their employees hard for long hours under harsh conditions. More frequently than workers in any other local industry, ironworkers called strikes. Excepting only cotton mill operatives, they ranked lowest in rates of homeownership and persisted least in the community. Clearly they were more willing to migrate to other, presumably better, jobs elsewhere than to accept unfavorable conditions permanently where they were.

Iron strikes notwithstanding, a half-century of industrialization brought Harrisburg minimal worker demonstrations, walkouts, and labor violence. This was neither because wages were higher or conditions better than elsewhere, nor because Harrisburg workers were more docile or cowed by their employers. Two principal factors appear to have been significant: the existence of ways to escape factory work, and the relatively easy hand of the major employers. There is no evidence that workers especially liked factory work or longed for careers as industrial drudges. In fact, it was quite the opposite. Among the various groups examined in detail, longtime residents of the city of whatever background successfully avoided the mills, developing contacts that led to other kinds of employment for themselves and their sons. Those who did not find better occupations in Harrisburg merely moved on, as the low persistence rates for the least attractive factory jobs attest. That course was easier than staying, trying to organize unions or political movements, and fighting.[15]

The relatively easy hand with which Harrisburg's employers dealt with their workers no doubt contributed to the lack of labor militance. As a group, the city's industrialists were not noted for great efficiency, for seeking new ways to get more out of their employees, or for introducing new machines to reduce the need for laborers. This was in part because they were not on the cutting edge with Andrew Carnegie, Henry Clay Frick, or Charles Schwab; in part because Frederick W. Taylor's scientific management still lay in the future. They did not encourage unions, but neither did they wage war against them. By closing down in times of trouble instead of confronting strikers with scab labor and armed might, they forestalled violence and avoided unnecessary provocation of their employeees.

How Harrisburg's entrepreneurs came to this policy is not known. Logic—a poor substitute when documentation is not available—can only suggest possible motives. Bringing in strikebreakers would have involved finding them and being willing to bear the costs of recruiting, housing, and protecting them. Unlike many Pennsylvania coal barons and ironmasters who had big-city ties, Harrisburg's entrepreneurs were self-contained and had fewer outside contacts. The former—for the most part absentee owners who did not live in the coal patches or factory towns where the troubles occurred—broke strikes by importing immigrants or southern blacks.[16] Harrisburg's entrepreneurs would have opposed importing outsiders almost as vigorously as the strikers would have. One of the advantages or handicaps (depending on point of view) of centering their lives where their factories were located was that they were reluctant to face down people with whom they rubbed shoulders daily or to unnecessarily disrupt life in "their" small city.[17] Overall, their employees gave them little trouble, rarely confronted them or questioned their authority, and did not appeal to outside groups for help in times of trouble. Rebellious workers had to be dealt with firmly, but embittering them and undermining the long-run welfare of the community made no sense. This was especially true so long as it was evident how the difficulties would end. Had strong unions challenged Harrisburg's entrepreneurs or long strikes crippled their interests, they might well have adopted other tactics. How widespread such policies were in other industrial centers, especially where owners and their families lived in the community, once again calls for additional research.

The employers' policy of not confronting workers or using local law officers to break strikes tended to keep Harrisburg's laborers politically inactive. Because they did not see government as being used against them, they saw no need to wrest it from the hands of their employers.[18] Similarly, the absence of significant numbers of immigrants forestalled the rise of boss-ridden, ethnic political machines. The elite's tolerance of, rather than hostility to, broadly representative municipal government further depoliticized labor. By manipulating discreetly behind the scenes, they managed to avoid political contests with militant workers. And when something important had to be done politically, the entrepreneurs used their intimate connections with the Cameron machine to acquire backdoor access to the state legislature. This allowed them to accomplish many objectives that could not be attained through the local political machinery. The tactic had the further advantage of shunting blame from themselves to the General Assembly.

The positive effects of industrialization were as important to Harrisburg as what was avoided. The coming of factories, for example, provided an economic base that could support a much larger community. Each stratum of the class structure became more diverse. Among the working classes, the proportion of unspecified laborers declined, both because they moved into new occupations and because new job titles were given to tasks once performed simply by "laborers." Growing prosperity among the middle and upper classes added to the number and variety of personal service positions. To casual laborers, servants, and craft workers were added at least an equal number of industrial workers and railroad employees. These included cleanup crews, haulers, machine tenders, factory artisans, master mechanics, and scores of new semi-skilled and skilled workers. Although jobs may have become more disciplined and regimented, they were also steadier and provided more regular pay.

The other classes also expanded in number and scope. Industrial foremen and recordkeepers, and railroad telegraphers, clerks, agents, and conductors, joined the ranks of white-collar and clerical workers. Added to the middle classes were small-scale industrialists and factory and railroad managers and administrators. At the top, company presidents and general managers joined the wealthy merchants, bankers, landowners, and professionals. Among all classes, industrialization was accompanied by increases in the percentages of persons who owned real estate and in the average value of those holdings.

Yet another significant change over time involved the succession of the sons and grandsons of the entrepreneurs to leadership positions. The younger men were more thoroughly educated and far more interested in civic, cultural, and philanthropic matters than their elders. Continuing control by the same families who had lived in the town for two or three generations kept much of the wealth produced at Harrisburg within the community. The city likewise benefited from the paternalistic dedication of the entrepreneurial families to the city's well-being. By contrast, absentee owners usually dealt with the communities in which their firms were located through resident managers. Wealth produced in those communities was drained away, and control of the community's economic base was almost entirely divorced from local political control.[19] The sons of the entrepreneurs, on the other hand, tended to be less innovative in business and less focused on industry. When the time came, they dutifully took over the enterprises started by their fathers. However, with the exception of the Baileys, who would try unsuccessfully to modernize their family mills, most presided over increasingly obsolete plants that eventually fell victim to national-scale competitors from the outside.

Toward a New Analytic Framework

Again, if the consequences of industrialization on Harrisburg had been unique, little more would need to be said. But they were not unique. Overall, developments in many if not most of the middle-size and smaller cities of the Northeast in the second half of the nineteenth century were essentially similar, however different they might be from one another in details. Albany, Reading, Trenton, and Wilmington illustrate the point once more (see Table 14.1). For example, in each of those cities industrialization created many new jobs and brought in large numbers of newcomers. The great majority of these were native-born American whites, although at Albany by 1880, immigrants did constitute a full quarter of the population. African Americans made up a little more than 10 percent of the populations of Albany and Wilmington, 5 percent at Trenton, and less than 1 percent at Reading. The percentage of the entire population engaged in manufacturing was larger in all four cities than it was at Harrisburg.

As at Harrisburg, the kinds of industry introduced in these places posed little threat to existing artisans and in fact often enhanced their job opportunities. Trenton's pottery mills, iron rope and cable manufactories, and rubber plants displaced few if any existing craft shops. Wilmington's car and shipbuilding industries, and the railroad maintenance and repair shops at Reading and Albany, not only did not interfere with existing crafts but they also created large numbers of additional jobs for carpenters, painters, blacksmiths, and machinists. The one notable exception was Reading, where wool-hat factories over time displaced what had once been the leading craft shop trade.

The degree to which labor organized before 1900 followed different

Table 14.1. Percentage of Foreign-Born, Blacks, and Persons Engaged in Manufacturing at Harrisburg, Albany, Reading, Trenton, and Wilmington, 1880

City	Population	% Foreign-Born	% Blacks	% Population in Manufacturing
Harrisburg	30,762	7.5	9.4	11.9
Albany	90,758	26.2	11.7	13.0
Reading	43,278	8.3	0.8	15.5
Trenton	29,910	19.1	4.6	15.5
Wilmington	42,478	13.4	12.9	18.5

Source: *1880 Census, Social Statistics of Cities*, Part 1, pp. 447, 458 (Albany); pp. 728, 731 (Trenton); pp. 762, 767 (Harrisburg); pp. 876, 880 (Reading); 903, 909 (Wilmington).

lines. Albany and Trenton had vigorous union traditions; conservative, Pennsylvania-German Reading and Wilmington—more like Harrisburg—did not. For example, Albany's workers supported the eight-hour day movement, established worker co-ops, and crusaded for an end to the prison labor contract system. Trenton, where in the pre–Civil War era many workers had belonged to unions, saw a larger number of workers enrolled in the Knights of Labor in the 1880s and later in various craft unions of the American Federation of Labor.

All the cities, both union and nonunion, endured occasional strikes and other worker demonstrations and movements. During the nineteenth century, however, none suffered protracted labor disputes or significant ethnic or class warfare. Workers involved themselves only sporadically in third-party political movements. The majority of Albany's workers, for example, seem to have accepted the notion of a community of interest with their employers.[20] Trenton, in spite of its well-known labor tradition, had few difficulties before the twentieth century. From 1884 until 1923 employers there tacitly recognized the pottery workers' union and bargained collectively with it regarding terms of employment.[21] The one serious labor confrontation at Reading, during the railroad strikes of 1877, stemmed in large part from the failure of public officials to act responsibly and from the vigor with which the line's president, Franklin B. Gowen, waged war on the demonstrators from his Philadelphia headquarters.[22] A Wilmington historian describes late-nineteenth-century labor relations there as "tranquil." Several of the factors that explain why would apply with equal force to Harrisburg and perhaps to the other three cities: (1) a large percentage of the factory workers were skilled, (2) the industries involved did not downgrade or threaten those workers' skills, (3) there was no large pool of immigrant laborers to inflame the situation, and (4) the employers, most of whom lived in the community, followed a policy of restraint in their disagreements with laborers.[23] Employers usually counted on their superior economic positions to outwait any strikers. Except for the absentee manager Gowen at Reading in 1877, resort to strikebreakers and force against workers was infrequent. Unions, though unwanted by employers, were grudgingly recognized when necessary rather than fought to extermination.

If nothing else, the experiences of Harrisburg and these other cities suggest that a fuller understanding of both the process and the impact of industrialization on American cities and on their residents requires more research and a broadened framework for analysis. For example, there must be more studies of second-stage manufacturing centers, with a focus on how they developed and the impact on the people involved. If Harrisburg was at all typical, it appears that industrialization at such places was in several ways fundamentally different from what happened at primary-stage centers and produced very different consequences. This and other recent studies demonstrate that it is clearly inadequate to say that a community "industrialized" and to assume that the process and its consequences are self-evident. The process, its consequences, or both may vary considerably from one community to the next. At the same time, to conclude that the unique character of each community's experience precludes meaningful generalizations is probably also incorrect. Future research needs to make allowance for a wider range of factors and situations, the more significant of which are:

- Whether the community was founded directly as an industrial center or evolved out of an existing commercial or preindustrial community
- Whether it developed around a single industry or emerged as a more diverse manufacturing center
- Whether the entrepreneurs who brought about the transformation arose out of the local elite or came from the outside
- Whether these leaders continued to live where their factories were located, took pride in the community, and assumed a paternalistic interest in its welfare for a generation or more, or simply exploited it as absentee owners through the agency of hired managers
- Whether the new manufacturing was innovative and driving first-stage industrialization, brought about by entrepreneurial pioneers, or second-stage, derivative industrialization designed to imitate the success of others
- Whether the new factories displaced, strengthened, or had little impact on existing craft production in the community
- Whether the newly created jobs went to local residents and migrants from the immediate vicinity, or to workers of very different racial, ethnic, or cultural backgrounds, or to those who came from distant regions or foreign countries
- Whether new ethnic working groups were large enough to form neighborhoods of their own, and whether they vied with one another and with longer-term residents of the community for homes and jobs
- Whether employers fostered labor hostility by driving workers harshly, treating them as outsiders, and otherwise alienating them, or whether they treated them with respect in spite of their inferior social and economic status
- Whether employers reacted to the rise of unions by waging warfare

against them, quietly ignoring them, or, when expedient, negotiating with them

- Whether employers reacted to strikes, stalemates with unions, or labor violence by closing down the mills and waiting out the troubles, or whether they met such problems with force, attempting to displace rebellious workers with strike breakers, or turned to government agencies for assistance to crush rebellions.

- Whether the employing class ignored the political life of the community or either openly or covertly managed it.

- Whether ethnic political machines took control of the management of local government

- Whether workers, frustrated by their employers' use of local government against them, organized politically to seize control at the ballot box and institute labor reforms.

Scholars have already sketched out the broad outlines of what industrialization has meant to American society. The full account of that central development in the life of the nation, however, remains to be explored and formulated. To date, too many studies have focused on the problem from the perspective of either the entrepreneurs or the workers, but not both. Primary-stage instead of second-stage industrial centers and major cities, rather than small or medium-size communities, have received the greatest attention. And despite a growing number of works on individual towns or cities, few have compared experiences across communities.[24] Until more research is done in these neglected areas, no comprehensive, systematic synthesis of the industrializing experience will be possible. If the findings of future scholars reveal that the process and impact at Harrisburg was common to that of many smaller and middle-size communities, it will go far to explain the lack of militance and radicalism in the American labor movement.

Epilogue

The burning of the State Capitol on February 2, 1897, and a stroke that crippled Colonel Henry McCormick that same day together symbolize the shift in Harrisburg's economy from manufacturing to government administration. Although McCormick lingered on for another three years, others had to take over his responsibilities. Because none of the rising generation of the family had much interest in ironmaking, less than three months after the colonel's stroke they were uniting their ironworks with that of the Baileys. Although the merged firm would last another half-century, iron manufacturing in Harrisburg was in retreat. Meanwhile, the legislature authorized an imposing new structure to replace the old capitol. The size, grandeur, and cost of the building, and the manner in which it loomed over the city, foreshadowed the greater role state government would thereafter play in the economic life of the community. The percentage of local residents engaged in manufacturing declined slowly at first, then more sharply in the years that followed. Those losses, however, were more than offset by gains in the governmental, banking, and service sectors.

How Harrisburg deindustrialized after 1900, and the impact of that process on its residents, would be a worthwhile research project, but it is beyond the scope of the present book. The remainder of this study can only trace briefly the decline, demise, or continuation of the various firms that dominated Harrisburg's industrial age at its peak. Although the founding entrepreneurs, and in most instances their descendants as well, have passed from the scene, a few of the mills, banks, and firms they founded remain in business. The venerable cotton mill, after shifting to silk production in the 1880s, limped along on slim profits for another half-century. During the Great Depression it was torn down to make way for a large YMCA building. By the early twentieth century the Pennsylvania Railroad, finding the city's industrial corridor too cramped, moved to new yards and repair shops at Enola, just across the Susquehanna. The greatly reduced use of those facilities today stands in marked contrast to the near-monopoly once exercised by railroads over the area's transportation. Even with abbreviated rail service, however, Harrisburg remains a major East Coast transportation center, boasting an international airport, several interstate highways passing near or through the area, and enormous trucking terminals in and around the city.

The once-dominant iron mills of the Baileys and the McCormicks faded more slowly. As the second generation of Baileys succeeded Colonel Henry McCormick, Charles L. Bailey, and Gilbert McCauley in 1901, the nation's iron and steel industry was in the throes of a profound restructuring. The United States Steel Corporation, founded that year with a capital of $1 billion, led the way. It brought under single control the properties of Andrew Carnegie, a number of large fabricating mills financed by the great banker J. P. Morgan, and several other major firms. The resulting combine, controlling approximately two-thirds of the nation's basic steel output, was wholly integrated, operated in all facets of steel manufacturing, and had facilities in all principal marketing regions. Apparently U.S. Steel briefly considered purchasing Central Iron & Steel but decided otherwise. If Central was to survive, it had to adapt.

Shortly before his death, McCauley advanced several schemes to save the firm. To secure its share of the market, he entered into a pooling arrangement with the other American plate manufacturers. The agreement allocated 8 percent of the market to Central. Because steel, the basic raw material for plate, was becoming scarce, McCauley urged the firm to erect its own plant.[1] When Edward and James Bailey came to power, they adopted his advice, in effect attempting to catch up with the rest of the industry. At the 1901 stockholder's meeting, Edward Bailey reported that neither United States Steel nor the two major independent producers—Jones & Laughlin and Sharon Steel—any longer had metal to sell to Central. The one other source, firms controlled by the Pennsylvania Railroad (such as Pennsylvania Steel), demanded $26 a ton and still could not supply all Central needed. Accordingly, the company voted to increase its capital stock from $1 million to $5 million. All but $45,000 worth was taken up by the firm's current few stockholders. Central promptly installed four fifty-ton open-hearth furnaces to produce steel. It got pig iron from its own Paxton furnaces (at this point, Lochiel Furnace belonged to Pennsylvania Steel) and from European sources. However, more trouble lay ahead. Iron ore was in short supply, and the company's primary supplier of coke, the Frick Company (now part of United States Steel), announced it could no longer spare coke for outsiders. Accordingly, in 1906 Central bought control of the Mohawk Ore Mine in Minnesota for $120,000, the Connellsville Basin Coke Company in West Virginia for $650,000, and fifty pressed-steel cars at a cost of $700,000 to haul coke to Harrisburg.[2]

By 1903, Central owned 66.2 acres of real estate, 4.5 miles of railroad track, and fifty new coal cars. On its land were a puddle mill, a nail plate mill, a nail factory, three plate mills, a universal mill, and two blast furnaces. Four open-hearth furnaces were under construction. Central employed 1,640 men at more than $533,400 in annual wages (about $325 per worker) and produced 133,100 tons of steel a year.[3] Although now ready to compete with the giants of the industry for a small share of the market, the effort was too little and had come too late. U.S. Steel's position in the industry was overwhelming. Whenever the market soured and independent companies slashed prices to capture customers, U.S. Steel simply undersold them.[4]

Meanwhile, in an apparent attempt to end internal friction, James Bailey in 1908 purchased the stock held by the McCormicks. He paid them only a small amount in cash, giving them his note for the balance, and used their former stock as collateral. The McCormick-Cameron bloc all promptly resigned from the board. Complete control by the Baileys was brief. Apparently the great expense of modernization impaired the value of the capital stock and prevented further payments to the McCormicks. At the creditors' request, receivers took over in 1912. By 1916 the McCormicks had recovered their stock, purchased the holdings of one of the New York allies of the Baileys, and thereafter controlled the company for the remainder of its existence.

In fact, however, the arrangement restored uneasy joint-ownership by the two families. The McCormicks and the Camerons held three seats on the board, and the Baileys had two. Robert H. Irons, who began at Central when he was sixteen and rose to the general superintendency, became the new president. Only during World War I and II did the firm earn solid profits. Between 1926 and 1928 it tore down the last of Harrisburg's anthracite blast furnaces, Paxton Nos. 1 and 2 and Lochiel (reacquired in 1917). All were obsolete. Vance McCormick, Colonel Henry's son, headed Central Iron & Steel from the death of President Irons in 1939 until the Barium Steel Corporation purchased the properties in 1946. During the 1960s the Phoenix Steel Company of Claymont, Delaware, bought up Central, but when it could not turn a profit it dismantled what remained of the mills. The last steel was poured in Harrisburg in 1975.[5]

The one continuing important producer of iron and steel in the Harrisburg area was and still is the large steelworks at Steelton. Pennsylvania Steel weathered the difficulties of the 1890s but lost its independence in 1915, when the Bethlehem Steel Corporation purchased the firm. Under the leadership of Charles M. Schwab, Bethlehem invested some $15 million in improvements at Steelton, just in time to reap high profits during World War I. With other steel-producers, the company faced the strictures of the early 1920s, the upturn after 1923, and the bleak depression decade of the 1930s. Another period of prosperity came with World War II and continued into the postwar era. Although suffering along with the other American steel producers in recent years, the Steelton plant in 1988 still provided jobs for 2,000 residents of the greater Harrisburg area.[6]

Only two of the city proper's major nineteenth-century industrial firms remain in business today: the successors to the Harrisburg Car Manufacturing Company and to Hickok's Eagle Works. The Harrisburg Pipe & Pipe Bending Company began repairing refrigerator cars in one of the shops of the failed car works. The venture proved profitable, and the firm expanded. Just ahead of World War I it undertook the manufacture of cylinders and automobile pistons. The war brought a flood of business, especially shells for the military and cylinders for a wide variety of military and civilian uses. In the 1920s and early 1930s the company's main product continued to be cylinders, but they were now used as containers for carbonated water chargers and for fire extinguishers. Because it had not made a foot of pipe since 1913, the company changed its name to the Harrisburg Steel Company in 1935. Not long after, it began producing bombs for various armsmakers. During World War II the company became a major producer of shells, bombs, and other war materiel for the U.S. government. Today, known as the Harsco Corporation, it is a major defense contractor—a manufacturer of primary metals, construction products, and fabricated metals. Its new headquarters are in Wormleysburg, directly across the river from Harrisburg, where it still operates a plant on the site of the old car works. The bulk of the firm's operations, however, are at other sites in eight states and Great Britain. Of its total work force of 12,000, only 2,600 are located in the Harrisburg area.[7]

The Hickok Manufacturing Company alone is controlled and managed by descendants of its founder. Grandson William Hickok III, after completing his education at Yale in 1895, returned to clerk in the office of the Eagle Works. The next year he became a member of the board of directors, and in 1906 president of the firm. His younger brother Ross followed the same route: Yale, work in the company office, then on to a high managerial position. After William died in 1933, Ross served as president until his own death in 1943.[8] At present, great-great-grandsons of the founder, bearing the family name and educated at Yale (with majors in history), continue to manage the firm. Although having no more employees than in the post–Civil War era, the company is still an important producer of ruling machines.

Having served as midwives at the birth of the city's first important industrial firms, Harrisburg's two major banks lived to help bury or transform them and now do business with their successors. After a series of mergers, the Harrisburg National Bank became the Commonwealth National Bank and since 1991 has belonged to Mellon Bank of Pittsburgh. The Dauphin Deposit Bank & Trust Company has preserved a variation of its original name. In the late nineteenth century the entrepreneurial families contolled both the major industries and the major banks. As manufacturing went into decline, the sons of the entrepreneurs gradually abandoned their factories, taking refuge in their more profitable banks. The McCormick family's 105-year domination of the Dauphin Deposit Bank ended with the death of Donald McCormick in 1945. The Reily-Bailey dynasty at the Harrisburg National and its successor ran until displaced in the late 1970s. In recent years, with deposits as the measure, the Dauphin Bank has outranked Commonwealth National. In 1990, with deposits of more than $1.9 billion, Dauphin ranked 201st in the United States. Commonwealth National, with a little more than $1.2 billion, ranked 275th.[9]

Just as Harrisburg's economic base changed, so did its population. The city's 30 percent growth rate per decade between 1900 and 1910 was half what it had been in the era of industrial development (see Appendix A). During the prosperous 1920s, the depression-torn 1930s, and the war and reconversion periods of the 1940s, the rate slackened to between 5 and 6 percent a decade. The city's population peaked at 89,544 in 1950, then dropped sharply and at an accelerating rate: 11 percent in the 1950s, 15 percent during the 1960s, and another 22 percent by 1980. In the 1980s the drain slowed, and preliminary census data for 1990 indicate a total population of 52,376, or 2 percent fewer than ten years before.

The racial and ethnic makeup of the community also changed. The decline in the proportion of foreign-born residents continued, sinking from 5 percent at the turn of the century to 3 percent in both 1980 and 1990. The one exception was between 1900 and 1910 when enough of the "New Immigrants" briefly swelled the total from 5 percent to 7 percent of the population. Immigration after 1920, without halting the overall decline, brought in new groups. Slavic and Mediterranean peoples, rather than Irish, Germans, or British, became dominant. Since

1950, Hispanic and Asian peoples have steadily increased their proportions in Harrisburg.

Although the number of African Americans grew steadily decade by decade throughout the city's history, their proportion of the total population declined gradually, from a high of 11 percent in 1850 to a low of 7 percent by 1920. During the 1920s, however, a reversal that has continued to the present set in. By 1950, blacks once more constituted 11 percent of the city's population. During the next three decades, as the population of the city proper declined, the number of African American residents increased by nearly 5,000 persons a decade. As in the post–Civil War period, many were from the South and again divided the black community between established older residents and impoverished, less well educated newcomers. By 1960 nearly one resident in five was black, by 1970 almost one in three, and by 1980 slightly fewer than 45 percent. Preliminary census figures for 1990 indicate that today African Americans constitute slightly more than half of the population.

Politically, the absence of ethnic-based boss domination did not prevent Harrisburg from having a "reform" movement early in the twentieth century. Inspired in part by the rising new capitol building, it took the form of the "City Beautiful" movement. Corruption and "bossism" were not the concerns of these reformers. They complained of a do-nothing government that failed to make the city attractive, clean up the riverfront, drain miasmic swampland, establish a system of parks, or provide clean, healthful drinking water. Mira Lloyd Dock, of an established and financially independent Harrisburg family, launched the movement in December 1900 when she addressed a meeting of the Board of Trade and its guests, calling for cleaning up the city and developing its natural beauty. A combination of entrepreneurial, business, professional, and old family elites provided leadership that crossed traditional party lines. They organized, raised private funds to finance the drafting of a comprehensive plan, and skillfully campaigned for passage of a city bond issue to pay for the improvements. With Vance C. McCormick as their candidate for mayor, and with support from virtually all important shapers of public opinion—the press, the churches, civic organizations, and business leaders—they won. For two decades the movement worked at improving the city practically as well as aesthetically, paving streets, building new bridges, putting in an adequate sewer system, creating parks, and beautifying the riverfront. By 1914, four bond issues had raised nearly $2.5 million for plan projects. Although the movement spent much of its force by World War I and its founders moved to other projects, by 1926 another $250,000 had been spent on related facets of the plan.[10]

Fig. 26. The New State Capitol, 1906. Although legislative sessions were held in Harrisburg from 1812 on, the Old Capitol was not completed until 1822. When it burned in 1897, this imposing domed structure was erected in its place.

Following Vance McCormick's term as mayor, the home base of the Camerons at last turned solidly Republican. Except for the years between 1911 and 1915, Republican mayors presided at City Hall until 1970. Briefly they continued the reform tradition. Democrats and Republicans alternately filled the mayor's office until 1982; since that year, only Democrats have held the post. At the state level the machine created by Simon Cameron remained in charge until 1921, but once the Camerons passed, the levers of power fell to non-Harrisburgers, and the city lost that special relationship with the General Assembly that had so favored its elite.

Present-day Harrisburg is now only the historic vestige of the nineteenth-century city of this study. The relevant entity now is the census bureau's "standard metropolitan area," which includes parts of four neighboring counties. Published federal census schedules show that as the city proper declined in population this larger entity grew from 345,071 residents in 1960 to 587,966 in 1990, an increase of 70 percent. The city population itself makes up an ever-smaller part of the whole: 23 percent in 1960, then 12 percent two decades later. As for immigrants, the largest national groups in the standard metropolitan area in 1980 were Italians (11.4 percent of the foreign-born) and Germans (10.6 percent), but the Slavic peoples combined outnumbered both, and East Asians (18 percent) and Hispanics (6.5 percent) were increasingly noticeable. Even so, the total foreign-born residents amounted to only 2 percent of the entire population, too small a group to wield much influence or to constitute a problem.[11]

Not so with race. Harrisburg resembles many large metropolitan areas in that African Americans constitute more than half the city's population but only a small fraction of the metropolitan area. This has created problems in both places. One problem is how to devise satisfactory working relationships with the white suburbanites who dominate the metropolitan area. The other problem, within the city, has to do with integration, civil rights, and racial tension.

Today, each weekday morning between 7:00 and 9:30, Harrisburg seems to come to life as many, if not most, of its work force hurry from homes in the suburbs to jobs in the city. Then, about 3:30 in the afternoon the process reverses and by 6:00 in the evening life at the heart of the city appears to be suspended. Harrisburg's economic base no longer rests on manufacturing but on government administration and services. Its largest employer, the Commonwealth of Pennsylvania, provides jobs for 23,500 persons living in the greater Harrisburg area, which includes most of Dauphin County and parts of Lebanon, Lancaster, York, Cumberland, and Perry counties. Medical and educational facilities rank high among the area's other major employers. So do military supply facilities and firms that provide other services. The six largest manufacturing companies combined (AMP Incorporated, Hershey Foods, Kinney Shoes, the Harsco Corporation, IBM, and Bethlehem Steel at Steelton) hire fewer employees (22,817) than the state government does.[12] Of the 53,264 residents in the city proper in 1980, some 21,131 (almost half of them women) were employed. Public administration provided work for more than 25 percent of them, 32 percent were in the professions or services, and 15 percent were in the wholesale or retail business; only 18 percent were in manufacturing and construction.[13]

Industrialization was but an interlude in Harrisburg's history. It came later than to many comparable places, forced residents to adapt to a new way of life, and then, within a century, ebbed away. As in many similar communities, outmoded local industries could not hold their own against newer and larger firms operating on a national scale for national markets. Albany, like Harrisburg, also began to slip in the 1880s, "left behind as the locus and scale of industry shifted" and its entrepreneurs failed to generate that type of manufacturing.[14] Trenton's industries persisted much longer, not seriously losing ground until the 1920s. In spite of heroic and imaginative efforts, however, local industries in the end lost out to national firms from the outside.[15] For Harrisburg, Albany, and Trenton, each the capital of a wealthy state, at least some of the economic damage was offset by the dramatic growth of government that created new occupations and income. Fortunately for Wilmington, Delaware, as its old industrial order began to fall into decline at the turn of the century, the DuPont Company chose that city to be the administrative and research center of its extensive chemical empire. Moreover, the DuPont family took a great interest in and contributed generously to many aspects of community life. The effect has been much the same as for the state capital cities, replacing industrial workers with upper and middle managers, professionals, and white-collar workers.[16] The city of Reading has thus far found no enduring substitute for iron manufacture and the railroad industry. Between 1900 and 1930 the city moved to light industries—manufacturing bicycles, hosiery, cigars, and pretzels; next came tourism, and empty mills were converted to factory outlet stores.[17] Meanwhile, the city has provided housing and services to large numbers of people who earn their living in the greater Philadelphia area.

Today many of the great American industrial centers that squeezed Harrisburg and its counterparts out of manufacturing face a similar experience as international industrial complexes both here and abroad challenge their existence. How they will survive the loss of livelihoods and the shifting to new economic activities that they forced on smaller cities earlier in the century remains to be seen.

Appendixes

LIST OF APPENDIXES

Appendix A

Percentages of Native Whites, Blacks, Foreign-Born, Irish-Born, and German-Born Residents, Harrisburg, 1790–1990

Year	Total Population	% American-Born Whites	% Blacks	% Foreign-Born	% Irish-Born[a]	% German-Born[a]
1790	875	n/a	3.0	n/a	n/a	n/a
1800	1,472	n/a	4.1	n/a	n/a	n/a
1810	2,287	n/a	2.6	n/a	n/a	n/a
1820	2,990	n/a	5.9	n/a	n/a	n/a
1830	4,312	n/a	11.4	n/a	n/a	n/a
1840	5,980	n/a	10.8	n/a	n/a	n/a
1850	7,834	78.2	11.3	10.5	5.4	4.5
1860	13,405	78.7	9.9	11.5	4.2	6.0
1870	23,104	78.1	9.8	12.1	3.8	5.4
1880	30,762	83.0	9.4	7.5	n/a	n/a
1890	39,385	84.4	9.1	6.4	1.4	3.0
1900	50,167	86.8	8.2	5.0	0.8	2.0
1910	64,186	86.5	7.1	6.4	0.5	1.3
1920	75,917	87.6	6.9	5.5	0.3	0.7
1930	80,339	87.4	7.9	4.6	1.2	2.6
1940	83,893	87.4	8.7	3.9	0.1	0.3
1950	89,544	85.4	11.3	3.2		
1960	79,697	77.9	18.9	3.1		
1970	68,061	66.6	30.7	2.7		
1980	53,264	53.9	43.6	2.5		
1990	52,376	n/a	50.1	n/a		

SOURCES: Compiled from published census data except for 1850, 1860, and 1870, which are based on computerized population census schedules.

n/a = not available.
[a]Irish and German percentages for 1920–90 were miniscule.

Appendix B
Rate of Population Growth by Decade, 1790–1990

Decade	% of Growth	Decade	% of Growth	Decade	% of Growth
1790–1800	68.2	1860–70	72.4	1930–40	4.4
1800–1810	55.4	1870–80	33.1	1940–50	6.7
1810–20	30.7	1880–90	24.8	1950–60	−11.0
1820–30	44.2	1890–1900	30.7	1960–70	−14.6
1830–40	38.7	1900–1910	27.9	1970–80	−21.7
1840–50	31.0	1910–20	18.3	1980–90	−1.7
1850–60	71.0	1920–30	5.8		

SOURCE: Calculated from Total Population column of Appendix A.

Appendix C
Changes in General Occupational Categories, 1850, 1860, 1870

	1850	1860	1870
Total listed occupations	2,336	4,203	7,443
Categories			
Major manufacturers, executives, managers	4.2%	5.5%	4.1%
Manufacturers: merchant and small	5.1	4.5	4.6
Professionals	3.5	2.5	2.1
Clerks, salespersons, semi-professionals	12.9	11.2	12.0
Workers	73.8	76.4	76.7
Miscellaneous	0.4	0.7	0.5

SOURCE: Computerized population census schedules, 1850, 1860, 1870.

Appendix D
Growth of Population and Work Force, 1850, 1860, 1870

	1850	1860	1870	Increase 1850–70
Total population	7,834	13,405	23,105	15,721
Percent of increase	—	71.0	72.4	194.9
Total listed occupations	2,336	4,203	7,443	5,107
Percent of increase	—	79.9	77.1	218.6
Percent of total population	29.8	31.4	32.2	—
Total workers (population schedules)	1,725	3,209	5,710	3,985
Percent of increase	—	86.0	77.9	231.0
Percent of total population	22.0	23.9	24.7	—
Percent of listed occupations	73.8	76.4	76.7	—
No. of industrial workers (population schedules)	206	661	1,481	1,275
Percent of increase	—	220.9	124.1	618.9
Percent of total population	2.6	4.9	6.4	—
Percent of listed occupations	8.8	15.7	19.9	—
Percent of total workers	11.9	20.6	25.9	—
No. of industrial workers (manufacturing schedules)	112	562	2,239	2,127
Percent of increase	—	401.8	298.4	1899.1
Percent of total population	1.4	4.2	9.7	—
Percent of listed occupations	4.8	13.4	30.1	—
Percent of total workers	6.5	17.5	39.2	—

SOURCE: Computerized population and manufacturing census schedules, 1850, 1860, 1870.

Appendix E
Craft, Industrial, and General Workers, 1850, 1860, 1870

	1850	1860	1870
Total Workers	1,725	3,209	5,710
Percent of increase		86.0	77.9
No. of Craft workers	853	1,239	1,684
Percent of increase		45.2	35.9
Percent of total workers	49.4	38.6	29.4
No. of industrial workers	206	661	1,481
Percent of increase		220.9	124.1
Percent of total workers	11.9	20.6	25.9
No. of general workers	666	1,309	2,545
Percent of increase		96.5	94.4
Percent of total workers	38.6	40.8	45.6
No. of "laborers" (unspecified)	495	547	1,411
Percent of general workers	74.3	41.8	55.4
No. of servants	126	747	1,115
Percent of general workers	18.9	57.1	43.8
No. of others	45	15	19
Percent of general workers	6.8	1.1	0.7

SOURCE: Computerized population census schedules, 1850, 1860, 1870.

Appendix F

Real Estate Holdings by Selected Categories, 1850, 1860, 1870

	1850		1860		1870	
Total value	$3,559,200		$8,529,300		$14,605,400	
By ranking	Landowners (No.)	% of Value	No.	% of Value	No.	% of Value
Top 1%	5	34.0	15	31.1	18	20.7
Top 5%	23	55.7	73	55.9	90	47.6
Top 10%	45	67.5	146	69.3	181	61.1
Top 20%	90	79.3%	292	81.4%	361	74.8%
2nd 20%	90	11.4	292	10.6	361	12.5
3rd 20%	90	5.5	292	4.9	361	6.1
4th 20%	90	3.0	292	3.2	361	3.8
5th 20%	91	1.4	292	1.6	361	2.0
Total landowners	451		1,460		1,805	
% of adults	10.6		20.7		14.4	
Landowners by gender						
Males	347	91.4	1,203	81.1	1,501	81.4
Females	104	8.5	257	18.9	304	18.6
Landowners by race and ethnicity						
White American-born	373	94.7	1,089	89.0	1,236	85.3
African American	30	0.6	109	1.2	69	0.7
Foreign-born	48	4.7	262	9.8	500	13.9
German-born	26	1.3	189	2.4	281	9.2
Irish-born*	16	2.1	103	6.8	152	4.3

SOURCE: Computerized population census schedules, 1850, 1860, 1870.

*To avoid distorting average Irish holdings, three exceptional holders in 1850 and 1860, and two in 1870, collectively possessing 73 percent, 64 percent, and 33 percent of the whole for those years, were excluded.

Appendix G

Real Estate Holdings of Persisters and Newcomers by Race and Ethnicity, 1850, 1860, 1870

	All Residents	Blacks	Irish*	Germans
1850				
Total real estate holders	451	30	16	28
Percent of all adults	11	6	5	10
Total real estate values	$3,559,200	$20,100	$30,600	$115,100
Average holding	7,892	670	1,913	1,714
Median holding	2,000	600	1,200	1,400
1860				
Persisters from 1850	n/a	15	31	38
Average holding	n/a	$1,738	$3,838	$3,942
New holders, 1860	n/a	91	84	133
Average holding	n/a	$784	$1,118	$1,429
Total real estate holders	1,460	106	115	171
Percent of all adults	21	15	25	26
Total real estate values	$8,529,350	$97,300	$212,900	$339,800
Average holding	5,842	918	1,815	1,987
Median holding	1,500	800	900	1,000
1870				
Persisters from 1850	n/a	7	10	33
Average holding	n/a	$2,286	$5,520	$10,858
Persisters from 1860	n/a	17	29	47
Average holding	n/a	$1,612	$2,355	$5,004
New holders, 1870	n/a	45	111	207
Average holding	n/a	$1,364	$2,702	$3,605
Total real estate holders	1,805	69	150	287
Percent of all adults	14	5	19	25
Total real estate values	$14,605,400	$104,800	$423,400	$1,349,000
Average holding	8,092	1,519	2,823	4,700
Median holding	2,500	1,000	1,500	2,500

SOURCE: Computerized population census schedules, 1850, 1860, 1870.

*To avoid distorting average Irish holdings, three exceptional holders in 1850 and 1860, and two in 1870, collectively possessing 73 percent, 64 percent, and 33 percent of the whole for those years, were excluded.

n/a = not available.

Appendix H

Birthplace, Ethnicity, and Race of Harrisburg Residents,
1850, 1860, 1870

	1850		1860		1870	
	No.	%	No.	%	No.	%
Total population:	7,834	—	13,405	—	23,105	—
Born in Pa.	6,550	84	11,332	85	18,205	79
Born in other states	460	6	538	4	2,107	9
Foreign-born	824	11	1,535	11	2,793	12
Total whites	6,948	89	12,079	90	20,834	90
Born in Pa.	5,754	83	10,135	84	17,067	82
Born in other states	362	5	409	3	974	5
Foreign-born	824	12	1,535	17	2,793	13
Total blacks	886	11	1,326	10	2,271	10
Born in Pa.	788	89	1,197	90	1,138	50
Born in other states	98	11	129	10	1,133	50
Total foreign-born	824	11	1,535	11	2,793	12
Irish	421	51	567	37	886	32
German	350	42	809	53	1,250	45
British	28	3	105	7	549	20
Other	25	3	54	4	108	4

SOURCE: Computerized population census schedules, 1850, 1860, 1870.

Appendix I

Industrial Workers by Birthplace, Race, Age-Group, Household
Status, Real Estate Ownership, and Persistence, 1860, 1870

Industrial Workers	Ctn Mill	Eagle Wks	Car Wks	Iron	RRs	Total
1860						
Total Employees	300	58	140	91	n/a	n/a
Percent in census	54	57	11	96	n/a	n/a
No. identified in census	161	46	16	87	159	469
Birthplace (percent)						
Pennsylvania	81	80	88	75	85	81
Other state	3	2	0	14	4	5
Foreign nation	16	17	13	12	11	13
Percent African American	0	0	0	1	0	0
Age (percent)						
Under 16	33	7	13	1	1	13
16–19	43	26	25	17	6	24
20–39	23	63	44	62	69	51
40+	1	4	19	20	24	13
Household status (percent)						
Head of household	8	44	50	53	69	42
Daughter or son	64	37	25	31	9	35
Tenant/other	28	20	25	16	21	23
Real estate ownership (percent)	4	15	44	28	38	22
Persistence (percent)						
In Harrisburg, 1850	19	25	41	13	20	21
In Harrisburg, 1870	19	25	50	25	32	28
Same occupation, 1870	3	14	17	13	22	12
1870						
Total Employees	280	78	442	992	n/a	n/a
Percent in census	35	58	20	41	n/a	n/a
No. identified in census	98	45	86	408	666	1,303
Birthplace (percent)						
Pennsylvania	87	82	72	56	83	74
Other state	6	0	14	17	6	10
Foreign nation	7	18	14	27	11	16
Percent African American	3	0	9	1	1	2
Age (percent)						
Under 16	20	4	1	4	0	3
16–19	37	22	11	17	5	12
20–39	36	53	69	61	74	66
40+	8	20	20	18	20	18

Appendix I *Continued*

Household status (percent)						
Head of household	22	51	69	58	71	62
Daughter or son	66	36	20	20	12	20
Tenant/other	12	13	12	22	17	18
Real estate ownership (percent)	5	20	20	16	19	17
Persistence (percent)						
In Harrisburg, 1860	17	40	44	7	18	15
Same occupation, 1860	4	18	11	3	4	4

SOURCES: Total employees from computerized manufacturing census schedules, 1860, 1870; all others from computerized population census schedules, 1850, 1860, 1870.

n/a = not available.

NOTES

INTRODUCTION

1. Lynn, for example, has been the focus of three recent book-length studies: Alan Dawley, *Class and Community: The Industrial Revolution in Lynn* (Cambridge, Mass., 1976); John Cumbler, *Working-Class Community in Industrial America: Work, Leisure, and Struggle in Two Industrial Cities [Lynn and Fall River, Mass.], 1880–1930* (Westport, Conn., 1979); and Paul G. Faler, *Mechanics and Manufacturers in the Early Industrial Revolution: Lynn, Massachusetts, 1780–1860* (Albany, N.Y., 1981). Numerous books on Lowell include John Coolidge, *Mill and Mansion: A Study of Architecture and Society in Lowell, Massachusetts, 1820–1865* (New York, 1942); Hannah Josephson, *The Golden Threads: New England's Mill Girls and Magnates* (New York, 1949); and Thomas Dublin, *Women at Work: The Transformation of Work and Community in Lowell, Massachusetts, 1826–1860* (New York, 1979). The scholarship on iron and steel, largely focused on the Pittsburgh region and approached in different ways, includes William T. Hogan, *Economic History of the Iron and Steel Industry of the United States*, 5 vols. (Lexington, Mass., 1971); Peter Temin, *Iron and Steel in Nineteenth-Century America: An Economic Inquiry* (Cambridge, Mass., 1964); David Brody, *Steelworkers in America: The Nonunion Era* (Cambridge, Mass., 1960); J. H. Bridge, *The Inside History of the Carnegie Steel Company* (New York, 1903); Harold C. Livesay, *Andrew Carnegie and the Rise of Big Business* (Boston, 1975); Francis G. Couvares, *The Remaking of Pittsburgh: Class and Culture in an Industrializing City, 1877–1919* (Albany, N.Y., 1984). Among studies of Detroit and the early automobile assembly lines are Alfred D. Chandler Jr., *Giant Enterprise: Ford, General Motors, and the Automobile Industry* (New York, 1964); John B. Rae, *The American Automobile* (Chicago, 1965); Sidney Fine, *The Automobile Under the Blue Eagle* (Ann Arbor, Mich., 1963); and Sidney Fine, *Sit-Down: The General Motors Strike of 1936–1937* (Ann Arbor, Mich., 1969). On the same subject, and more recent, are Stephen Meyer III, *The Five-Dollar Day: Labor Management and Social Control in the Ford Motor Company, 1908–1911* (Albany, N.Y., 1981); David Gartman, *Auto Slavery: The Labor Process in the American Automobile Industry, 1897–1950* (New Brunswick, N.J., 1986); and Joyce Shaw Peterson, *American Automobile Workers, 1900–1933* (Albany, N.Y., 1987).

2. Philip Scranton, *Proprietary Capitalism: The Textile Manufacture at Philadelphia, 1800–1885* (New York, 1983), by shifting away from the much-studied New England mills, provides a totally new perspective on that industry. Stuart M. Blumin, *The Emergence of the Middle Class: Social Experience in the American City, 1760–1900* (New York, 1989), pp. 299–300, notes that both Lynn and Lowell, because they were satellites of a major metropolitan area, were able to evolve as atypically specialized one-industry towns. Pittsburgh and Detroit were set apart from other "typical" American communities by their size. More than other major cities, they were also close to being dominated by single industries.

3. Between 1830 and 1900 the number of towns and cities in the United States with 10,000

to 100,000 residents increased from 22 to 402. The economies of most of these, especially in the Northeast, rested on a railroad or industrial base. The number of smaller towns (5,000 to 10,000 residents) grew from 33 to 465 in the same period, and many of them boasted at least one factory that was their single largest employer. U.S. Bureau of the Census, *Historical Statistics of the United States* (Washington, D.C., 1960), p. 14 (hereafter cited as Census Bureau, *Historical Statistics*).

4. Studies of nineteenth-century industrial communities that I found particularly useful included Clyde and Sally Griffen, *Natives and Newcomers: The Ordering of Opportunity in Mid-Nineteenth-Century Poughkeepsie* (Cambridge, Mass., 1978); Susan E. Hirsch, *Roots of the American Working Class: The Industrialization of Crafts in Newark, 1800–1860* (Philadelphia, 1978); Carol E. Hoffecker, *Corporate Capital: Wilmington in the Twentieth Century* (Philadelphia, 1983); Michael B. Katz, *The People of Hamilton, Canada West: Family and Class in a Mid-Nineteenth-Century City* (Cambridge, Mass., 1975); Steven J. Ross, *Workers on the Edge: Work, Leisure, and Politics in Industrializing Cincinnati, 1788–1890* (New York, 1985); Stephan Thernstrom, *Poverty and Progress: Social Mobility in a Nineteenth-Century City [Newburyport, Mass.]* (Cambridge, Mass., 1964); and Daniel J. Walkowitz, *Worker City, Company Town: Iron- and Cotton-Worker Protest in Troy and Cohoes, New York, 1855–1884* (Urbana, Ill., 1978). Studies of Pennsylvania communities in the nineteenth century, excluding several each of Philadelphia and Pittsburgh, include Harold W. Aurand, *Population Change and Social Continuity: Ten Years in a Coal Town [Hazleton, Pa., 1880–1890]* (Selinsgrove, Pa., 1986); Edward J. Davies II, *The Anthracite Aristocracy: Leadership and Social Change in the Hard Coal Regions of Northeastern Pennsylvania, 1800–1930* [esp. Wilkes-Barre and Pottsville] (DeKalb, Ill., 1985); James J. Farley, "The Frankford Arsenal, 1816–1870: Industrial and Technological Change," Ph.D. diss. Temple University, 1991; Burton W. Folsom Jr., *Urban Capitalists, Entrepreneurs, and City Growth in Pennsylvania's Lackawanna and Lehigh Regions, 1800–1920* (Baltimore, 1981), centered on Scranton and Wilkes-Barre; two studies by Anthony F. C. Wallace, *Rockdale: The Growth of an American Village in the Early Industrial Revolution* (New York, 1972) and *St. Clair: A Nineteenth-Century Coal Town's Experience with a Disaster-Prone Industry* (New York, 1987); Michael P. Weber, *Social Change in an Industrial Town: Patterns of Progress in Warren, Pennsylvania, from Civil War to World War I* (University Park, Pa., 1976); and two by Thomas R. Winpenny, *Bending Is Not Breaking: Adaptation and Persistence Among Nineteenth-Century Lancaster Artisans* (Lanham, Md., 1990) and *Industrial Progress and Human Welfare: The Rise of the Factory System in Nineteenth-Century Lancaster* (Washington, D.C., 1982). For an illustrated study of middle-size northeastern industrial cities focused on historical preservation, see Mary Proctor and Bill Matuszeski, *Gritty Cities* (Philadelphia, 1978).

5. See David Ward, *Cities and Immigrants: A Geography of Change in Nineteenth-Century America* (New York, 1971), pp. 19–46; Brian Greenberg, *Worker and Community: Response to Industrialization in a Nineteenth-Century American City, Albany, New York, 1850–1884* (Albany, N.Y., 1985), pp. 9–14; and Allan R. Pred, *The Spatial Dynamics of U.S. Urban-Industrial Growth, 1800–1914: Interpretive and Theoretical Essays* (Cambridge, Mass., 1966), pp. 16–24.

6. Jeffrey G. Williamson and Joseph A. Swanson, "The Growth of Cities in the American Northeast, 1820–1870," *Explorations in Entrepreneurial History*, 2d ser., Supplement, vol. 4, no. 1 (1966).

7. With few exceptions, only the men of Harrisburg have been considered. Obituary and published biographical sketches of even prominent women centered on fathers, husbands, and sons. A woman who married could be traced in directories or census records only if her husband's name was known.

Part I: On the Eve of Industrialization

1. Edwin Whitefield, "View of Harrisburg, Pennsylvania, from the West, 1846," plate 28, and "Johnstown, Pa.," c. 1846, plate 29, in *Pennsylvania Prints from the Collection of John C. O'Connor and Ralph M. Yeager* (University Park, Pa., 1980). For information on Whitefield, see John W. Reps, *Views and Viewmakers of Urban America* (Columbia, Mo., 1984), pp. 215–16.

2. "Plan [Map] of the Borough of Harrisburg," from the original survey by J. C. Sidney, published by Sam'l. Moody (n.p., 1850) (hereafter "Map of Harrisburg, 1850"). On file at the Division of Archives and Manuscripts, Pennsylvania Historical and Museum Commission (hereafter cited as PHMC).

3. William Henry Egle, *History of the Counties of Dauphin and Lebanon* (Philadelphia, 1883) (hereafter cited as Egle's *Dauphin County*), pp. 319–22.

4. Ibid., pp. 19–23; George P. Donehoo, *Harrisburg and Dauphin County* (Dayton, Ohio, 1925), pp. 49–51. The "Proprietaries" was the term used for the office of the proprietor(s) of Pennsylvania.

5. *Pennsylvania Archives*, 2d ser., vol. 19 (1893): 749; 3d ser., vol. 1 (1894): 45, vol. 8 (1896): 107–9. Donehoo, *Harrisburg and Dauphin County*, p. 51, discusses the Blunston licenses and cites the manuscript sources.

6. Egle's *Dauphin County*, pp. 293–95.

7. Ibid., pp. 295–97. A photograph of the first courthouse is in Richard H. Steinmetz Sr. and Robert D. Hoffsommer, *This Was Harrisburg: A Photographic History* (Harrisburg, 1976), p. 45.

8. For histories of the various denominations and drawings of many of the churches, see Egle's *Dauphin County*, pp. 329–49. Several of the same drawings were used to decorate the borders of Sidney's "Map of Harrisburg, 1850." Although the 1850 map lists the church on Front between Walnut and Locust as the United Brethren Church, it was the site of the Baptist church built in 1830–31. That building may have been sold to the United Brethren, who occupied "a small church" on Front from 1850 to about 1859. The Baptists, meanwhile, were building a new church at Second and Pine streets. See ibid., pp. 345–46.

9. Ibid., pp. 15–16; Donehoo, *Harrisburg and Dauphin County*, pp. 52–56.

10. Egle's *Dauphin County*, pp. 344, 348–49.

11. Ibid., pp. 312–16. Most histories of the community state that Harris set aside the four acres to induce the Commonwealth to locate its capital at Harrisburg. Theophile Cazenove, a French visitor to the community, learned, probably from Harris's brother-in-law Major John Hanna, that the four acres were set aside to get the Assembly to incorporate the borough. See Cazenove's *Record of the Journey of Theophile Cazenove Through New Jersey and Pennsylvania*, extracts of which are in Asa Earl Martin and Hiram Herr Shenk, *Pennsylvania History Told by Contemporaries* (New York, 1925) (hereafter cited as Cazenove's *Journey*), pp. 344–52; see p. 348.

12. Population figures are from published federal population census data. Beginning in 1850, the federal manuscript population census schedules devoted one line to each man, woman, and child located by the census taker. Names were entered by household in order of visitation. Included were each person's name, age, gender, race, the occupation of males over age sixteen (and sometimes of females and males under sixteen), the value of real estate owned, the place of birth, whether the person had attended school in the previous year, and whether the person was literate. By computerizing the manuscript population census data for Harrisburg in 1850, 1860, and 1870, I was able to use the data for a variety of comparisons and analyses. The observations on occupations are based on this data. (This computerized population census data is hereafter referred to as "computerized population census schedules.")

13. *View of Harrisburg, Penn.* [*1855*], J. T. Williams, Publisher, Historical Society of Dauphin County, Harrisburg. The lithograph as been reproduced on the inner side of both covers of Michael Barton, *Life by the Moving Road: An Illustrated History of Greater Harrisburg* (Woodland Hills, Calif., 1983).

14. Computerized population census schedules, 1850.

15. A census taker for a single ward of Reading, Pennsylvania, in 1850 gave not only the state or nation of birth for each person, but also the town or township. Of the total, a little more than 40 percent were born in Reading, and another 23 percent in the rest of Berks County. John Modell, "The Peopling of a Working-Class Ward: Reading, Pa., 1850," *Journal of Social History* 5 (Fall 1971): 71–95.

16. Computerized population census schedules. Comparable data for the general population of the United States in the 1980 census shows that less than 15 percent were under age ten and half were under thirty and that more than a quarter of the population was fifty or older. Indeed, more than 11 percent were over age sixty-five.

Chapter 1: Economic Foundations of the Community

1. Robert G. Crist, "Dauphin County—A Reprise," introduction to the reprint of the 1875 *Combination Atlas Map of Dauphin County, Pennsylvania* (Harrisburg, 1985) (hereafter cited as *Dauphin County Atlas*), p. 5; Philip S. Klein and Ari Hoogenboom, *A History of Pennsylvania* (New York, 1973), pp. 178–80. George Swetnam, *Pennsylvania Transportation* (Gettysburg, Pa., 1968), gives a brief general account of transportation in the Commonwealth.

2. Klein and Hoogenboom, *History of Pennsylvania*, pp. 53–69, 89–91; Tim H. Blessing, "The Development of a Dysfunctional Society: The Systems of Pennsylvania's Juniata Valley, 1740–1985" (Ph.D. diss., The Pennsylvania State University, 1989), pp. 43–55. Peter C. Mancall, *Valley of Opportunity, Economic Culture Along the Upper Susquehanna, 1700–1800* (Ithaca, N.Y., 1991), gives an excellent and detailed account of the development of much of Harrisburg's frontier hinterland.

3. Quoted in Donehoo, *Harrisburg and Dauphin County*, pp. 58–59.

4. Egle's *Dauphin County*, pp. 112–13, 118–19, 291–95, 297, 299, 303; Cazenove's *Journey*, pp. 347–48.

5. Cazenove's *Journey*, p. 348.

6. George H. Morgan, *Annals of Harrisburg*, 2d ed. (Harrisburg, 1906), p. 385; Egle's *Dauphin County*, p. 319; Brenda Barrett, "Historic Cultural Resources Reconaissance: Harrisburg Local Flood Protection Study, Dauphin County, Pennsylvania" (Report for the U.S. Army Corps of Engineers, 1977), p. 4.

7. F. Cuming, *Sketches of a Tour, 1807–1809*, reprinted in Reuben G. Thwaites, *Early Western Travels, 1748–1846* (Cleveland, Ohio, 1904), 4:37.

8. James Weston Livingood, *The Philadelphia-Baltimore Trade Rivalry, 1780–1860* (1947; reprint, New York, 1970), pp. 28–29.

9. Ibid., pp. 3–5. Of eighty-six turnpike roads patented by the Pennsylvania Assembly between 1792 and 1821, the Commonwealth subscribed to the capital stock of fifty-six companies, totaling $1,861,542. This amounted to 26 percent of all turnpike capital and 44 percent of the capital of the firms aided. Pennsylvania Senate *Journal, 1821*, pp. 670–71.

10. Livingood, *Philadelphia-Baltimore Trade Rivalry*, pp. 5–21.

11. Ibid., pp. 44–47, 53; *Laws of Pennsylvania, 1792–1814*, passim; Swetnam, *Pennsylvania Transportation*, pp. 44–46. For the Harrisburg to Pittsburgh routes, the freight they carried, and the costs, see Robert M. Blackson, "The Panic of 1819 in Pennsylvania" (Ph.D. diss.,

The Pennsylvania State University, 1978), pp. 180–87, 199; and Catherine Elizabeth Reiser, *Pittsburgh's Commercial Development, 1800–1850* (Harrisburg, Pa., 1951), pp. 74–83.

12. Morgan, *Annals*, pp. 356–57; *Laws of Pennsylvania, 1808–1809*, pp. 152–60; ibid., *1810–1811*, p. 258; ibid., *1811–1812*, p. 15; ibid., *1816–1817*, p. 52. The minutes of the Bridge Company's board of directors for 1812–1934 are in MG-112 at the Historical Society of Dauphin County, Harrisburg, Pa. (hereafter cited as HSDC).

13. Joseph Wallace to Daniel I. Rupp, October 22, 1845, in Rupp's *History and Topography of Dauphin, Cumberland, . . . Counties* (Lancaster, Pa., 1846) (hereafter cited as Rupp's *Dauphin County*), pp. 270–72.

14. Ibid., p. 270. The last toll was collected on May 15, 1957 (Steinmetz and Hoffsommer, *This Was Harrisburg*, p. 66).

15. Egle's *Dauphin County*, pp. 501–2; *Laws of Pennsylvania, 1792–1814*, passim.

16. Egle's *Dauphin County*, pp. 483–85.

17. Ibid., pp. 498–99; Luther Reily Kelker, *History of Dauphin County, Pennsylvania*, 3 vols. (New York, 1907) (hereafter cited as Kelker's *Dauphin County*), 3:24–25; *Biographical Annals of Cumberland County, Pennsylvania* (Chicago, 1905), pp. 829–30.

18. In the early years, the family spelled the name "Colder." *Biographical Encyclopedia of Pennsylvania in the Nineteenth Century* (Philadelphia, 1874), p. 316; Kelker's *Dauphin County*, 3:610; Obituary, *Telegraph*, March 5, 1861; J. Allen Barrett, "Old Stagecoach Travel and Stops in Dauphin County," *Dauphin County Historical Review* 3 (December 1854).

19. Nicholas B. Wainwright, *History of the Philadelphia National Bank* (Philadelphia, 1953), pp. 24–25; Helen Bruce Wallace, *A Century of Banking* (Harrisburg, 1914), pp. 12–13.

20. Belden L. Daniels, *Pennsylvania Birthplace of Banking in America* (Harrisburg, 1976), pp. 147–55.

21. Wallace, *Century of Banking*, p. 16.

22. Ibid., pp. 17–18; Charter of Harrisburg Bank (1814), listing the shareholders, in the bank's Minutes Books. The records of the Harrisburg Bank (later Harrisburg National Bank), which are in the custody of its successor institution, the Commonwealth National Bank, are used with permission.

23. The observations on the management of banks are based on the minutes books of Harrisburg banks. For the Harrisburg Bank's early headquarters, see Wallace, *Century of Banking*, pp. 18–19.

24. Wallace, *Century of Banking*, p. 28.

25. Harrisburg Bank, Minutes Books, November 25 and 30, 1814; October 3, 1815; March 6 and August 28, 1816.

26. Wallace, *Century of Banking*, pp. 15, 28; Harrisburg Bank, Minutes Books, June 13, November 28 and November 30, 1814; October 13, 1815; March 6 and August 28, 1816.

27. Harrisburg Bank, Minutes Books, July 6, October 23, and December 23, 1814; July 20, November 7, and December 11, 1814; Wallace, *Century of Banking*, pp. 21–22.

28. Harrisburg Bank, Minutes Books, January 1, July 19, and August 30, 1815.

29. Ibid., October 12 and November 7, 1814; Wallace, *Century of Banking*, pp. 23, 40. Because Harrisburg was the state capital, no doubt the bank discounted notes for legislators from all parts of the Commonwealth.

30. Egle's *Dauphin County*, pp. 492–93.

31. Wainwright, *Philadelphia National Bank*, p. 48; Wallace, *Century of Banking*, p. 13.

32. Blackson, "Panic of 1819 in Pa.," pp. 306–7, 344.

33. Robert McCullough and Walter Leuba, *The Pennsylvania Main Line Canal* (York, Pa., 1973), pp. 18–40.

34. Egle's *Dauphin County*, p. 472; Obituary, *Harrisburg Patriot*, August 16, 1864.

35. Except where otherwise indicated, this sketch is based on Lee F. Crippen, *Simon Cam-*

eron: Ante-Bellum Years (Oxford, Ohio, 1942), pp. 1–14; Erwin Stanley Bradley, *Simon Cameron: Lincoln's Secretary of War* (Philadelphia, 1966), pp. 19–54; and James B. McNair, *Simon Cameron's Adventure in Iron, 1837–1846* (Los Angeles, 1949), pp. 4–11.

36. *Majority and Minority Reports of the Joint Committee of the Senate and House of Representatives Relative to an Investigation into Any Corrupt Means . . . Employed by the Banks . . . for the Purpose of Influencing the . . . Legislature . . .* (Harrisburg, 1842), p. 93.

37. Cameron to former Governor John A. Schulze, November 18, 1831, Society Small Collection, Historical Society of Pennsylvania, Philadelphia (hereafter cited as HSP).

38. McNair, *Simon Cameron's Adventure*, pp. 11–29; Cameron to E. D., W. E., and Simon Gratz, 1838–42, Simon Gratz Collection, HSP (the quotations are from Cameron to W. E. Gratz, December 11, 1838, and March 12, 1841).

39. For example, see Cameron to Simon Gratz, February 12, 1838, HSP.

40. *Laws of Pennsylvania, 1834–1835,* April 15, 1835.

41. Computerized population census schedules.

42. Marian Inglewood, *Then and Now in Harrisburg* (Harrisburg, 1925), p. 71.

43. *Laws of Pennsylvania, 1830–31*, pp. 373–74; Paul J. Westhaeffer, *History of the Cumberland Valley Railroad, 1835–1919* (Washington, D.C., 1979), pp. 5–9.

44. *Laws of Pennsylvania, 1831–1832*, pp. 590–91; ibid., *1834–1835*, p. 58; ibid., *1835–1836*, pp. 385–86; ibid., *1836–1837*, pp. 249–50.

45. Egle's *Dauphin County*, pp. 321–22; William Bender Wilson, *History of the Pennsylvania Railroad Company*, 2 vols. (Philadelphia, 1899), 1:368–94.

46. Harrisburg Bank, Minutes Books, 1814–38; biographical sketches in Egle's *Dauphin County;* H. Napey, *The Harrisburg Business Directory and Stranger's Guide* (Harrisburg, 1842).

47. Harrisburg Bank, Minutes Books; Wallace, *Century of Banking*, pp. 29, 31–32.

48. *Laws of Pennsylvania, 1833–1834*, no. 183; Bern Sharfman, *Dauphin Deposit: The First 150 Years* (Harrisburg, 1985), pp. 11–13. The Minutes Book of the Harrisburg Savings Institution (later the Dauphin Deposit Bank) is in MG-110 at HSDC; those of the Dauphin Deposit are in the custody of the bank itself.

49. Robert G. Crist, manuscript history of the Dauphin Deposit Bank, used with permission of the author.

50. Dauphin Deposit Bank, Special Accounts Book. Used with permission.

51. Dauphin Deposit, Minutes Book; Pennsylvania Auditor General, *Annual Report on Banks, 1840* (Harrisburg, 1840).

Chapter 2: The Maturing Infrastructure

1. Egle's *Dauphin County*, pp. 462–64. For the quotation, see *Harrisburg Morning Herald*, June 27, 1854.

2. *Harrisburg Telegraph*, January 7 and 14, 1846; H. W. Schotter, *The Growth and Development of the Pennsylvania Railroad Company* (Philadelphia, 1927), p. 118.

3. John F. Stover, *American Railroads* (Chicago, 1961), pp. 28, 41–43.

4. *Harrisburg Democratic Union*, September 5, 1849.

5. *Morning Herald*, March 14, 1854.

6. McCullough and Leuba, *Pennsylvania Main Line Canal*, p. 146.

7. For the canal, see Harvey H. Segal, "Canals and Economic Development," in *Canals*

and American Economic Development, ed. Carter Goodrich (New York, 1961), p. 242; for the Pennsylvania Railroad, see Henry V. Poor, *History of the Railroads and Canals of the United States of America* (New York, 1860), pp. 473–74; Pennsylvania Railroad annual reports, 1852–58.

8. Wilson, *History of the Pa. Railroad*, 1:230; *Laws of Pennsylvania, 1831–1832, 1832–1833*, passim.

9. Wilson, *History of the Pa. Railroad*, 1:241–44; Norbert C. Soldon, "James Donald Cameron: Pennsylvania Politician" (Master's thesis, The Pennsylvania State University, 1959), pp. 8–9.

10. James Arthur Ward, *J. Edgar Thomson: Master of the Pennsylvania* (Westport, Conn., 1980), pp. 116–17.

11. Bradley, *Cameron*, p. 55; Crippen, *Cameron*, p. 113.

12. Wilson, *History of the Pennsylvania Railroad*, 1:243–44; Samuel Richey Kamm, *The Civil War Career of Thomas A. Scott* (Philadelphia, 1940), pp. 9–11; Northern Central Railroad annual reports, 1856–65.

13. Soldon, "Cameron," pp. 8–9.

14. I have used the later names of consolidating companies rather than the original names of the many short lines in the area. Crist, "A Reprise," *Dauphin County Atlas*, pp. 9–10; Egle's *Dauphin County*, p. 322.

15. Computerized population census schedules.

16. Swetnam, *Pennsylvania Transportation*, p. 56.

17. Egle's *Dauphin County*, p. 312.

18. Ibid., p. 518.

19. *Laws of Pennsylvania, 1842*, p. 261. The provision would have been added only at the instigation of someone from the Middletown Bank, and, as Cameron told a legislative committee that same session, he had attended nearly every session of the legislature since 1817 (see also Chapter 1, n. 36).

20. See Sharfman, *Dauphin Deposit*, pp. 23–31, 49–59; Harrisburg Bank, Minutes Books, December 11, 1844; Dauphin Deposit, Minutes Books, January 11, May 3, and October 25, 1845; inventory of James M. Haldeman estate, Dauphin County Courthouse.

21. Office of the County Clerk, Dauphin County Courthouse.

22. Harrisburg Bank, Minutes Books.

23. Ibid.

24. Ibid., June 5, 1850; February 12, 1851; September 8, 1852.

25. Ibid., July 3 and December 4, 1850; February 25, 1852.

26. Ibid., September 23, 1857; March 10, June 16, August 18, and November 2, 1858.

27. Harrisburg Bank, Minutes Books.

28. Wallace, *Century of Banking*, pp. 30–32; Harrisburg Bank, Minutes Books, 1852–59.

29. U.S. House of Representatives, Select Committee on Government Contracts, *Report and Testimony* 37th Cong., 2d sess., Report 2, serial no. 1143, and Testimony, p. 1387; *Biographical Encyclopedia*, pp. 316–18.

30. For Eby, see *Biographical Encyclopedia*, p. 404. For the other partners, see Committee on Government Contracts, *Testimony*, pp. 1394, 1396–1401.

31. Egle's *Dauphin County*, pp. 326–29, 462–63; Rupp's *Dauphin County*, p. 269.

32. Egle's *Dauphin County*, p. 463; Harrisburg Gas Company, *A Century of Public Service* (Harrisburg, 1949?); *Telegraph*, September 19, 1849; January 15 and 22, 1851; *Morning Herald*, May 26, 1855.

33. Egle's *Dauphin County*, pp. 462–64.

34. Biddle to Ayres (Ayres File), Ayres to Biddle, vols. 97 and 98, passim, Nicholas Biddle Papers, Library of Congress, Manuscripts Division, Washington, D.C.

35. Egle's *Dauphin County,* pp. 43–64.

Chapter 3: Harrisburg Industrializes

1. George H. Morgan, *Industries of Harrisburg* (Harrisburg, 1874), p. 6, observed: "Twenty-five years ago Harrisburg was proverbially one of the 'deadest' towns in the country." He dated its awakening from the founding of the cotton mill in 1849.

2. *Telegraph,* February 21, 1849.

3. *Harrisburg Keystone,* September 12, 1848; January 2, 1849.

4. *Telegraph,* February 21, 1849.

5. *Keystone,* March 6, 1849.

6. Harrisburg Bank, Minutes Books, entries for June 25 and July 2, 1851.

7. *Telegraph,* January 11, 1854; see also June 20 and December 19, 1849; *Democratic Union,* June 20 and July 4 and 18, 1849; *Harrisburg Whig State Journal,* July 22, 1851; June 30 and September 8, 1853. The comments appear to have been based on conversations with persons connected with the company.

8. Winpenny, *Industrial Progress and Human Welfare,* pp. 25–37.

9. *Democratic Union,* January 3 and March 7, 1849.

10. *Telegraph* and *Democratic Union,* June 20, 1849.

11. *Telegraph,* July 11, 1849. The data on the wealth of the leaders is from the computerized population census schedules.

12. Papers of the Harrisburg Cotton Mill Company, MG-158, Minutes Book, HSDC; *Democratic Union,* July 18 and August 1, 1849; *Keystone,* December 11, 1849.

13. *Democratic Union,* July 18, 1849.

14. The Rawn Collection, MG-62, HSDC, Journal of C. C. Rawn, entry for March 28, 1851; bound charter filed in cotton mill Minutes Book, HSDC; also see minutes for August 23, 1851; April 7, 1852.

15. The shareholder lists are in a separate booklet kept with the cotton mill Minutes Book.

16. Quotation from the *Telegraph,* September 17, 1851. See also ibid., November 19, 1851, and cotton mill Minutes Book, November 3, 1851; April 7 and December 11, 1852. The information on Buehler is from Egle's *Dauphin County,* p. 427, and the computerized population census schedules for 1850.

17. Cotton mill Minutes Book, May 28, 1852, loose sheet dated June 1, 1852.

18. Cotton mill Minutes Book, passim; *Whig State Journal,* October 20, 1853; *Telegraph,* October 26 and November 24, 1853; *Democratic Union,* November 2, 1853.

19. Cotton mill Minutes Book; Pennsylvania Auditor General, *Annual Reports on Banks, 1854–1856;* Haldeman probate records, Dauphin County Courthouse.

20. Horace Andrew Keefer, *Early Iron Industries of Dauphin County* (Harrisburg, 1927), p. 25; Lenore Embick Flower, "Central Iron & Steel Company, Harrisburg, Pennsylvania," manuscript in possession of Milton E. Flower, Carlisle, Pa., pp. 6–7; used with permission. Lenore Flower undertook this project at the instance of Vance C. McCormick and had full access to company records, most of which are apparently no longer extant. She suspended the study when McCormick died in 1946.

21. *Biographical Annals of Cumberland County,* pp. 829–30. Harold C. Livesay and Patrick G. Porter, "Iron on the Susquehanna: New Cumberland Forge," *Pennsylvania History* 37 (July 1970): 261–68, detail Haldeman's early iron operations and related businesses.

22. Flower, "Central Iron & Steel," p. 7; Evan Miller, "The Iron Furnaces of Dauphin County" (a portion of his unpublished compilations on Harrisburg), deposited with the HSDC; Egle's *Dauphin County*, p. 577; and data from computerized federal manuscript manufacturing census schedules for 1850. The manuscript manufacturing schedules of 1850, 1860, 1870, and 1880 are a major source of information on manufacturing at Harrisburg in those years. To facilitate their use, the city's manufacturing schedules were put on computer (as the population schedules were; see Part I Introduction: On the Eve of Industrialization, n.12) and are hereafter cited as "computerized manufacturing census schedules." With census-to-census variations in what was defined as manufacturing, the exact questions asked, and the order in which responses were recorded, the schedules supposedly reported on all manufacturing establishments producing at least $500 worth of goods a year. The schedules indicated the name of the manufacturer (person, persons, or firm); the nature of the business; the amount of capital invested; the amount and cost of raw materials used that year; the quantity and value of goods produced during the year; the type of power used and number of horsepower generated by steam; the average number of employees (sometimes broken down to show the number of men, women, and youngsters employed); the average monthly (1850 and 1860) or annual (1870 and 1880) payroll (sometimes broken down by gender); the number of months operated during the year (1870); and the kinds and number of machines used in production. Users of this data should be aware of the imprecision and lack of uniformity, both within and among the decennial censuses, and that at least in Harrisburg even major firms were sometimes not reported.

23. Keefer, *Early Iron Industries of Dauphin County*, pp. 25–26; Flower, "Central Iron & Steel," p. 10; Miller, "Iron Furnaces," no. 15; Harrisburg Bank, Minutes Books, October 17, December 19, and December 29, 1849; computerized manufacturing census schedules, 1850.

24. Frederick Moore Binder, *Coal Age Empire: Pennsylvania Coal and Its Utilization to 1860* (Harrisburg, 1974), p. 66. According to Arthur C. Bining, *Pennsylvania Iron Manufacture in the Eighteenth Century*, 2d ed. (Harrisburg, 1973), pp. 47, 161, the large furnace at Cornwall, about twenty-five miles east of Harrisburg, supplied the capital city's forges with part of their iron. However, there were numerous other charcoal furnaces within a dozen miles of Harrisburg: to the south, Stubbs Furnace (1796–1890) and Christiana (later called Cameron) Furnace (1840–1904), both located above the mouth of Swatara Creek near Middletown; to the east, Manada Furnace (1835–75) on Manada Creek; to the north, Victoria Furnace (1830–57) on Clark's Creek northeast of the borough of Dauphin, Emmeline Furnace (1837–53) at the mouth of the same creek, and Montebello Furnace (1836–50s) at Duncannon on the Susquehanna River in Perry County; to the west (across the Susquehanna), Haldeman Furnace (1806–44) at the mouth of the Yellow Breeches Creek. Some of these were later converted into or replaced by anthracite furnaces. See Keefer, *Early Iron Industries of Dauphin County*, pp. 8–19; Miller, "Iron Furnaces," no. 10; and "Iron Production in Pennsylvania, 1850," *Journal of the Franklin Institute* 51 (January 1851): 69–72, with accompanying charts.

25. According to Binder, *Coal Age Empire*, p. 65, it required 200 bushels of charcoal (at 5 cents each), costing $10, to produce one ton of pig iron, or two tons of anthracite, costing a total of $5. The charts accompanying "Iron Production in Pennsylvania, 1850" (see preceding note) show that the largest annual output of the state's 59 cold-blast charcoal furnaces, employing about 60 men each, averaged 1,200 tons. For 66 hot-blast charcoal furnaces the comparable output was 1,500 tons with approximately 72 employees each. The 56 anthracite furnaces on average produced more than 3,950 tons with about 88 men each. Porter Furnace reported to this source an output of 3,614 tons with 71 employees.

26. For the market in the greater Harrisburg area, see various letters in the Duncannon Iron Works Collection, Special Collections, Pattee Library of The Pennsylvania State University. For more general conditions, see Paul F. Paskoff, *Industrial Evolution: Organization, Structure, and Growth of the Pennsylvania Iron Industry, 1750–1860* (Baltimore, 1983), p. 106. Paskoff

argues that the iron industry in general evolved gradually and did not go through a sudden-takeoff period of growth.

27. Harrisburg Bank, Minutes Books, 1840s and 1850s.

28. Egle's *Dauphin County*, pp. 472, 528; Miller, "Iron Furnaces," no. 16; chart of anthracite blast furnaces accompanying "Iron Production in Pennsylvania, 1850."

29. Darwin H. Stapleton, "The Diffusion of Anthracite Iron Technology: The Case of Lancaster County," *Pennsylvania History* 45 (April 1978): 151.

30. Computerized manufacturing census schedules, 1850, 1870; Harrisburg Bank, Minutes Books, June 1848–August 1860. The credit reports are from the R. G. Dun & Company Collection, Baker Library, Harvard University Graduate School of Business Administration, and used with permission (hereafter referred to as Dun & Co.). Only short summaries of the reports on persons or firms were preserved. They are arranged by the state where the subject was located and by volume, page, and date of report. For reports on Porter, see Dun & Co., Pa. 54:83 (March 25 and December 29, 1858; January 17 and July 21, 1859; September 26, 1860).

31. J. P. Lesley, *The Iron Manufacturer's Guide* (New York, 1859), p. 16; Harrisburg Bank, Minutes Books, May 30, 1855, and July 15, 1857; McCormick Estate Trust Papers, Dauphin Deposit Bank & Trust Company, Harrisburg (hereafter cited as McCormick Estate), used with permission.

32. For biographical data on the Bailey brothers, see Flower, "Central Iron & Steel," p. 47; *The Manufactories and Manufacturers of Pennsylvania of the Nineteenth Century* (Philadelphia, 1875), pp. 299–300 (hereafter cited as *Manufacturers of Pennsylvania*); Egle's *Dauphin County*, pp. 599–600.

33. Flower, "Central Iron & Steel, pp. 44–45; Lesley, *Iron Manufacturer's Guide*, p. 237; computerized manufacturing census schedules, 1860; Dun & Co., Pa. 54:23 (January 30, 1857).

34. Harrisburg Bank, Minutes Books, July 15, 1857; computerized manufacturing census schedules, 1860.

35. Dun & Co., Pa. 54:3 (November 2 and 26, 1858; January 12 and 25 and July 21, 1859; January 24 and December 31, 1863).

36. Trust Records, McCormick Estate; Flower, "Central Iron & Steel," p. 7; computerized manufacturing census schedules, 1850, 1860. The price seems exceedingly low, considering the plant's capital and output in 1850 and the fact that it included twenty-five acres of land, six lots in the village of Fairview, a slitting mill, a rolling mill, and a nail factory. However, the plant may have deteriorated, and if Pratt was in serious distress and McCormick was the only interested buyer, it may have gone for that price.

37. Egle's *Dauphin County*, pp. 584–85; obituary of Henry McCormick, *Iron Age*, July 19, 1900.

38. Wilson, *History of the Pennsylvania Railroad*, 1:32, 55, 58; Pennsylvania Railroad, *Annual Report, 1861*, p. 46; ibid., *1869*, p. 49; *Telegraph*, January 23 and February 8, 1860. Shelton Stromquist, *A Generation of Boomers: The Pattern of Railroad Labor Conflict in Nineteenth-Century America* (Urbana, Ill., 1987), pp. 142–45, discusses railroad division towns. By the 1870s and 1880s, in the west, they were built at 200- to 300-mile intervals; earlier, in the east, they were not as far apart. For an account of railroad repair and maintenance facilities as an industry and its impact on a town's economy, see Spyridon George Patton, "Some Impacts of the Reading Railroad on the Industrialization of Reading, Pa., 1838–1910" (Ph.D. diss., University of Pittsburgh, 1978), pp. 46–70.

39. *Since 1853: An Informal History of the Harrisburg Steel Corporation and Its Predecessor Company* (Harrisburg, 1953), pp. 8–10. Except for the sketch of Fleming and the history of the firm after 1884, the anonymous author of this work relied almost wholly on William T. Hildrup, *History and Organization of the Harrisburg Car Manufacturing Company* (hereafter

cited as *Harrisburg Car Co.*) (Harrisburg, 1884). See also *Biographical Encyclopedia*, pp. 378–79; and *Encyclopedia of Contemporary Pennsylvania*, 2:219–21.

40. Not just Fleming was thinking along these lines. According to the *Telegraph*, February 12, 1852, J. R. Jones (a machine-shop operator and owner of the Novelty Iron Works) and J. C. Bucher (who owned land on the opposite side of the railway track from Jones in South Harrisburg) had formed a co-partnership to build railroad cars and were planning to erect one of the largest buildings in the state, but apparently that project fell through.

41. *Jones Wister's Reminiscences* (Philadelphia, 1920), p. 174; obituary of Wm. Calder Jr., *Telegraph*, July 19, 1880.

42. Hildrup, *Harrisburg Car Co.*, p. 64.

43. *Since 1853*, pp. 10–11; Hildrup, *Harrisburg Car Co.*, pp. 3, 52–53.

44. Hildrup, *Harrisburg Car Co.*, pp. 54–55.

45. Charles Nelson Hickok, comp., *The Hickok Genealogy* (Rutland, Vt., 1938), pp. 186–87; *Manufacturers of Pennsylvania*, pp. 266–68; *Biographical Encyclopedia*, p. 284; Obituary, *Telegraph*, May 26, 1891. The unpublished journal (apparently no longer extant) is quoted in a promotional booklet, *Between the Lines, 1844–1944*, p. 25, published by the W. O. Hickok Manufacturing Company.

46. *Between the Lines*, p. 25.

47. *Whig State Journal*, July 22, 1851; *Telegraph*, April 28, 1852; *Morning Herald*, May 18, 1855.

48. Dun & Co., Pa. 54:23 (May 21, 1850; March 14 and December 22, 1851).

49. For the variety of products and a picture of the cider press, see *Between the Lines*, pp. 9–10; *Morning Herald*, May 23, 1855. The cider press is described in *Manufacturers of Pennsylvania*, p. 268.

50. The records of the Eagle Works are in the offices of the W. O. Hickok Manufacturing Company, Harrisburg, and are used with permission.

51. Ibid., computerized population census schedules, 1850, 1860, 1870.

52. Computerized manufacturing census schedules, 1860; Dun & Co., Pa. 54:23 (February 1 and 12, 1855) and 54:3 (August 21, 1857; February 27 and November 26, 1858; June 14 and July 21, 1859; February 21 and September 26, 1860; July 31, 1861; June 19, 1862).

Chapter 4: The Civil War Interlude

1. Harrisburg's experience paralleled that of Philadelphia. According to J. Matthew Gallman, *Mastering Wartime: A Social History of Philadelphia During the Civil War* (New York, 1990), pp. 266–98, the start of the war there touched off a brief economic slump followed by a general recovery as contracts for war materiel poured in.

2. Janet Mae Book, *Northern Rendezvous: Harrisburg During the Civil War* (Harrisburg, 1951), relates the community's wartime experience chiefly from local newspaper files; Robert G. Crist, *Confederate Invasion: 1863* (Camp Hill, Pa., 1963), discusses the defense of Harrisburg at the time of Gettysburg.

3. Fred A. Shannon, *The Organization and Administration of the Union Army, 1861–1865*, 2 vols. (Cleveland, Ohio, 1928), 1:207; William J. Miller, *The Training of an Army: Camp Curtin and the North's Civil War* (Shippensburg, Pa., 1990).

4. Book, *Northern Rendezvous*, pp. 56–58. According to Miller, *Training of an Army*, p. 237, more troops passed through Curtin than any other camp in the north.

5. For Cameron's prewar political career, see Crippen, *Cameron*, pp. 63–244, and Bradley, *Cameron*, pp. 76–135.

6. Bradley, *Cameron*, pp. 136–74; quotation from p. 145.

7. Ibid., pp. 196–215. The charges of corruption were only part of the reason for Cameron's removal. For example, without discussing the matter with Lincoln or the Cabinet, in his first annual report he called for arming and using former slaves to help quell the rebellion.

8. Kamm, *Civil War Career of Thomas Scott*, pp. 22–23, 49, 130–32; Committee on Government Contracts, *Report* (see Chapter 2, n. 29), p. xv. Because Scott was to assist Cameron only temporarily, he retained his position with the Pennsylvania Railroad Company at a higher salary than he received from the government. Named assistant secretary of war in August 1861, he suspended his railroad salary until he returned to private life a year later.

9. Quoted in John P. Usher, *President Lincoln's Cabinet* (Omaha, Nebr., 1925), p. 19.

10. Committee on Government Contracts, *Report*, p. xv; Schotter, *Growth and Development of the Pa. Railroad*, p. 85; Ward, *Thomson*, p. 117.

11. For detailed treatment of the struggle among the railroad companies on which the following account is based, see Kamm, *Scott*, pp. 9–11, 24–82; and Thomas Weber, *The Northern Railroads in the Civil War* (New York, 1952), pp. 27–37.

12. Kamm, *Scott*, p. 26.

13. Ibid., pp. 24–25, 35, 63–66.

14. Ibid., pp. 58–63.

15. Committee on Government Contracts, *Report*, pp. xiv–xxv; for a defense of Scott's role in this controversy, see Kamm, *Scott*, pp 66–82.

16. Committee on Government Contracts, *Report*, p. lx.

17. For Calder and Hildrup, see *Biographical Encyclopedia*, pp. 316–18, 334–36. The self-serving quality of the entries and level of detail suggest the information was supplied by the subjects themselves.

18. Committee on Government Contracts, *Testimony*, pp. 155–59; ibid., *Report*, p. lx.

19. Ibid., *Testimony*, p. 159.

20. Ibid., pp. 1387–1401; *Biographical Encyclopedia*, pp. 316–18.

21. Computerized population census schedules.

22. Committee on Government Contracts, *Testimony*, p. 148.

23. Ibid., p. 163.

24. Ibid., p. 161.

25. Ibid., pp. 163–64.

26. Hildrup, *Harrisburg Car Co.*, p. 56. Jones Wister (*Jones Wister's Reminiscences*, p. 37), ironmaster at the Duncannon works north of Harrisburg, stated, "The first year industry was paralyzed," but by 1862 demand for iron increased sharply.

27. Hildrup, *Harrisburg Car Co.*, p. 8. Reconstructed wholesale and consumer price indexes show the former (1910–14: 100) rising from 89 in 1861 to 185 by 1865; consumer prices (1860: 100) in the same period rose from 101 to 175 (Census Bureau, *Historical Statistics*, pp. 115, 127).

28. Hildrup, *Harrisburg Car Co.*, p. 56.

29. *Biographical Encyclopedia*, pp. 334–36; Crist, *Confederate Invasion*, p. 15.

30. Harrisburg City Council, Minutes Books, Municipal Building, Harrisburg.

31. Crist, *Confederate Invasion*, p. 26. It can be argued, of course, that virtually all businesses charged for services to the war effort.

32. In the five years between 1849 and 1854, the Harrisburg and Dauphin Deposit banks had grown 36 percent and more than 200 percent respectively. The next five years, thanks in part to the Panic of 1857, their assets declined 12 percent and 30 percent respectively (Pennsylvania Auditor General, *Annual Reports on Banks*, 1849–69).

33. Harrisburg Bank and Dauphin Deposit Bank minutes books, 1861–65; *Dauphin County Atlas*, p. 24.

34. This was one of the yardsticks Thomas M. Cochran used in his "Did the Civil War Retard Industrialization?" *Mississippi Valley Historical Review* 48 (September 1961): 197–210.

35. Pennsylvania Railroad, *Annual Report, 1862*, p. 4.

36. Calculated from data in Poor, *History of the Railroads and Canals*, p. 474; Pennsylvania Railroad annual reports, 1860–65; H. W. Schotter, *Growth and Development of the Pennsylvania Railroad*, p. 115.

37. Calculated from data in Northern Central Railroad annual reports.

38. Calculated from data in Cumberland Valley Railroad annual reports; Pennsylvania Auditor General, *Annual Reports on Railroads, Canals, etc.*, 1862–65, 1869, 1874. See also Westhaeffer, *History of the Cumberland Valley Railroad*, pp. 65–84.

39. Pennsylvania Railroad, *Annual Report, 1864*, pp. 12–14; ibid., *1865*, p. 11; ibid., *1867*, pp. 25–26; George H. Burgess and Miles C. Kennedy, *Centennial History of the Pennsylvania Railroad Company* (Philadelphia, 1949), pp. 292–93.

40. Weber, *Northern Railroads*, p. 37.

41. Harrisburg cotton mill Minutes Book. The value of stock is in the inventory of the Forster estate, Dauphin County Courthouse.

42. Cotton mill Minutes Book.

43. *Jones Wister's Reminiscences*, p. 137.

44. Flower, "Central Iron & Steel," pp. 44–47.

45. McCormick Estate Trust Papers, real estate records, Dauphin Deposit Bank & Trust Archives, Harrisburg; computerized manufacturing census schedules.

46. Calder to Saunders, April 16, 1861, Cameron Papers, Library of Congress.

47. Committee on Government Contracts, *Testimony*, p. 158.

48. Hildrup, *Harrisburg Car Co.*, pp. 5–6, 53–56.

49. Dun & Co., Pa. 54:3 (June 9, 1863; June 1, 1864); 54:62 (March 16, 1866).

Chapter 5: Expansion and Consolidation

1. Computerized manufacturing census schedules, 1870. Maps of Harrisburg in city directories between 1870 and 1900 show no new structures in the Pennsylvania Railroad yards.

2. *Telegraph*, February 14, 1864; Flower, "Central Iron & Steel," pp. 18–19; *A Biographical Album of Prominent Pennsylvanians*, 3d ser. (Philadelphia, 1874), p. 201.

3. *Biographical Album*, pp. 202–3; Dun & Co., Pa. 54:16 (November 18, 1871; January 20, 1872); 37 (October 14, 1865); 81 (November 18, 1871; April 11, 1873; April 6, 1877).

4. Accounts of the founding of the Pennsylvania Steel Company will be found in a typewritten summary, "Steelton Plant Bethlehem Steel Company, Steelton, Pennsylvania," December 1957 (hereafter cited as "Steelton Plant History"), compiled from Bethlehem Steel Company Archives, Bethlehem Steel Corporation Collection, Hagley Museum and Library, Wilmington, Delaware; Elting E. Morison, *Men, Machines, and Modern Times* (Cambridge, Mass., 1966), pp. 162–66; Jeanne McHugh, *Alexander Holley and the Makers of Steel* (Baltimore, 1980), pp. 211–21; and Burgess and Kennedy, *Centennial History of the Pennsylvania Railroad*, pp. 292–93. The quotation is from "Steelton Plant History," p. 2. The Pennsylvania Railroad's investment is reported in Burgess and Kennedy, *Centennial History*, p. 293.

5. McHugh, *Holley*, pp. 211–12.

6. "Steelton Plant History," p. 4. For the alleged roles of Calder and Cameron, see their respective obituaries in the *Telegraph*, July 19, 1880, and the *Patriot*, August 31, 1918.

7. "Steelton Plant History," pp. 5, 6, 8, 19. See also John B. Yetter, *Steelton, Pennsylvania: Stop–Look–Listen* (Harrisburg, 1979), pp. 1–2.

8. Harrisburg National Bank, Minutes Book, November 1, 1865.

9. *Telegraph,* November 21, 25, and 28, 1865; *Patriot,* November 29, 1865. Harrisburg residents in fact contributed only $24,577.50, about $5,000 short of the actual cost of the land ("Steelton Plant History," p. 5).

10. Dun & Co., Pa. 55:365 (October 3, 1882).

11. Temin, *Iron and Steel,* pp. 126–32; Joseph Frazier Wall, *Andrew Carnegie* (New York, 1970), pp. 329–33; Morrison, *Men, Machines, and Modern Times,* pp. 169–96.

12. Pennsylvania Secretary of the Interior, *Annual Report, 1875–1876,* p. 663; "Steelton Plant History," pp. 7–12; Morrison, *Men, Machines, and Modern Times,* pp. 165–69; McHugh, *Holley,* pp. 212–14; Harrisburg Board of Trade, *Industrial and Commercial Resources* (Harrisburg, 1887), p. 26 (hereafter cited as Board of Trade, *Resources*); Dun & Co., Pa. 54:220 (August 19, 1879); 55:365 (June 4, 1880; October 3, 1882). For the new plant near Baltimore, see Mark Reutter, *Sparrows Point* (New York, 1988).

13. Based on U.S. totals (converted from long tons into regular tons) in Census Bureau, *Historical Statistics,* pp. 416–17, and Pennsylvania Steel totals in Board of Trade, *Resources,* p. 27.

14. *Dauphin County Atlas,* pp. 28, 30; Pennsylvania Secretary of the Interior, *Annual Report, 1878–1879,* pp. 68–69.

15. Guillard Dock Journal, Dock Family Papers, MG-43, PHMC; *Dauphin County Atlas,* p. 29; Dun & Co., Pa. 54:213 (February 3 and August 30, 1876; September 1, 1879; March 10, 1880).

16. Flower, "Central Iron & Steel," pp. 15, 29, 46–47; *Dauphin County Atlas,* p. vii; computerized manufacturing census schedules, 1870.

17. Flower, "Central Iron & Steel," pp. 44–50.

18. Bound records of the McCormick Estate. Because no inventory was made of his estate and no listing of securities appears among the extant records of the trust, it is not possible to measure James McCormick's wealth precisely or to know the full extent of his business connections.

19. The agreement, dated February 1, 1870, was registered with the Dauphin County clerk. Under a supplemental agreement in 1874, the widow, Eliza McCormick, accepted more than $400,000 in a trust fund in lieu of her quarter of the estate. She received an annual income of $25,000 from the trust but could not draw against its principal. Agreements in Trust Papers, McCormick Estate.

20. McCormick Company records, Dauphin Deposit Archives.

21. Computerized manufacturing census schedules, 1870; *Dauphin County Atlas,* pp. vii–viii; Board of Trade, *Resources,* pp. 29–30. The figures for 1875 may be inflated, especially for the nail works. The Panic of 1873 was in progress, and the firm did not produce as many nails again until 1885. The computerized manufacturing census schedules for 1880, though extant and apparently missing no pages, inexplicably have no entries for Lochiel, for any of the McCormick iron properties, for the Pennsylvania Steel Company, or for the cotton works. Published totals that year for Harrisburg are larger than the sum of the census data.

22. Flower, "Central Iron & Steel," pp. 51–59.

23. Data on Lochiel's output is in Pennsylvania Bureau of Statistics, *Annual Report, 1873–87.* For Pennsylvania Steel, see Board of Trade, *Resources,* p. 27. For later history of Lochiel, see Dun & Co., Pa. 56:257 (December 3, 1885); "Steelton Plant History," p. 11.

24. This account is based on Hildrup, *Harrisburg Car Co.,* pp. 7–31. See also *Since 1853,* pp. 17–39, based on company records for the period after 1885.

25. Hildrup, *Harrisburg Car Co.,* pp. 12–16, describes the facility in full.

26. The quotation in the previous paragraph is from *Since 1853,* p. 29.

27. Ibid., p. 38.

28. *Biographical Encyclopedia,* pp. 316–18; *Manufacturers of Pennsylvania,* p. 227; Obituary, *Telegraph,* July 19, 1880.

29. Morgan, *Industries of Harrisburg,* p. 62.

30. *Between the Lines,* p. 13; Sylvester K. Stevens, *Pennsylvania Titans of Industry,* 3 vols. (New York, 1948), 2:340–41.

31. Morgan, *Industries of Harrisburg,* p. 65.

32. *Dauphin County Atlas,* pp. 23–24.

33. Stockholder lists are filed with the minutes in the years indicated. See Harrisburg National Bank, Minutes Books.

34. John W. Jordan, *Encyclopedia of Pennsylvania Biography,* 28 vols. (New York, 1918–52), 4:1143.

35. Wallace, *A Century of Banking,* p. 37.

36. Based on a list of directors chosen each year, according to the Harrisburg National Bank Minutes Books.

37. Ibid., February 10, 1864.

38. Ibid., March 8, 1876.

39. Ibid., September 14 and 21, 1870; April 10, 1878.

40. Ibid., January 8, 1889; John Thom Holdsworth, *Financing an Empire: History of Banking in Pennsylvania,* 4 vols. (Chicago, 1928), 3:237–41.

41. Board of Trade, *Resources,* pp. 72–78.

42. Harrisburg National Bank, Minutes Books, March 9, 1892; September 28, October 25, November 7, and December 18, 1894; February 20 and October 2, 1895.

43. Wallace, *A Century of Banking,* pp. 30–32.

44. Sharfman, *Dauphin Deposit,* pp. 26–33.

45. Board of Trade, *Resources,* pp. 72, 77.

46. Dun & Co., Pa. 54:180. Apparently the Calder from Lancaster was not related to the Calders of Harrisburg.

47. Board of Trade, *Resources,* pp. 34–35; Kelker's *Dauphin Co.,* 2:636; Morgan, *Annals of Harrisburg,* p. 461.

48. *Since 1853,* pp. 38–39.

49. Ibid., pp. 40–83.

50. Reports of visits by Archibald Johnston of the Bethlehem Steel Corporation to the Paxton Mill, April 1896, and to Central's Universal Mill, May 12, 1896, Box 19J, Archibald Johnston Collection, Hagley Museum and Library Archives, Wilmington, Delaware.

51. Flower, "Central Iron & Steel, pp. 34–35, 60–64, 69–73.

52. Hildrup, *Harrisburg Car Co.,* p. 24.

53. Steinmetz and Hoffsommer, *This Was Harrisburg,* p. 87; Barrett, "Harrisburg Flood Protection Study," p. 42.

54. "Steelton Plant History," pp. 10–12; *Patriot,* April 22, 1893; Dock Journal, entry for April 23, 1893.

55. Hickok Manufacturing Company, Corporate Minutes Books, used with permission; will (1946, no. 339) and inventory of Hickok's estate (B:252), Dauphin County Courthouse.

Chapter 6: The Process of Industrialization

1. For the four cities, see Greenberg, *Worker and Community,* esp. pp. 11–18; John T. Cumbler, *A Social History of Economic Decline: Business, Politics, and Work in Trenton* (New

Brunswick, N.J., 1989), esp. pp. 1–26; Raymond W. Albright, *Two Centuries of Reading, Pennsylvania, 1748–1948* (Reading, Pa., 1948); and Carol E. Hoffecker's two works, *Wilmington, Delaware: Portrait of an Industrial City, 1830–1910* (Charlottesville, Va., 1974) and *Corporate Capital* (on Wilmington), esp. pp. 12–16. See also Proctor and Matuszeski, *Gritty Cities*, pp. 169–74 (for Reading), 187–93 (for Trenton), and 245–51 (for Wilmington).

2. Hoffecker, *Corporate Capital*, p. 4; for the term "civic capitalism," see Cumbler, *Social History of Economic Decline*, pp. 3, 5–6.

3. For the waterpower problem, see Barrett, "Historic Cultural Resources of Harrisburg," pp. 7–9, 34, 37.

4. There is no precise and generally accepted definition for factories. Alfred D. Chandler Jr., *The Visible Hand: The Managerial Revolution in American Business*, p. 51, speaks of the factory as a "large industrial establishment, with its battery of machines, foundries, or furnaces," relying on a "central source of power and heat" and operated by "a large number of workers who had no other source of income than their wages." In similarly imprecise terms in 1880, Carroll D. Wright, the first U.S. commissioner of labor, defined factories as "several workmen" involved in "an association of separate occupations conducted in one establishment" where raw materials were "converted into finished goods by consecutive, harmonious processes, carried along by a central power" ("The Factory System of the United States," in *1880 Census, Manufactures, Part 1*, pp. 533, 548). A substantial number of workers and the use of power-driven machinery seem to be the essential factors.

5. Computerized manufacturing census schedules, 1850.

6. Industrialists who had not previously helped build the infrastructure were Charles Bailey, David Fleming, William T. Hildrup, and William O. Hickok.

7. Based on a study of Sidney's "Map of Harrisburg, 1850" (see n. 2 in Part I Introduction); Hother Hage's "Map of the City of Harrisburg, 1860–61" (Philadelphia, n.d.); and H. W. Hopkins and L. Cunningham's "Map of the City of Harrisburg" (Philadelphia, 1871). Copies of all these maps are in the PHMC.

8. Steam-driven looms were widely used in New England after the Civil War, especially in the New Bedford area.

9. Paskoff, *Industrial Evolution*, chap. 6, esp. pp. 106–7; charts accompanying "Manufacture of Iron in Pennsylvania," *Journal of the Franklin Institute* 51 (January 1851): 69–72; Lesley, *Iron Manufacturers Guide*, esp. pp. 16, 237–38.

10. *Jones Wister's Reminiscences*, pp. 173–74.

11. Flower, "Central Iron & Steel," pp. 23 and 23a, discusses Henry McCormick's experiments. For examples of the Baileys catching up with rather than leading the industry, see pp. 52, 60.

12. Computerized manufacturing census schedules, 1860.

13. McCormick's West Fairview Nail Works and the Pennsylvania Steel Company, though in the same size category and clearly a part of the same larger community and economic sphere, have not been included because they were not in the city proper.

14. Computerized manufacturing census schedules, 1870.

15. Ibid., 1850, 1860, 1870. For problems with the 1880 computerized manufacturing census schedules, see Chapter 5, n. 21.

16. In 1880 "All Others" included 25 percent of all capital investment, 41 percent of all output, and 10 percent of all employees. In 1890 the percentages were, respectively, 47, 48, and 56; in 1900 they were 19, 22, and 30.

17. The published census on manufacturing in Harrisburg for 1900 (*1900 Census, Manufacturing*, Part 2, pp. 778–79), for example, shows only five shoe factories, three brick and tile yards, six foundry and machine shops, and seven iron and steel mills, which could have employed more than 25 workers each. Together they accounted for 3,314 of the 8,217 Harris-

burg residents engaged in manufacturing. To that number should be added the employees of a few of the nineteen printing establishments (377 employees), twenty-five cigarmaking shops (618 employees), and four lumber-planing mills (88 employees). However, most of those probably had fewer than 25 employees. Finally, from the "All Others" category, the employees of the silk mill, of the successor to the car works, of a men's clothing factory, of three meat packers, and of the typewriter manufacturers, each of whom probably had more than twenty-five employees, should be added.

18. Flower, "Central Iron & Steel," pp. 76–81, discusses the complaints against Carnegie and U.S. Steel.

19. Flower (ibid., p. 15) indicates that, as early as 1878, ores from Spain and Africa were brought to Harrisburg more cheaply than it would cost to mine them locally. For Carnegie's acquisition of coke and ore, see Livesay, *Andrew Carnegie and Rise of Big Business*, pp. 147–55.

20. Morrison, *Men, Machines, and Modern Times*, p. 169; computerized manufacturing census schedules, 1870; Board of Trade, *Resources*, p. 26.

21. "Steelton Plant History," pp. 6–12.

22. Computerized manufacturing census schedules, 1860, 1870; Flower, "Central Iron & Steel," p. 69.

23. Flower, "Central Iron & Steel," pp. 62–63, 79–82; Report of visit to Central Iron Works, May 12, 1896, Box 19J, Archibald Johnston Collection.

24. See Chapter 7, below.

25. Sidney's "Map of Harrisburg, 1850."

26. Hopkins and Cunningham's *"Map of Harrisburg"* (1871).

27. I am indebted to Clyde McGeary of Camp Hill, Pennsylvania, who generously allowed me to use his copy no. 1307, *Atlas of the City of Harrisburg, Dauphin County, Pennsylvania, made from plans, deeds, and surveys*, Harrisburg Title Company, publisher (Philadelphia, 1901), now on deposit at the Rare Books Room of Pattee Library, The Pennsylvania State University, University Park, Pa.

Part II: A Generation Later

1. Thomas Hunter, "Harrisburg, Pa., 1879: Viewed from Fort Washington," published by C. J. Corbin (n.d.). A copy of this is on deposit at the HSDC.

2. The additions of 1860, made when the legislature conferred city status on Harrisburg, approximately tripled the town's 1850 area; the "Allison Hill" and South Harrisburg additions of 1868–69 doubled its 1860 area. In 1850 Harrisburg fronted the river from a half-block below Paxton Street in the south to Herr Street in the north, in 1860 from Hanna Street to Maclay, and in 1870 from Poplar Street to Maclay. Paxton Creek (roughly modern Cameron Street) was the town's eastern border in 1850, Thirteenth Street was that border in 1860, and a half-block east of Eighteenth Street was the eastern border in 1870. See Steinmetz and Hoffsommer, *This Was Harrisburg*, pp. 61–63; Egle's *Dauphin County*, map p. 297 and pp. 322–24. The Pennsylvania State Archives has the three maps used: Sidney's "Map of Harrisburg, 1850"; Hage's "Map of Harrisburg, 1860–61"; and H. W. Hopkins and L. Cunningham's "Map of Harrisburg" (1871). Population growth was determined from computerized population census schedules, 1850, 1860, 1870.

3. Sketches of Harris's grave and the new courthouse appear on the border of Hunter's lithograph (see n. 1, above). See also Egle's *Dauphin County*, pp. 293–94.

4. Drawings of several of the churches appear on the border of Hunter's lithograph (see n. 1, above), and both sketches and brief histories are in Egle's *Dauphin County*, pp. 329–49.

For membership of the various entrepreneurs, I have relied on their obituaries and on George B. Stewart, ed., *Centennial Memorial English Presbyterian Congregation [Market Square]* (Harrisburg, 1894), pp. 416–18; and George P. Donehoo, *A Brief Sketch of the History of the Presbyterian Church of Harrisburg, Pennsylvania [Pine Street]* (Harrisburg, 1926), pp. 15–21.

5. Incorporated in May 1861, the line from Camp Curtin was not completed until the company reorganized with better funding in 1864. The new line was sold at sheriff's auction in 1870. Another group of incorporators took over in 1873 and added the line to Steelton in 1875. See Richard H. Steinmetz and Harold E. Cox, *Street Railways of Harrisburg* (Forty Fort, Pa., 1988).

6. Erik H. Monkkonen, *America Becomes Urban: The Development of U.S. Cities and Towns, 1780–1980* (Berkeley and Los Angeles, 1988), pp. 160–61.

7. Charles R. Boak, Memoirs on growing up in Harrisburg, typescript, n.d., copy on deposit at the HSDC.

8. Direct observation. The Harrisburg Cemetery Association's pamphlet entitled *Walking Tour of Harrisburg Cemetery* has a map showing the location of the plots of prominent families. According to the *Patriot,* May 29, 1875, blacks were buried in the Harris Free Cemetery and in grounds belonging to the Wesley Union AME Church.

9. Part of the increased value of real estate was undoubtedly due to wartime inflation. However, assuming that land values paralleled rises in the general cost of living from 1860 to 1870, the gains were still impressive. The reconstructed cost-of-living index of the Federal Reserve Bank of New York, using 1913 = 100, showed 1860 as 61, 1865 as 102, and 1870 as 91. Had Harrisburg real estate values reflected only index changes, they would have amounted to $6.0 million in 1865 and $5.4 million in 1870. See Census Bureau, *Historical Statistics,* p. 127.

10. For a discussion of this problem, see Clyde Griffen, "Occupational Mobility in Nineteenth-Century America: Problems and Possibilities," *Journal of Social History* 5 (Spring 1972): 310–19.

Tables in this and similar studies based on data from the manuscript population and manufacturing census schedules create the illusion of greater precision than is justified by the nature of the data used. Conclusions based on such information cannot be exact and are at best close approximations. To remind the reader of this, the text of this book frequently uses approximate major fractions such as "nearly half," "more than a third," and "three-quarters" rather than the mathematically more refined percentages of the tables, which seem more precise.

11. This very significant difference between the population and manufacturing schedules has three principal causes: census takers for the two schedules were gathering data in different ways, the manufacturing schedules listed all employees whether they were residents of Harrisburg or not, and the manufacturing census in 1850 included no railroad workers—giving it a lower base from which to start.

12. Egle's *Dauphin County,* p. 211.

Chapter 7: The Entrepreneurs and Other Elites

1. Four useful studies of elites that have relevance for Harrisburg are E. Digby Baltzell, *Philadelphia Gentlemen* (New York, 1958); John N. Ingham, *The Iron Barons: A Social Analysis of an American Urban Elite, 1874–1965* (Westport, Conn., 1978), which looks at the iron and steel manufacturers in Philadelphia, Pittsburgh, Cleveland, Youngstown, and Wheeling; Folsom, *Urban Capitalists,* who compares the business elites of Scranton and Wilkes-Barre (and to a lesser degree other northeastern Pennsylvania industrial centers); and Davies, *Anthra-*

cite Aristocracy, whose study of Wilkes-Barre, Pottstown, and other anthracite communities, though geographically overlapping Folsom's work, offers a number of fresh insights.

2. Crippen, *Cameron,* p. 3. Sources of biographical data on the entrepreneurial families are too widely scattered to be cited in full. For the entrepreneurs, see Chapters 1, 2, and 3, above.

3. Harrisburg's entrepreneurs compare in many ways with those in Ingham's *Iron Barons,* pp. 79–82. Because the town had much less of an English heritage, however, proportions of Scots-Irish and Presbyterians were even greater (63 percent and 75 percent to 32 percent and 43 percent) than among Ingham's ironmasters, while proportions of English and Welsh stock and of Episcopalians were much smaller (21 percent and 6 percent to 53 percent and 23 percent).

4. Of Harrisburg's entrepreneurs, 21 percent arose from poverty, while 57 percent came from elite families. Of Ingham's ironmasters, only 4 percent came from poor families and 66 percent from wealthy backgrounds (ibid., p. 79).

5. *Harrisburg, Pa., Directory,* J. A. Spofford and H. Napey, publishers (Harrisburg, 1845); James Gopsill, *Gopsill's Directory of Lancaster, Harrisburg, Lebanon, and York, 1863–1864* (Jersey City, N.J., 1863); W. Harry Boyd, *Boyd's Harrisburg City Directory, 1887* (Harrisburg, 1887); Hopkins and Cunningham's *"Map of Harrisburg"* (1871); *Dauphin County Atlas.*

6. For detailed analyses of similar elites in larger and more complex settings, see Baltzell, *Philadelphia Gentlemen,* pp. 173–363; and Ingham, *Iron Barons,* pp. 84–219. Folsom's *Urban Capitalists,* pp. 68–83, contrasts Scranton's relatively "open" social system with Wilkes-Barre's "closed" system. Harrisburg more closely resembled Wilkes-Barre. Davies, *Anthracite Aristocracy,* pp. 41–59, confirms the similarities of Wilkes-Barre and Harrisburg. There were also interesting differences: Wilkes-Barre's elite was British rather than Scots-Irish, and far more Episcopalian than Presbyterian. Although law was the preferred profession of elite males in both communities, Harrisburg's elite included several physicians.

7. I told this to an officer of the successor bank where I was doing research, and he said he believed it was still the practice. This was confirmed by a descendant of the Bailey family long connected with both the bank and the church.

8. *Laws of Pennsylvania, 1845,* pp. 26–28.

9. Robert Grant Crist, *Harrisburg Hospital: The First 100 Years* (Harrisburg, 1973), pp. 1–5, 81–84.

10. Anthony Arms, *The Years Speak Volumes* (Camp Hill, Pa., 1976), pp. 19–23, 45–51.

11. For information on the sons of the first group of entrepreneurs, see William Henry Egle, *Pennsylvania Genealogies: Scotch-Irish and German* (Harrisburg, 1886): Ayres, pp. 45–48; Elder, pp. 159, 161, 164–65; Forster, pp. 214–16. For George Harris, see Obituary, *Patriot,* August 14, 1882; for James Calder, see Kelker's *Dauphin County,* 3:611–13; for the sons of Simon Cameron, see Bradley, *Cameron,* pp. 51–52, and obituary of Brua Cameron in the *Telegraph,* January 14, 1864.

12. Obituaries of David Harris, *Telegraph,* March 15, 1880, and *Patriot,* March 16, 1880, and of Thomas Jefferson Harris, *Telegraph* and *Patriot,* August 12, 1878.

13. John G. Haldeman will be discussed below. For Jacob S. Haldeman, see Obituary, *Patriot,* November 28, 1889. For Richard J. Haldeman, see *Biographical Directory of the American Congress, 1774–1971* (Washington, D.C., 1971), p. 987; *Biographical Annals of Cumberland County,* p. 831; and Kelker's *Dauphin County,* 3:25–26.

14. *Across the Continent in 1865* (a portion of the diary of Henry McCormick) (Harrisburg, 1937), esp. pp. 8, 34; inventory of the Estate of John Haldeman, H-12, Dauphin County Courthouse.

15. It is possible, though I found no evidence, that Edwin and William O. Hickok Jr., and Theodore G. Calder also went to college. Yale was also a favorite college for Pittsburgh, Youngstown, and Cleveland ironmasters and their sons. See Ingham, *Iron Barons,* pp. 94–95, 148–52, 170–72, 194–95, 209.

16. The movement of the sons of Harrisburg entrepreneurs into the leadership of industrial firms and banks was similar to the patterns at Wilkes-Barre. See Davies, *Anthracite Aristocracy*, p. 75. On the other hand, according to Folsom, *Urban Capitalists*, pp. 96–112, few sons of entrepreneurs there succeeded their fathers as business leaders. Fundamental differences between the communities, not the superiority of Harrisburg's sons or the shortcomings of Scranton's, probably account for this. Harrisburg had a lengthy preindustrial history and traditions that the city of Scranton, founded much later, did not have. Scranton, far more than Harrisburg, was largely an iron and steel town. Its founders' policy of openness to attract capital and industry was at the expense of control and stability, as Folsom notes. Later, critical economic decisions affecting the town were made elsewhere. The leadership was also badly fragmented, in part because of the community's large and diverse ethnic groups. Finally, with the restructuring of the iron and steel industry, though Harrisburg's firms slowly faded away, Scranton's leading firm simply moved away.

17. Jordan, *Encyclopedia*, 18:160–62; Obituary, *Patriot*, April 13, 1925.

18. Dun & Co., Pa. 54:62 (March 2, 1869); Pa. 54:153 (August 16, 1870); Obituary, *Patriot*, February 18, 1886.

19. Obituary, *Patriot*, October 4, 1948.

20. Lewis R. Hamersly, *Who's Who in Pennsylvania* (New York, 1904); Obituary, *Patriot*, February 12, 1923.

21. For Edward Bailey, see *Edward Bailey: In Memoriam* (Harrisburg, 1938); Jordan, *Encyclopedia*, 15:254–55; and obituaries in the *Telegraph* and the *Patriot*, October 18, 1938. For Charles L. Bailey Jr., see Jordan, *Encyclopedia*, 2:444. For William J. Calder, see Kelker's *Dauphin County*, 3:614–15, and Obituary, *Patriot*, March 1, 1911. For James M. Cameron, see Obituary, *Patriot*, October 27, 1949. For Maurice Eby, see Jordan, *Encyclopedia*, 15:379–81, and Obituary, *Patriot*, April 6, 1914. For William H. Eby, see *Patriot*, October 1, 1886, and Obituary, March 16, 1906. For Charles Fleming, see *Patriot*, March 19, 1883; for George Fleming, see *Patriot*, January 17, 1898; and for David Fleming see *Telegraph*, January 14, 1890. For William O. Hickok Jr., see Kelker's *Dauphin County*, 3:605–6; Jordan, *Encyclopedia*, 18:269–70, and Obituary, *Patriot*, October 25, 1881. For William Hildrup Jr., see Jordan, *Encyclopedia*, 10:276–77, and Obituary, *Patriot*, May 15, 1920. For the sons of James Mc-Cormick Jr., see their obituaries: William, *Patriot*, February 12, 1923; Robert, *Patriot*, May 5, 1925; Henry, *Patriot*, January 6, 1939; James, *Telegraph*, October 1, 1943; and Donald, *Patriot*, May 16, 1945.

22. Obituary of Henry B. McCormick, *Telegraph*, December 27, 1941. For Vance C., see Jordan, *Encyclopedia*, 2:688–89, and the insightful sketch by Paul B. Beers, *Profiles from the Susquehanna Valley* (Harrisburg, 1973), pp. 169–76.

23. Quoted in *Diary and Letters of Rutherford Birchard Hayes*, ed. Charles Richard Williams (Columbus, Ohio, 1925), 4:591.

24. Beers, *Profiles from the Susquehanna Valley*, pp. 17–23, portrays the role of Front Street in late nineteenth- and early twentieth-century Harrisburg. James Boyd's novel, *Roll River* (New York, 1935), set in Midian (a thinly disguised Harrisburg), deals perceptively with two generations of the Front Street set between the Civil War and the Great Depression. Boyd, whose well-off family lived a block off Front on Pine Street, had relatives who lived on Front Street. He worked briefly for the *Patriot*, and taught at the Harrisburg Academy. John O'Hara's Harrisburg novel, *A Rage to Live* (New York, 1949), though better literature, deals chiefly with the post–World War I era. In spite of its Front Street setting and characters, it is less useful in understanding their role in Harrisburg.

25. Handwritten biographical notes by Ross A. Hickok for the period 1876–1918, Hickok Papers, MG-146, PHMC, box 7, genealogical folder 6.

26. Ibid.

27. Morton J. Horwitz, *The Transformation of American Law, 1780–1860* (Cambridge, Mass., 1977), esp. pp. 140–59.

28. James M. Sellers to Simon Cameron, November 7, 1855, and Stephen Miller to Cameron, November 16, 1855; George Bergner to Cameron, August 8, 1856, and Miller to Cameron, August 28, 1856, microfilm of Cameron Papers, HSDC. See also *Dauphin County Atlas,* p. iv.

29. Haldeman Collection, Hagley Museum; Hickok Papers.

30. For example, Stormquist, *Generation of Boomers,* pp. 174–87, discusses the difference between towns that were entirely railroad towns and towns where railroads shared economic power with industries and a variety of commerce.

Chapter 8: Recruiting and Persistence of Industrial Workers

1. I am indebted to Steven J. Ross for the term "factory artisan," which he defines and discusses in *Workers on the Edge,* pp. 95–118.

2. For example, see above, Chapter 3.

3. Industrialization drew population to the larger industrial centers of Newark (N.J.) and Cincinnati at higher rates. Both industrialized earlier and more heavily than Harrisburg. In the decades between 1820 and 1860, Newark's population grew 68 percent, 58 percent, 125 percent, and 85 percent. Cincinnati expanded 149 percent in the 1840s and 40 percent in the 1850s. See Hirsch, *Roots of the American Working Class,* p. 17; Ross, *Workers on the Edge,* p. 74. Published census returns for several intermediate cities of Pennsylvania show that between 1850 and 1890 Harrisburg's population grew 403 percent, Lancaster's grew 159 percent, and Reading's 234 percent. Three cities industrializing even later—Altoona, Scranton, and Williamsport—expanded 745, 379, and 716 percent respectively between 1860 and 1890. See *1890 Census, Population,* Part 1, pp. 370–73.

4. Those not found may have listed no occupation or designated it so that was not recognized as industrial, may have lived outside the city limits, or may simply have been missed by the census taker. Illegible handwriting, misspelled or differently spelled names, or other such factors no doubt prevented the identification of some persons from one census to the next.

5. The percentages of foreign-born persons in Newark and Cincinnati in 1860 stood at 46 percent and 37 percent respectively. For Pennsylvania industrial cities, which were as high or higher, see Chapter 10, n. 2.

6. Although Dublin, *Women at Work,* pp. 35–40, discussed why farmers' daughters accepted jobs in the Lowell mills, many of their reasons would seem to apply equally well to young men there and elsewhere who were also leaving the farm. Few were driven by economic necessity, but instead left what they regarded as the isolation and dreariness of farm life for the perceived excitement of urban life. They wanted to escape dependence on family and "make it on their own," and to take advantage of the supposed greater economic and social opportunities of the city. See also Dublin, "Rural-Urban Migrants in Industrial New England: The Case of Lynn, Massachusetts, in the Mid-Nineteenth Century," *Journal of American History* 73 (December 1986): 623–44. For a discussion of how the employment of youngsters in factories was perceived favorably by working families rather than seen as exploitation, see Farley, "Frankford Arsenal," chap. 6.

7. *Morning Herald,* September 4, 1854.

8. *Telegraph,* August 24, 1853. For the longtime persistence of this complaint, see also ibid., April 17, 1850, and October 21, 1865.

9. The grouping of single-firm industries under the heading "All Others" in the manufactur-

ing censuses of 1880, 1890, and 1900, which obscured data on capital, costs, and output of individual firms (see Chapter 6), similarly affected information on the labor forces of those same companies. At Harrisburg those years, "All Others" included employees of the cotton mill and the car works as well as those of many lesser firms. In 1880 the category embraced 38 percent of all workers in manufacturing, in 1890 some 56 percent, and a decade later 30 percent. When those not in the "All Others" category are separated into industrial workers and craft shop employees, industrial workers in 1880 were found on average to have earned 14 percent more than craft workers; in 1890 they earned about 7 percent more, and a decade later 24 percent more.

10. Hildrup, *Harrisburg Car Co.*, p. 4.

11. For the labor system of the Philadelphia textile industry, see Scranton, *Proprietary Capitalism*, esp. pp. 42–71. For the labor systems north and south of Boston, see George Rogers Taylor, *The Transportation Revolution, 1815–1860* (New York, 1958), pp. 270–77. For Lowell, see Dublin, *Women at Work*.

12. *Telegraph*, January 11, 1854.

13. Because the cotton mill was located on the banks of the Susquehanna at almost the exact midpoint of the city, it is unlikely that any substantial number of employees came from outside Harrisburg. Although the 1860 census listed fifty-three cotton mill employees younger than age sixteen, it should be remembered that the census forms did not call for the occupations of anyone in that age category. Consequently, many of those not located were probably child employees under sixteen years of age who had no listed occupation in the census. The shortfall of older male employees may be because some were listed as unspecified laborers in the census and could not be identified as cotton mill employees. That would not account for the large number of unlisted female cotton mill employees, however, because no females were listed as laborers in 1860. The even larger number of unidentified cotton mill employees in 1870 is a greater mystery, because all occupations, regardless of gender or age, were supposed to be listed. Respondents must simply have not given that information.

14. Calculated from computerized manufacturing census schedules for 1850 and 1860.

15. Only five of the eight names on the payroll in 1860 could be found in the census of 1870.

16. The census of 1880 gives no data on real estate holdings.

17. See Hildrup, *Harrisburg Car Co.*, pp. 4–5.

18. Simon and J. Donald Cameron's control and management of the Northern Central in the early 1860s effectively ended when that line became a subsidiary of the Pennsylvania Railroad during the Civil War.

19. Modell, "Peopling of a Working-Class Ward," pp. 71–95.

Chapter 9: Craft Workers

1. Winpenny, *Bending Is Not Breaking*, esp. chap. 1.

2. Peter Seibert, executive director of the Historical Society of Dauphin County, who is studying Harrisburg craft shops, says a few very good artisans lived there. He suggests that the lack of a strong tradition may be due to three distinct cultures—English, Scots-Irish, and German—meeting and overlapping in the county. The Boyds (English woodworkers), the Beattys (Scots-Irish weavers, silversmiths, and clockmakers), and the Boas family (German metalsmiths), representing these cultures, all settled in the town between 1770 and 1790. (Seibert to author, April 24, 1991.)

3. Bruce Laurie, *Artisans into Workers: Labor in Nineteenth-Century America* (New York, 1989), pp. 15–16.

4. W. J. Rorabaugh, *The Craft Apprentice from Franklin to the Machine Age in America* (New York, 1986).

5. For persistence in a single craft, I have used city directories, census schedules, and biographical sketches. For Muench, see computerized manufacturing census schedules, 1850, 1860; the others who shifted trades were traced in the computerized population census schedules for 1850, 1860, and 1870.

6. Dawley's *Class and Community* offers a particularly forceful example.

7. Ross, *Workers on the Edge,* pp. xvii, 97–100. Although our ultimate conclusions about the impact of industrialization are different, Hirsch, *Roots of American Working Class,* pp. 19–36, deftly traces the often very drawn out process by which craft production was turned to factory production. Winpenny's *Bending Is Not Breaking* examines in depth the persistence of crafts in Lancaster. Farley, "Frankford Arsenal," chap. 7, points out that industrialization of gun production continued to utilize the skills of artisans and frequently created new machine-related skills.

8. Hirsch, *Roots of the American Working Class,* pp. 24–27.

9. The estimates of artisans employed at the car works were based on the proportions of each hired in 1873 applied to the total number of car works employees listed in the manuscript manufacturing census schedules for 1860 and 1870. The numbers in the repair shops of the Pennsylvania Railroad are directly from the manuscript manufacturing census schedules for 1870.

10. Griffen, "Occupational Mobility in the Nineteenth Century," p. 317.

11. For this study, confectioners were included with bakers, tobacconists with cigarmakers, and chairmakers with cabinetmakers.

12. Most persistence studies of American urban centers show half or fewer of the population remaining from one census to the next. An example of a smaller city is Griffen and Griffen, *Natives and Newcomers,* pp. 15–18. For a large city, see Stephan Thernstrom, *The Other Bostonians* (Cambridge, Mass., 1973), p. 222; the table on that page gives comparative persistence data for several antebellum cities, and rarely do they exceed 50 percent.

13. These percentages of persistence are roughly comparable to those of the same occupations at Poughkeepsie, New York, in the same censuses. See Clyde Griffen, "Workers Divided: The Effect of Craft and Ethnic Differences in Poughkeepsie, New York, 1850–1880," in *Nineteenth-Century Cities: Essays in the New Urban History,* ed. Stephan Thernstrom and Richard Sennett (New Haven, 1969), p. 76.

14. This and the material following is based on the computerized manufacturing census schedules for 1850, 1860, 1870, and 1880.

15. Charles R. Boak memoirs at HSDC (see above, Part II Introduction: A Generation Later, n. 7).

16. Ibid., pp. 19–21; the quotation is from p. 21. The best treatment of the industry is Patricia A. Cooper, *Once a Cigar Maker* (Urbana, 1987).

17. Boak memoirs, pp. 3–7.

18. Ibid., p. 7.

19. Ibid., pp. 21–22.

20. Cooper, *Once a Cigar Maker,* pp. 198–209, discusses cigarmaking in southeastern Pennsylvania, where shops or factories ranged from "buckeyes" (one-man shops) to sizable factories. Many were branches of Philadelphia and New York firms, located to take advantage of nonunion labor, low wages, and inducements from local communities to come.

21. Boak memoirs, p. 110.

22. Ibid.

23. Because they were on piecework, cigarmakers had a long tradition of working the hours they pleased. Only a few large shops enforced strict hours. As piece rates were lowered,

however, journeymen cigarmakers had to put in long hours at maximum speed just to earn a living. Cooper, *Once a Cigar Maker,* pp. 41–42, 208.

24. Ibid., pp. 314–15, discusses how displaced cigarmakers adjusted to unemployment in the 1930s.

Chapter 10: Ethnic Minorities

1. For the notion that migrating peoples found familiar frameworks on which to rebuild their lives in new communities, see Katz, *People of Hamilton,* pp. 17–18. Aurand, *Population Change and Social Continuity,* pp. 105–15, indicates that even immigrants moving from one American community to another identified with replicated institutions, such as churches, mutual aid societies, labor unions, and jobs.

2. Published census data for 1850–1900. In 1850, when Harrisburg had 11 percent blacks, Philadelphia was second with 9 percent. In 1870, when Harrisburg's proportion of blacks was 10 percent, the Pennsylvania cities nearest were Norristown, Wilkes-Barre, and Williamsport, each with about 4 percent. Philadelphia had 3 percent, Pittsburgh and Allegheny City each had 2 percent. The percentages of foreign-born residents of Pennsylvania cities with populations in excess of 10,000 were Scranton, 45 percent; Erie, 35; Pittsburgh, 32; Allegheny, 29; Philadelphia, 27; Wilkes-Barre, 26; Pottsville, 22; Lancaster, 17; Williamsport, 16; Allentown and Norristown, 15; Altoona, 14; Harrisburg and Easton, 12; Reading, 11; and York, 10. See *1870 Census,* vol. 1, *Population,* pp. 243–57.

3. Blacks outnumbered immigrants five to three in 1900, while the combined German and Irish populations sank from 94 percent of the foreign-born to 56 percent. The 401 Irish only slightly outnumbered the British (367) and the Russians (356).

4. *Laws of Pennsylvania, 1780,* pp. 67–73.

5. Arthur Zilversmit, *The First Emancipation: The Abolition of Slavery in the North* (Chicago, 1967). Zilversmit stated that there were a few slaves in Pennsylvania on the eve of the war. Published federal census returns show none after 1840. Much of the material dealing with Harrisburg's blacks in this chapter is based on material published in greater detail and in a somewhat different context in Gerald Eggert, "Two Steps Forward, a Step-and-a-Half Back: Harrisburg's African American Community in the Nineteenth Century," *Pennsylvania History* 58 (January 1991): 1–36, and is used with permission. For Harrisburg, see manuscript population census schedules, 1790–1860. At the state level, 36 percent of Pennsylvania blacks were slaves in 1790, 10 percent were slaves in 1800, a little more than 3 percent in 1810, and thereafter 1 percent or fewer through 1840. See Edward R. Turner, *The Negro in Pennsylvania* (1911; reprint, New York, 1969), p. 253.

6. Eggert, "Two Steps Forward," table 1, p. 3.

7. In 1810 some 77 percent of Dauphin County blacks lived outside Harrisburg; by 1840, only 32 percent did. The black population of the county outside Harrisburg increased from only 194 to 311 in the same thirty years.

8. Leonard Curry, *The Free Black in Urban America, 1800–1850* (Chicago, 1981), pp. 239–40.

9. Eggert, "Two Steps Forward," p. 4. The quotation is from a newspaper advertisement in the *Oracle of Dauphin,* cited by Mary D. Houts in her "Black Harrisburg's Resistance to Slavery," *Pennsylvania Heritage* 4 (December 1977): 11. In establishing their own institutions, Harrisburg blacks lagged from one to two decades behind Philadelphia blacks. See Gary B. Nash, *Forging Freedom: The Formation of Philadelphia's Black Community, 1720–1840* (Cambridge, Mass., 1988), pp. 66–133.

10. Eggert, "Two Steps Forward," pp. 4–6. Houts, "Black Harrisburg's Resistance," pp. 10–11, supplies the quotations from borough ordinances.

11. Houts, "Black Harrisburg's Resistance," pp. 9–13; Eggert, "The Impact of the Fugitive Slave Law on Harrisburg: A Case Study," *Pennsylvania Magazine of History and Biography* 109 (October 1985): 540–43, 545.

12. Eggert, "Impact of the Fugitive Slave Law," pp. 543–45.

13. Ibid., pp. 546–69. For the text of the law, see U.S., *Statutes at Large,* 9:462–65.

14. Computerized population census schedules, 1850. The published census showed 892 blacks and mulattoes, but the census taker erred. By too freely using ditto marks in the race column, he showed the socially prominent Dr. Christian Seiler and his family as black.

15. Persons living in the same household and having the same surname were classified as family members. Those with different surnames, including some who might have been family members—such as stepchildren or married female relatives (daughters, mothers, sisters) who had returned to the household—were classified as "singles."

16. The two black physicians were variously listed in the population census schedules for 1850, 1860, and 1870 as "doctor," "I. Doctor" (perhaps meaning Indian doctor), and "druggist."

17. Houts, "Black Harrisburg's Resistance," p. 11.

18. Egle's *Dauphin County,* pp. 472, 520.

19. *Telegraph,* October 30, 1850.

20. For a detailed account of Miller's work as editor of the local papers, see Eggert, " 'Seeing Sam': The Know-Nothing Episode in Harrisburg," *Pennsylvania Magazine of History and Biography* 111 (July 1987): 305–40.

21. Computerized population census schedules 1850. The very few foreign females with American husbands have been counted as American rather than foreign families. For example, the four Irish wives who had American husbands were assumed to be controlled by the husband's nationality. One of the husbands was William Calder Sr., the stagecoach entrepreneur; another was Daniel Boas, a leading lumber merchant. Both families were protestant and town leaders.

22. Thernstrom, *Poverty and Progress,* pp. 156–57, discusses this with reference to the Irish of Newburyport, Massachusetts. He concludes that, although Irish parents acquired real estate, relatively few of their children advanced above occupations as laborers.

23. Examples of blacks going south to assist their race are anecdotal. T. Morris Chester of Harrisburg was one. Some Philadelphia blacks who did are noted in Roger Lane, *William Dorsey's Philadelphia and Ours: On the History and Future of the Black City in America* (New York, 1991), pp. 44–46 and passim. See also Pauli Murray, *Proud Shoes: The Story of an American Family* (New York, 1984).

24. For the purging and other abuses suffered by blacks in Harrisburg during the war and immediately after, see Eggert, "Two Steps Foward," pp. 12–16.

25. Ibid., p. 22.

26. Computerized population census schedules, 1850, 1860, 1870.

27. W.E.B. Du Bois, *The Philadelphia Negro* (New York, 1899), p. 269, argued that recent emancipation and pauperism went hand in hand in postwar Philadelphia. Elizabeth Hafkin Pleck, *Black Migration and Poverty: Boston, 1865–1900* (New York, 1979), pp. 44–67, demonstrates that Boston's black newcomers from Virginia came from urban areas, already had experience with urban work, and were only 32 percent illiterate. I believe the adult illiteracy rate of more than 61 percent for Harrisburg's black newcomers indicates a much larger number of recent fieldhands there.

28. Eggert, "Two Steps Foward," pp. 23–24.

29. Pauline Allen, "Freed Slave Began Hill Development," *Patriot,* February 9, 1982.

30. Editor R.J.M. Blackett provides a biographical sketch of Chester in his *Thomas Morris Chester: Black Civil War Correspondent* (Baton Rouge, La., 1989), pp. 14–34.

31. Michael J. Nestleroth, "The Black Community of Harrisburg, 1880–1910: A Study of the Black Community of a Small Northern City" (college term paper, Penn State, Harrisburg, 1973), pp. 6–8 (copy on file at HSDC). (Although this is an undergraduate study, the research appears to have been done with care.)

32. Ibid., pp. 16–17. A sample of all black males instead of only married ones would probably have produced less-impressive gains.

33. Ibid., pp. 5, 18, 32–33.

34. Ibid., p. 13; Egle's *Dauphin County*, p. 323.

35. *1890 Census, Population*, Part 1, p. 571; *1900 Census, Population*, Part 1, p. 638.

36. Nestleroth, "The Black Community of Harrisburg," pp. 30–31. For the law and practice regarding blacks in Pennsylvania schools, see Ira V. Brown, *The Negro in Pennsylvania History* (University Park, Pa., 1970), pp. 52–54. For Dr. Day, see the sketch prepared for the Harrisburg school board, Genealogical File, State Library, Harrisburg, and his obituaries, *Telegraph*, December 3, 1900, and *Patriot*, December 4, 1900.

37. For example, see Du Bois, *Philadelphia Negro*; Pleck, *Black Migration and Poverty*; Roger Lane, *Roots of Violence in Black Philadelphia, 1860–1900* (Cambridge, Mass, 1986); David M. Katzman, *Before the Ghetto: Black Detroit in the Nineteenth Century* (Urbana, Ill., 1973); and Kenneth L. Kusmer, *A Ghetto Takes Shape: Black Cleveland, 1870–1930*.

38. The migration to Harrisburg between 1865 and 1870 seems to have been an exception rather than the rule. See Herbert G. Gutman, *The Black Family in Slavery and Freedom, 1750–1925* (New York, 1975), p. 433.

39. Recent studies of cities that had large influxes of blacks after 1914 include Peter Gottlieb, *Making Their Own Way: Southern Blacks' Migration to Pittsburgh, 1916–1930* (Urbana, Ill., 1987); Kusmer, *A Ghetto Takes Shape*; and Joe William Trotter Jr., *Black Milwaukee: The Making of an Industrial Proletariat, 1915–1945* (Urbana, Ill., 1958).

40. George C. Wright, *Life Behind a Veil: Blacks in Louisville, Kentucky, 1865–1930* (Baton Rouge, La., 1985), argues that gains made by blacks in Louisville, Kentucky (in some respects more impressive than the gains at Harrisburg) were essentially token devices of whites for co-opting black leaders and placating the remainder to keep them "in their place." Kenneth L. Kusmer, "The Black Urban Experience in American History," in *The State of Afro-American History*, ed. Darlene Clark Hine (Baton Rouge, 1986), pp. 106–8, notes that blacks in northern cities shifted to an accommodationist stance before World War I, when segregation and hostility to blacks were increasing. Once their communites grew larger after the war, black professionals became less dependent on whites, and black militance increased.

Chapter 11: Labor Relations in the Early Industrial Era

1. *Morning Herald*, July 6, 1854. The festivities this particular year were better organized and more elaborate then usual.

2. Quotations from the *Telegraph*, February 7, 1852; see also February 21, 1852, and January 15, 1853.

3. *Patriot*, July 15, 1867.

4. Ibid., August 8, 10, and 15, 1867. Notice of the Lochiel Iron Works Library Association picnic appeared in the August 12, 1867, *Telegraph*.

5. Hildrup, *Harrisburg Car Co.*, pp. 4–5.

6. Ibid., pp. 54–55.

7. *Biographical Encyclopedia*, pp. 334–36; *Patriot*, January 5, 1882.

8. The annual figures in Hildrup's history of the Harrisburg Car Works and in his annual reports to the state differ, perhaps because they were calculated from different dates during the year. Those in the history appear to be based on company annual reports.

9. According to reports to the state by Wister Brothers and the McCormick-owned Paxton furnaces in 1876, McCormick paid between 8 and 23 percent higher wages for the same jobs. McCormick also paid differing wages for essentially similar tasks at his Harrisburg furnaces and his nail works in West Fairview. Data from Pennsylvania Secretary of Internal Affairs, *Annual Report, 1875–1876*, pp. 652–53; ibid., *1878–1879*, pp. 100–101.

10. For examples, see *Telegraph*, June 20 and July 25, 1849, and September 25, 1850; *Whig State Journal*, June 24, 1851; and *Telegraph*, June 25, 1851, and April 4, 1855.

11. *Telegraph*, April 14 and 28, 1852. According to the *Morning Herald*, April 19, 1854, the refusal of the proprietor of the *Patriot* to discharge a worker who had "ratted" led all the other employees to stop operations.

12. *Whig State Journal*, March 24, 1853.

13. The letter from Oscar and Fenn's reply appear in the *Telegraph*, March 23, 1853; that of "Z" is in the April 2 issue. The only other expressions of republican sentiment I found in Harrisburg in the first decade of industrialization came during the debate over general incorporation laws (see Chapter 3). The lack of a strong spirit of republicanism at Harrisburg may have been the result of weak crafts and the absence of a respected craft tradition, or it may have been that local industries did not displace a significant number of artisans. Such a spirit may even have existed, but without a labor movement or press received little expression.

14. *Laws of Pennsylvania, 1848*, pp. 278–79; "An Act for the Relief of the Heirs of James Caldwell, Deceased, and Relative to the Hours of Labor in Manufacturing Establishments," ibid., *1849*, pp. 671–72.

15. *Telegraph*, April 17, 1853.

16. Ibid., August 24, 1854.

17. Ibid., October 21, 1865.

18. Ibid., October 26, 1853.

19. Ibid., October 20, 1853. The information on Morgan and Barr comes from computerized population census schedules for 1850, 1860, and 1870 and from Harrisburg directories for 1839, 1842, 1843, and 1845. The *Telegraph*, April 28, 1852, listed Morgan as one of those on strike. The two also later belonged to the Harrisburg Guard of Liberty, an offshoot of the Know-Nothing movement.

20. *Democratic Union*, November 2, 1853.

21. Sessions Docket Book, vol. 9, p. 35, Dauphin County Courthouse.

22. Cited in Gerald Eggert, *Steelmasters and Labor Reform, 1886–1923* (Pittsburgh, Pa., 1981), p. 6.

23. According to the *Patriot*, July 1, 1871, NLU President Richard Trevelick was in Harrisburg on June 30. On July 3 it reported that 100,000 Pennsylvanians and 1,000 residents of Harrisburg belonged to the National Labor Union. For the number of heats during the week and on Saturdays, see *Patriot*, April 11, 1871.

24. *Patriot*, April 11 and 12, 1871. For the company's other difficulties, see Chapter 5.

25. *Patriot*, April 17, 1871. According to the editor, the delay in printing the letter, dated April 12, was due to "unavoidable circumstances."

26. Pennsylvania Secretary of Internal Affairs, *Annual Report, 1880–1881*, p. 310; *Patriot*, July 1 and 2, 1874.

27. Pennsylvania Secretary of Internal Affairs, *Annual Report, 1880–1881*, p. 315; *Patriot*, May 10 and 28, 1875; *Telegraph*, May 10, 1875.

28. David Montgomery, "Strikes in Nineteenth-Century America," *Social Science History* 4

(1980): 81–104, argues that between 1845 and 1865, work rules, conditions, and hours seldom were the explicit issue. Instead, workers were fighting to defend themselves against a steadily rising cost of living (ibid., p. 88).

Chapter 12: From the Riots of 1877 to Century's End

1. The best and fullest account of the 1877 disorders is Robert V. Bruce, *1877: Year of Violence* (Chicago, 1959).

2. Ibid., pp. 93–158.

3. Ibid., pp. 59–63; quotation from p. 59.

4. Testimony of Dauphin County Sheriff William J. Jennings, "Report of the Committee Appointed to Investigate the Railroad Riots in July, 1877," in *Pennsylvania Legislative Documents Comprising the Department and Other Reports, Session of 1878* (Harrisburg, 1878), Document 29, 5:650 (hereafter cited as "Report on Railroad Riots").

5. *Telegraph,* July 23, 1877; Testimony of Mayor Patterson, "Report on Railroad Riots," p. 640. The computerized population census schedules for 1870 listed Patterson's occupation as "railroad clerk."

6. *Telegraph,* July 23, 1877; Patterson testimony, quoted in "Report on Railroad Riots," p. 643. The mayor, testifying in March 1878, confused the time, relating this as an incident taking place on Monday night rather than Saturday.

7. *Telegraph,* July 23, 1877.

8. Report of General J. K. Seigfried to Adjutant General James W. Latta, November 1, 1877, reprinted in George B. Stichter, "The Schuylkill County Soldiery in the Industrial Disturbances in 1877," *Publications of the Historical Society of Schuylkill County,* 1:194–207 (hereafter cited as Seigfried Report). See also *Telegraph,* July 23, 1877.

9. Testimony of Harrisburg attorney David Mumma, "Report on Railroad Riots," p. 657.

10. *Telegraph,* July 23, 1877; testimony of Patterson, "Report on Railroad Riots," pp. 644–45.

11. *Telegraph,* July 23, 1877. The newspaper also spelled his name "Talbert," and Mayor Patterson referred to him as "Torbett." At Tolbert's request, he was allowed to "correct" the newspaper's account. His version appeared on July 25. The most notable difference was that the revised account omitted a statement that the government's mortgage on the railroad companies had been purchased with money "wrung from the sweat of honest laboring men." Let the state take over the roads, he had declared, and "the men will be treated like men." This was loudly cheered by the crowd.

12. *Telegraph,* July 23, 1877; "Report on Railroad Riots," pp. 640–41.

13. *Telegraph,* July 23, 1877. *Patriot,* July 20, 1877.

14. *Telegraph,* July 23, 1877.

15. Seigfried Report, p. 196.

16. Testimony of A. J. Herr, "Report on Railroad Riots," pp. 636–39; and of tollgate-keeper Thomas Reckford, pp. 651–52. The quotations are by Reckford.

17. Seigfried Report, p. 201. Jones Wister in his *Reminiscences* (p. 182) had only contempt for those who surrendered their arms at Marysville: Once disarmed by the crowd, they "sneaked behind [their captors] like so many whipped dogs" and were finally "sent home as useless baggage."

18. *Patriot,* July 24, 1877. See also Patterson testimony and Mumma testimony, "Report on Railroad Riots," pp. 641, 644, 656–61.

19. *Telegraph,* July 24, 1877.

20. "Report on Railroad Riots," pp. 644–45, 648.

21. Ibid., pp. 642, 646.

22. Patterson estimated the crowd at 600 to 1,000; Jennings thought there were between 200 and 300 (ibid., pp. 642, 647).

23. Ibid., p. 647.

24. Ibid., pp. 642–43, 646. Ironmaster Jones Wister related a different version of the affair (*Jones Wister's Reminiscences*, pp. 182–83). In his account the heroes were the town's businessmen, not public officials. As the crisis drew near, many of the "leading citizens" called on Henry McCormick to head a party of volunteers. Meeting in secret at McCormick's home, plans were made to assemble at the sound of the town bell. McCormick, who had been a colonel during the Civil War, "was to command with the title of general." The group arranged to have all firearms and ammunition removed from gun shops and hardware stores. When it appeared that a mob was about to set the offices of the *Telegraph* afire at 8:00 in the evening, the town bell sounded the alarm. The volunteers assembled and divided into two companies, one headed by McCormick, the other by Wister. The two groups "covered Market Street from house to house, armed with clubs and pistols, and marched down to the depot, where several thousand strikers had gathered. Some blows were struck . . . but there were no fighting men in the crowd." The "determined advance" had "completely awed them, and they melted away like snow in the spring." The volunteers formed themselves into "a law-and-order" company and "marched through the town all night, requiring each man on the street to state his reasons for being out or to go to his home." They then wired for U.S. Army forces that "could not be frightened off" by mobs. Thereafter, regular soldiers guarded trains in the city, and the blockade was broken. Neither Wister nor McCormick was mentioned in any contemporary newspaper account or in the testimony gathered from Patterson, Jennings, or any other witnesses before the legislative committee that inquired into the riots.

25. Bruce, *1877: Year of Violence*.

26. "Report on Railroad Riots," pp. 644–45.

27. Ibid., p. 649.

28. Ibid., p. 657.

29. Monkkonen, *America Becomes Urban*, pp. 194–97; see Monkkonen's citations for evidence.

30. None of the Harrisburg accounts mentions women in the crowds. With a population of 23,104 in 1870 and 30,762 in 1880, Harrisburg probably had about 28,000 residents in 1877. If 47 percent were male and a third of those were either younger than thirteen or older than sixty-nine (as was the case in 1870), the total number of males between the ages of thirteen and sixty-nine in 1877 would have been about 8,800.

31. Events in nearby Reading showed the consequences of less-responsible leadership. In the absence of the mayor there, neither the police nor the sheriff took action when a mob began destroying railroad property. Franklin Gowen, president of the Philadelphia & Reading and recent prosecutor of the Molly Maguires, requested state militiamen, which were sent. Violence followed, in which eleven were killed and many were injured. Ronald L. Filippelli, "The Railroad Strike of 1877 in Reading," *Historical Review of Berks County* 38 (Spring 1972): 48–51, 62–71; Bruce, *1877: Year of Violence*, pp. 188–94.

32. "The Statistics of Strikes," in Pennsylvania Secretary of Internal Affairs, *Annual Report, 1880–1881*, p. 383; ibid., *1887*, sec. 12. The data was supplied by U.S. Commissioner of Labor Carroll D. Wright and supposedly included only strikes between 1881 and 1886 that lasted seven days or longer.

33. *Patriot*, January 2, 1882.

34. Ibid., January 17, 1882.

35. *Morning Call*, July 28 and 29, 1886. According to the July 28, 1886, *Harrisburg*

Independent, "all through the panic the fires burned brightly" at Old Hot Pot, "and men were kept in work when hundreds of similar establishments were idle."
36. *Patriot,* July 29 and 30, 1886.
37. *Harrisburg Independent,* July 29, 1886.
38. "Statistics of Strikes," sec. 12; *Patriot,* August 30, September 1 and 6, and November 1, 1886.
39. "Statistics of Strikes," sec. 12.
40. Ibid.; *Patriot,* August 6, 1887.
41. *Patriot,* October 8–11 and 15, 1895.

Chapter 13: Industrialization and Harrisburg Politics

1. Monkkonen, *America Becomes Urban,* pp. 89–130, has an excellent discussion of the political evolution of urban government in the United States.
2. The borough charter of 1808 is in *Laws of Pennsylvania, 1808,* pp. 20–24. The quotation is from the charter preface.
3. *Laws of Pennsylvania, 1837,* p. 126.
4. Borough charter, 1808, sec. 2.
5. The city charter of 1860 is in *Laws of Pennsylvania, 1860,* pp. 175–200.
6. *Laws of Pennsylvania, 1867,* pp. 423–25; ibid., *1868,* pp. 1136–44.
7. Ibid., *1874,* pp. 230–70, esp. secs. 3, 4, 19, 30.
8. Borough charter, 1808, secs. 6, 7, 11. For the changes, see *Laws of Pennsylvania, 1824,* pp. 270–71; ibid., *1839,* p. 183.
9. City charter, 1860, secs. 8, 9, 11, 15.
10. Based on extensive reading of local newspapers and the council Minutes, vols. 1 and 2.
11. Egle's *Dauphin County,* pp. 365–68.
12. Ibid., pp. 352–55. See above, Chapter 7.
13. See the *Telegraph,* the *Patriot,* and other Harrisburg newspapers for 1850–67.
14. See my articles in the *Pennsylvania Magazine of History and Biography:* "Impact of the Fugitive Slave Law on Harrisburg," 109 (October 1985): 537ff.; " 'Seeing Sam': The Know-Nothing Episode in Harrisburg," 111 (July 1987): 305ff.
15. The terms and political affiliations of the burgesses, mayors, and council members have been compiled from newspaper reports of election returns, city council minutes, and city directories.
16. The occupations of council members are based on computerized population census schedules and city directories.
17. Computerized population census schedules, 1850, 1860, 1870.
18. City directories after the 1860s do not indicate race. According to Beers, *Profiles from the Susquehanna Valley,* p. 32, no African American held office at city hall until one became city controller in 1967. Two years later the election to council of the first black and the first Jewish woman broke three barriers for that body, one racial, one based on gender, and one based on religion.
19. Hoffecker, *Corporate Capital,* pp. 5–7, summarizes some of the literature on these issues. Hoffecker cites sociologist Floyd Hunter, for example, who found in Atlanta in the 1950s that the city's economic leaders exercised influence commensurate with their financial powers because of their economic stake in the community. Even when lesser businessmen held impressive positions, they were perceived as messenger boys of those at the top. Robert O. Schulze, discussing Ypsilanti, Michigan, also in the 1950s, found that where absentee owners of the principal businesses ran their plants through resident hired managers, the managers avoided local politics.

20. Erwin Stanley Bradley, *The Triumph of Militant Republicanism: A Study of Pennsylvania and Presidential Politics, 1860–1872* (Philadelphia, 1964), deals effectively with the operations of the machine. James A. Kehl, *Boss Rule in the Gilded Age: Matt Quay of Pennsylvania* (Pittsburgh, 1981), discusses the Cameron machine, especially under Quay's leadership.

21. Paul B. Beers, "Boies Penrose, 1860–1921," in *Pennsylvania Kingmakers,* ed. Robert G. Crist (University Park, Pa., 1985), p. 40. The argument that follows is based on circumstantial evidence added to any direct information that is available. I have considered what measures affected Harrisburg, who likely would have introduced them, who controlled the state machine, and who benefited.

22. According to Klein and Hoogenboom, *History of Pennsylvania,* p. 318, between 1866 and 1873, with 8,700 of 9,230 legislative acts falling into that category, special legislation had become a scandal and would be specifically forbidden by the new Pennsylvania Constitution of 1873.

23. Letter from "Anti-Freehold Law" to editor, *Telegraph,* February 9, 1860.

24. "A Committee of Council" to editor, ibid., February 19, 1860. I found no ties between Fisher and the rising industrial entrepreneurs.

25. Letter to editor, ibid., March 8, 1860.

26. "Harrisburg" to editor, and "Fair Play to All" to editor, ibid., February 25 and March 10, 1860.

27. Letter to editor, *Patriot,* February 22, 1860.

28. An editorial in the *Telegraph,* March 1, 1867, attributed authorship of the proposal to former Mayor Augustus L. Roumfort and noted objections in the *Patriot* of that date. No copy of the March 1, 1867, *Patriot* is available.

29. *Telegraph,* March 16, 1867; *Patriot,* March 21, 1868.

30. *Telegraph,* March 21, 1868.

31. Ibid., April 10; *Patriot,* April 22, 1868

32. *Patriot,* October 12 and 14, 1869.

33. Bradley, *Triumph of Republicanism,* pp. 357–61.

34. Article III, sec. 7. A copy of the constitution is in *Laws of Pennsylvania, 1874,* pp. 1–42.

35. *Laws of Pennsylvania, 1874,* pp. 230–70, esp. secs. 1, 3, 4, 16, 19, 20, 30.

36. Klein and Hoogenboom, *History of Pennsylvania,* p. 318.

37. Bradley, *Triumph of Republicanism,* pp. 253–54, 260–61, 274–75.

38. Election returns as reported in the *Patriot,* October 14, 1869; October 12, 1870; October 14, 1872; February 17, 1874; February 16, 1876; and February 20, 1879.

39. *Patriot,* October 12, 1871. On Day, see Chapter 10, above.

40. *Telegraph,* February 7, 1882; *Patriot,* February 8, 1882.

41. *Patriot,* February 16 and 22, 1882.

42. Ibid., February 20, 1884.

43. Shelton Stromquist, *Generation of Boomers,* pp. 142–87, contrasts railroad cities (created by and wholly dependent on that single enterprise) with older market cities that added railroads to existing multisector economies. Strikes in the former often aroused the entire community against the company that dominated them. In market cities, by contrast, other sectors served as a moderating influence. The same analysis would apply with equal force to almost any community dominated by a single industry.

44. Harold W. Aurand, *From the Molly Maguires to the United Mine Workers* (Philadelphia, 1971), pp. 96–114; Bruce, *1887: Year of Violence,* pp. 296–99.

45. Election returns, *Patriot,* November 7, 1877. The Greenback movement was not exclusively labor-oriented. How many of the votes were cast because of the strikes, how many were in favor of inflationist doctrines, and how many were for other reasons cannot be known.

46. Ibid., February 20, 1878.

47. The percentages for the various cities were compiled from election data in John A.

Smull, *Smull's Legislative Handbook* (Harrisburg, 1879), pp. 316–466. For the difficulties at Reading, see Filippelli, "Railroad Strike of 1877 in Reading," pp. 48–51, 62–71; Bruce, *1877: Year of Violence*, pp. 189–94. For Altoona and Wilkes-Barre, see ibid., pp. 185–87, 295, 298–99. Michael Nash, *Conflict and Accommodation: Coal Miners, Steel Workers, and Socialism, 1890–1920* (Westport, Conn., 1982), demonstrates for other communities the relationship between strike defeats and increases in third-party voting.

48. Election returns, *Patriot*, February 20, 1879; February 18, 1880; February 22, 1882.

49. Monkkonen, *America Becomes Urban*, pp. 117–20, briefly discusses Robert A. Dahl's four-stage theory. The two nineteenth-century stages were succeeded in the twentieth century by reformer regimes that displaced the corrupt ethnic-based bosses and were followed in turn by professional politicians and bureaucrats.

50. Council Minutes and Ordinance Books.

Chapter 14: The Impact by Century's End

1. During the onset of industrialization between 1850 and 1870, Harrisburg's population had grown an average of 9.7 percent a year; by contrast, between 1870 and 1900 the average was 3.9 percent. For the other changes, see A. C. Stamm, *The Progress of Harrisburg* (Harrisburg, 1935), pp. 4–9.

2. For these and the following physical changes by 1900, I have relied on the detailed *Atlas of the City of Harrisburg* (1901) compiled by the Harrisburg Title Company.

3. Calculated from *1900 Census, Population*, Part 2, pp. 566, 568.

4. *1890 Census, Farms and Homes, Proprietorship and Indebtedness*, p. 367; *1900 Census, Population*, Part 2, p. 709.

5. Barry Bluestone and Bennett Harrison, *The Deindustrialization of America* (New York, 1982), pp. 49–81. On the steel industry, see John Strohmeyer, *Crisis in Bethlehem* (Bethesda, Md., 1986), and John P. Hoerr, *And the Wolf Finally Came: The Decline of the American Steel Industry* (Pittsburgh, 1988), esp. pp. 8–9, 78–81. On autoworkers, see David Halberstam, *The Reckoning* (New York, 1986), pp. 486–89.

6. For a more realistic appraisal of the "fragile affluence" enjoyed before the 1980s, see Gregory Pappas, *The Magic City: Unemployment in a Working-Class Community* (Ithaca, N.Y., 1989), pp. 13–34.

7. The seminal modern treatment is E. P. Thompson, *The Making of the English Working Class* (New York, 1963). Among the many scholars dealing with the American experience are David Montgomery, *Beyond Equality: Labor and the Radical Republicans, 1862–1872* (New York, 1967); Dawley, *Class and Community;* Hirsch, *Roots of the American Working Class;* and Ross, *Workers on the Edge.*

8. Hirsch, *Roots of the Working Class*, pp. 3–13, has an especially idealized view of the preindustrial world of work. For example, her statements that "independence and skill were the artisan's chief virtues, and these were possessed by all mechanics" (p. 7) and that "mechanics and farmers alike labored from sun up to sun down, but work was not an onerous burden to either" (p. 9) both overstate the case. The latter could only have been written by someone who has not worked as either a farmer or mechanic. According to Laurie, *Artisans into Workers*, p. 36, as early as the 1780s, masters were slighting "their moral and educative obligations to apprentices." Even at its supposed golden age in late medieval Europe, the apprenticeship system was not a harmonious order. See Robert Darnton, *The Great Cat Massacre and Other Episodes in French Cultural History* (New York, 1984), pp. 75–104. Peter Laslett, *The World We Have Lost Further Explored* (New York, 1984), pp. 5–8, who admiringly describes that era's family-work system as highly satisfying emotionally, a "circle of affection," and a "love-relationship," admits that it also could be a "scene of hatred" in which tension was "incessant and unrelieved."

9. Rorabaugh, *The Craft Apprentice,* pp. 3–130; the quotation is from p. 16. Once Rorabaugh turns to the rise of industry, it is almost as if he had not read the first half of his own book. After establishing that the apprenticeship ideal had really never worked in America, he complains (pp. 130, 209) that the industrial system, unlike the apprentice system, failed to provide youths with mentoring and guidance through their difficult adolescent years. At pp. 167–72 he suggests that the factory system's neglect of the young, and the lack of promising trades, contributed to the rising restlessness and alienation of working-class adolescents, to their drinking sprees, to membership in urban street-gangs, and to participation in such movements as Know-Nothingism and the California gold rush. See Bernard Elbaum, "Why Apprenticeship Persisted in Britain, but Not in the United States," *Journal of Economic History* 49 (June 1989): 337–49.

10. Computerized population census schedules, 1860.

11. Laurie, *Artisans into Workers,* p. 36.

12. Gary J. Kornblith, "The Craftsman as Industrialist: Jonas Chickering and the Transformation of American Piano-Making," *Business History Review* 59 (Autumn 1985): 349–68, uses the term "artisan entrepreneur" for masters who successfully adapted to industrial techniques without entirely abandoning craft standards. See also Chapter 9, above.

13. For examples of the role of ethnic and religious rivalry in dividing workers and the communities in which they lived, see David Montgomery, "The Shuttle and the Cross: Weavers and Artisans in the Kensington Riots of 1844," *Journal of Social History* 5 (Summer 1972): 411–46; Walkowitz, *Worker City, Company Town,* pp. 255–60; Richard J. Oestreicher, *Solidarity and Fragmentation: Working People and Class Consciousness in Detroit, 1875–1900* (Urbana, Ill., 1986). For the calculated use of ethnic rivalries to increase output and divide workers, see Livesay, *Carnegie,* p. 134.

14. Brody, *Steelworkers in America,* pp. 1–26, discusses the constant drives for greater efficiency and output in the leading steel mills. His evidence is drawn largely from Carnegie Steel, U.S. Steel, and Bethlehem Steel. See also Livesay, *Carnegie,* pp. 109–28.

15. Stephan Thernstrom and Peter R. Knights, "Men in Motion," *Journal of Interdisciplinary History* 1 (Autumn 1970): 7–37, argue that transiency left blue-collar workers alienated but "invisible and politically impotent" and minimized their chances for organizing effectively. For low persistence and high geographic mobility generally, see Monkkonen, *America Becomes Urban,* pp. 194–97.

16. By the 1870s the principal owners of the anthracite coal companies, for example, lived in Philadelphia and New York. See Wallace, *St. Clair,* pp. 55–70; and Aurand, *From Molly Maguires to UMW,* pp. 15–19. Pittsburgh's iron and steel masters lived either in fashionable districts well removed from the mills or in New York City. Carnegie, for example, lived at Homewood, fifteen miles northeast of Pittsburgh (his principal mills were miles south of the city), and after 1887 divided his time between his mansion in New York City and Skibo Castle in Scotland. With few exceptions, violence in labor disputes grew out of employers using force to break up picket lines and demonstrations, to displace strikers, or to protect strikebreakers. Until Homestead, for example, Carnegie closed down during labor troubles and waited out the strikers. The use of Pinkerton detectives in 1892 to clear the way for strikebreakers produced a pitched battle and bloodshed. Livesay, *Carnegie,* pp. 134–44. Although the industries in Wilmington, Delaware, were locally owned, the employers, who had easy access to labor pools in Philadelphia and Camden, brought in strikebreakers. Hoffecker, *Wilmington,* p. 126.

17. For the advantages of a community of resident entrepreneurs over absentee entrepreneurs, see Hoffecker, *Corporate Capital,* pp. 5–8.

18. Pennsylvania's coal and iron and steel industries were notorious for their use of local and state police power in putting down unions and strikes. Among other things, Pennsylvania law authorized the "Coal and Iron Police," nominated by the corporations and appointed by the governor, to act with the authority of regular police in protecting company property. Company officials frequently held commissions in this quasi-state body. If local government

proved inadequate or uncooperative in labor disputes, corporate officials used their influence with the state to secure any additional force needed. Aurand, *From Molly Maguires to UMW*, pp. 25–26, 96–114. An example of how use of the government against worker interests had political repercussions is the collapse of law and order in Scranton during the 1877 railroad strikes. State and federal troops occupied the city for three months, and in the violence that followed, three people were killed and twenty-five were wounded. Although other factors were involved, Scranton voters elected Terence V. Powderly, the Greenback-Labor candidate, as mayor in February 1878. Shortly after taking office, Powderly suspended the entire police force and appointed new personnel, all of whom were members of the Greenback-Labor party or belonged to the Knights of Labor. See Vincent J. Falzone, "Terence V. Powderly: Politician and Progressive Mayor of Scranton, 1878–1884," *Pennsylvania History* 41 (July 1974): 289–309.

19. Hoffecker, *Corporate Capital*, pp. 6–7.
20. Greenberg, *Worker and Community*, p. 85.
21. Cumbler, *Social History of Economic Decline*, pp. 31–34, 94–100.
22. Bruce, *1877: Year of Violence*, pp. 188–94.
23. Hoffecker, *Wilmington*, pp. 115–27.
24. Studies dealing with both entrepreneurs and workers include Josephson's *Golden Threads* and Wallace's *St. Clair* and *Rockdale*. Recent comparative studies include Folsom's *Urban Capitalists;* Walkowitz's *Worker City, Company Town;* and Cumbler's *Working-Class Community*.

Epilogue

1. Flower, "Central Iron & Steel," pp. 73–74. For a good brief account of the formation of U.S. Steel, see Joseph F. Wall, *Andrew Carnegie* (New York, 1970), pp. 765–93.
2. Flower, "Central Iron & Steel," pp. 74–83.
3. Ibid., p. 81.
4. Eggert, *Steelmasters and Labor Reform*, pp. 61–76.
5. Ibid., pp. 84–100; Barrett, "Harrisburg Flood Protection Study," 49; Paul Beers to author, February 28, 1992.
6. "Steelton Plant History," pp. 10–21; *Patriot-News*, August 21, 1988.
7. *Since 1853*, pp. 40–83; *Harsco Corporation Annual Report, 1985;* interview with Gerald F. Gilbert Jr., senior vice president and secretary, November 12, 1986.
8. *Between the Lines*, pp. 16–17.
9. *Moody's Bank and Financial Manual, 1990* (New York, 1990), pp. a4–a5.
10. William H. Wilson, *The City Beautiful Movement* (Baltimore, 1989). Chapter 6 of Wilson's book is devoted to Harrisburg's experience and provides an excellent account of how the city's elite managed to sell to the electorate a program that combined their interests with those of the community as a whole. For an example of propaganda circulated by the movement in Harrisburg, see J. Horace McFarland, *The Awakening of Harrisburg* [Harrisburg, 1906?]. See also George P. Donehoo, *Harrisburg the City: Beautiful, Romantic, and Historic* (Harrisburg, 1927), pp. 176–83.
11. *1980 Census, Population*, vol. 1, chap. C, pt. 40: *Pennsylvania*, sec. 1, p. 46.
12. *Patriot-News*, August 21, 1988.
13. Tri-County Regional Planning Commission, *Commercial Industrial Development Handbook, 1983*, pp. 72, 96. Based on U.S. census data.
14. Greenberg, *Worker and Community*, pp. 18–19.
15. Deindustrialization at Trenton is the subject of Cumbler's *Social History of Economic Decline*.
16. Hoffecker, *Corporate Capital*, pp. 4–5.
17. Procter and Matuszeski, *Gritty Cities*, p. 171.

SELECTED BIBLIOGRAPHY

Many sources used for this study contributed only small bits of information. For that reason this bibliography includes only sources of major importance. Also not listed here are newspapers, computerized manuscript or published federal census materials, and Harrisburg city directories—which provided much of the basic data used. Full citations of those sources are in the notes.

Primary Sources by Depository

Baker Library, Harvard University Graduate School of Business Administration, Boston, Mass.:
 R. G. Dun & Co. Collection
Commonwealth National Bank Archives, Harrisburg, Pa.:
 Harrisburg Bank and Harrisburg National Bank, Minutes of the Board of Directors
Dauphin County Courthouse, Harrisburg, Pa.:
 Probate Records
Dauphin Deposit Bank & Trust Co. Archives, Harrisburg, Pa.:
 Dauphin Deposit Bank: Minutes Books, Special Accounts Book, State of the Institution Books (weekly balance sheets)
 McCormick Estate Trust Papers
 Minutes Books of the McCormick Company
Flower, Milton E., Professor, Carlisle, Pa.:
 Lenore Embick Flower, "Central Iron and Steel Company, Harrisburg, Pennsylvania" (Manuscript)
Hagley Museum and Library Archives, Wilmington, Del.:
 Haldeman Family Papers
 Archibald Johnston Collection
 "Steelton Plant, Bethlehem Steel Company, Steelton, Pennsylvania," Bethlehem Steel Corporation Collection
Harrisburg City Clerk's Office, Municipal Building:
 Minutes of City Council, Ordinance Books
Hickok Manufacturing Co. Archives, Harrisburg, Pa.:
 Time Books, Dividend Books, Minutes Books
Historical Society of Dauphin County, Harrisburg, Pa.:
 Charles R. Boak Collection
 Simon Cameron Papers
 Harrisburg Bridge Co. Minutes Books
 Harrisburg Cotton Manufacturing Company Minutes Book
 Harrisburg Savings Institution (original name of Dauphin Deposit Bank) Minutes Book
 McCormick Family Papers
 C. C. Rawn diaries

Historical Society of Pennsylvania, Philadelphia:
    Simon Cameron items, Society Small Collection
    Samuel Morse Felton Papers
    Simon Gratz Autograph Collection
    Papers of Wayne MacVeagh
Library of Congress, Manuscripts Division, Washington, D.C.:
    Nicholas Biddle Papers
    Simon Cameron Papers
National Archives, Washington, D.C.:
    Records of the Bureau of Refugees, Freedmen, and Abandoned Lands, for the
        District of Columbia, RG 105 (microfilm)
Pattee Library of the Pennsylvania State University Library, Special Collections:
    Duncannon Iron Works Collection
    Douglas Dinsmore, "Archive Report: Duncannon Iron Works," with calendar
        of letters
Pennsylvania State Archives, Historical and Museum Commission, Harrisburg, Pa.:
    Dauphin County Tax Records, RG 47
    Dock Family Papers, MG 43
    Elder Family Papers, MG 45
    Haldeman-Wright Papers, MG 64
    Ross A. Hickok Papers, MG 146

Published Primary Sources

Combination Atlas Map of Dauphin County, Pennsylvania. Philadelphia, 1875.
    Reprinted by Dauphin County Historical Society, Wilkes-Barre, Pa., 1985,
    with "Reprise" by Robert G. Crist.
"Documents Relating to the Manufacture of Iron in Pennsylvania." Journal of the
    Franklin Institute 51 (January 1851): 69–72 and charts.
Harrisburg Board of Trade Statistical Committee. Industrial and Commercial Re-
    sources of the City of Harrisburg. Harrisburg, 1887.
Harrisburg Title Company. Atlas of the City of Harrisburg, Dauphin County,
    Penna., Made from Plans, Deeds, and Survey. Harrisburg, 1901.
Hildrup, William T. History and Organization of the Harrisburg Car Manufacturing
    Company. Harrisburg, 1884. Copy in State Library, Harrisburg.
Pennsylvania Auditor General. Annual Reports on Banks. Harrisburg, 1814–73.
Pennsylvania Auditor General. Annual Reports on Railroads, Canals, and Telegraph
    Companies. Harrisburg, 1860–75.
Pennsylvania Bureau of Statistics. Annual Reports. Harrisburg, 1872–1900.
Pennsylvania Legislature. "Report of the Committee Appointed to Investigate the
    Railroad Riots in July 1877." Documents Comprising the Department and
    Other Reports, Session of 1878. Harrisburg, 1878.
U.S. House of Representatives. Report and Testimony Before Select Committee on
    Government Contracts. 37th Cong., 2d sess., report 2, serial no. 1143.

Secondary Works on Harrisburg, Its People, and Its Institutions

Barton, Michael. Life by the Moving Road: An Illustrated History of Greater Harris-
    burg. Woodland Hills, Calif., 1983.

Beers, Paul B. *Profiles from the Susquehanna Valley.* Harrisburg, 1973.
Book, Janet. *Northern Rendezvous: Harrisburg During the Civil War.* Harrisburg, 1951.
Bradley, Erwin Stanley. *Simon Cameron: Lincoln's Secretary of War.* Philadelphia, 1966.
Crippen, Lee F. *Simon Cameron: Ante-Bellum Years.* Oxford, Ohio, 1942.
Donehoo, George P. *Harrisburg: The City Beautiful, Romantic, and Historic.* Harrisburg, 1927.
Eggert, Gerald G. "The Impact of the Fugitive Slave Law on Harrisburg: A Case Study." *Pennsylvania Magazine of History and Biography* 109 (October 1985): 537–69.
———. " 'Seeing Sam': The Know-Nothing Episode in Harrisburg." *Pennsylvania Magazine of History and Biography* 111 (July 1987): 305–40.
———. "Two Steps Forward, a Step-and-a-Half Back: Harrisburg's African American Community in the Nineteenth Century." *Pennsylvania History* 58 (January 1991): 1–36.
Egle, William Henry. *History of the Counties of Dauphin and Lebanon in the Commonwealth of Pennsylvania: Biographical and Genealogical.* Philadelphia, 1883.
Hickok Manufacturing Company. *Between the Lines, 1844–1944: An Informal History of the W. O. Hickok Manufacturing Company.* Harrisburg, 1944.
Kelker, Luther Reily. *History of Dauphin County, Pennsylvania.* 3 vols. New York, 1907.
McNair, James B. *Simon Cameron's Adventure in Iron, 1837–1846.* Los Angeles, 1949.
Morgan, George H., comp. *Annals of Harrisburg, Comprising Memoirs, Incidents, and Statistics from the Period of Its First Settlement.* Revised and enlarged by L. Frances Morgan Black. Harrisburg, 1906.
Nestleroth, Michael J. "The Black Community of Harrisburg, 1880–1910: A Study of the Black Community of a Small Northern City." College term paper, Penn State, Harrisburg, 1973, on file at the Historical Society of Dauphin County, Harrisburg.
Sharfman, Bern. *Dauphin Deposit: The First 150 Years.* Harrisburg, 1985.
Since 1853: An Informal Story of the Harrisburg Steel Corporation and Its Predecessor Companies. N.p., [1944].
Soldon, Norbert C. "James Donald Cameron: Pennsylvania Politician." Master's thesis, Penn State, 1959.
Steinmetz, Richard H. Sr., and Robert D. Hoffsommer. *This Was Harrisburg: A Photographic History.* Harrisburg, 1976.
Wallace, Helen Bruce, comp. *A Century of Banking* [A history of the Harrisburg Bank/Harrisburg National Bank]. Harrisburg, 1914.
Westhaeffer, Paul J. *History of the Cumberland Valley Railroad, 1835–1919.* Washington, D.C., 1979.

General Secondary Works

Binder, Frederick Moore. *Coal Age Empire: Pennsylvania Coal and Its Utilization to 1860.* Harrisburg, 1974.
A Biographical Album of Prominent Pennsylvanians. 3d ser. Philadelphia, 1890.

Biographical Encyclopedia of Pennsylvania of the Nineteenth Century. Philadelphia, 1874.

Bradley, Erwin Stanley. *The Triumph of Militant Republicanism: A Study of Pennsylvania and Presidential Politics, 1860–1872.* Philadelphia, 1964.

Brody, David. *Steelworkers in America: The Nonunion Era.* Cambridge, Mass., 1960.

Bruce, Robert V. *1877: Year of Violence.* Chicago, 1959.

Burgess, George H., and Miles C. Kennedy. *Centennial History of the Pennsylvania Railroad Company.* Philadelphia, 1949.

Daniels, Belden L. *Pennsylvania: Birthplace of Banking in America.* Harrisburg, 1976.

Encyclopedia of Contemporary Biography of Pennsylvania. 2 vols. New York, 1889–90.

Holdsworth, John Thom. *Financing an Empire: History of Banking in Pennsylvania.* 4 vols. Chicago, 1928.

Jordan, John W. *Encyclopedia of Pennsylvania Biography.* 28 vols. New York, 1918–52.

Kamm, Samuel Richey. *The Civil War Career of Thomas A. Scott.* Philadelphia, 1940.

Klein, Philip S., and Ari Hoogenboom. *A History of Pennsylvania.* New York, 1973.

Lane, Roger. *Roots of Violence in Black Philadelphia, 1860–1900.* Cambridge, Mass., 1986.

Lesley, J. P. *The Iron Manufacturer's Guide to the Furnaces, Forges, and Rolling Mills of the United States.* New York, 1859.

Livesay, Harold C. *Andrew Carnegie and the Rise of Big Business.* Boston, 1975.

Livingood, James Weston. *The Philadelphia-Baltimore Trade Rivalry, 1780–1860.* 1947. Reprint, New York, 1970.

McCullough, Robert, and Walter Leuba. *The Pennsylvania Main Line Canal.* York, Pa., 1973.

McHugh, Jeanne. *Alexander Holley and the Makers of Steel.* Baltimore, 1980.

The Manufactories and Manufacturers of Pennsylvania of the Nineteenth Century. Philadelphia, 1875.

Monkkonen, Eric H. *America Becomes Urban: The Development of U.S. Cities and Towns, 1780–1980.* Berkeley and Los Angeles, 1988.

Morison, Elting E. *Men, Machines, and Modern Times.* Cambridge, Mass., 1966.

Paskoff, Paul F. *Industrial Evolution: Organization, Structure, and Growth of the Pennsylvania Iron Industry, 1750–1860.* Baltimore, 1983.

Pred, Allan R. *The Spatial Dynamics of U.S. Urban-Industrial Growth, 1800–1914: Interpretive and Theoretical Essays.* Cambridge, Mass., 1966.

Rorabaugh, W. J. *The Craft Apprentice from Franklin to the Machine Age in America.* New York, 1986.

Schotter, H. W. *The Growth and Development of the Pennsylvania Railroad Company.* Philadelphia, 1927.

Temin, Peter. *Iron and Steel in Nineteenth-Century America: An Economic Inquiry.* Cambridge, Mass., 1964.

Turner, Edward Raymond. *The Negro in Pennsylvania: Slavery–Servitude–Freedom, 1639–1861.* 1911. Reprint, New York, 1969.

Wainwright, Nicholas B. *History of the Philadelphia National Bank.* Philadelphia, 1953.

Ward, David. *Cities and Immigrants: A Geography of Change in Nineteenth-Century America.* New York, 1971.
Ward, James Arthur. *J. Edgar Thomson: Master of the Pennsylvania.* Westport, Conn., 1980.
Weber, Thomas. *The Northern Railroads in the Civil War.* New York, 1952.
Williamson, Jeffrey G., and Joseph A. Swanson. "The Growth of Cities in the American Northeast, 1820–1870." *Explorations in Entrepreneurial History,* 2d ser., vol. 4, no. 1, Supplement, 1966.
Wilson, William H. *The City Beautiful Movement.* Baltimore, 1989.
Wright, Richard R. Jr. *The Negro in Pennsylvania: A Study in Economic History.* 1912. Reprint, New York, 1969.

Community Studies

Aurand, Harold W. *Population Change and Social Continuity: Ten Years in a Coal Town [Hazleton, Pa., 1880–1890].* Selinsgrove, Pa., 1986.
Bodnar, John. *Immigration and Industrialization: Ethnicity in an American Mill Town [Steelton].* Pittsburgh, 1977.
Couvares, Francis G. *The Remaking of Pittsburgh: Class and Culture in an Industrializing City, 1877–1919.* Albany, N.Y., 1984.
Cumbler, John T. *A Social History of Economic Decline: Business, Politics, and Work in Trenton.* New Brunswick, N.J., 1989.
Dawley, Alan. *Class and Community: The Industrial Revolution in Lynn.* Cambridge, Mass., 1976.
Folsom, Burton W. Jr. *Urban Capitalists, Entrepreneurs, and City Growth in Pennsylvania's Lackawanna and Lehigh Regions, 1800–1920.* Baltimore, 1981.
Greenberg, Brian. *Worker and Community: Response to Industrialization in a Nineteenth-Century American City, Albany, New York, 1850–1884.* Albany, N.Y., 1985.
Griffen, Clyde, and Sally Griffen. *Natives and Newcomers: The Ordering of Opportunity in Mid-Nineteenth Century Poughkeepsie.* Cambridge, Mass., 1978.
Hirsch, Susan E. *Roots of the American Working Class: The Industrialization of Crafts in Newark, 1800–1860.* Philadelphia, 1978.
Hoffecker, Carol E. *Corporate Capital: Wilmington in the Twentieth Century.* Philadelphia, 1983.
————. *Wilmington, Delaware: Portrait of an Industrial City, 1830–1910.* Charlottesville, Va., 1974.
Katz, Michael B. *People of Hamilton, Canada West: Family and Class in a Mid-Nineteenth-Century City.* Cambridge, Mass., 1975.
Laurie, Bruce. *Working People of Philadelphia, 1800–1850.* Philadelphia, 1980.
Nash, Gary B. *Forging Freedom: The Formation of Philadelphia's Black Community, 1720–1840.* Cambridge, Mass., 1988.
Proctor, Mary, and Bill Matuszeski. *Gritty Cities.* Philadelphia, 1978.
Ross, Stephen J. *Workers on the Edge: Work, Leisure, and Politics in Industrializing Cincinnati, 1788–1890.* New York, 1985.
Scranton, Philip. *Proprietary Capitalism: The Textile Manufacture at Philadelphia, 1800–1885.* New York, 1983.
Thernstrom, Stephan. *Poverty and Progress: Social Mobility in a Nineteenth-Century City [Newburyport, Mass.].* Cambridge, Mass., 1964.

Walkowitz, Daniel J. *Worker City, Company Town: Iron- and Cotton-Worker Protest in Troy and Cohoes, New York, 1855–1884.* Urbana, Ill., 1978.

Wallace, Anthony F. C. *Rockdale: The Growth of an American Village in the Early Industrial Revolution.* New York, 1972.

———. *St. Clair: A Nineteenth-Century Coal Town's Experience with a Disaster-Prone Industry.* New York, 1987.

Weber, Michael P. *Social Change in an Industrial Town: Patterns of Progress in Warren, Pennsylvania, from Civil War to World War I.* University Park, Pa., 1976.

Winpenny, Thomas R. *Bending Is Not Breaking: Adaptation and Persistence Among Nineteenth-Century Lancaster Artisans.* Lanham, Md., 1990.

———. *Industrial Progress and Human Welfare: The Rise of the Factory System in Nineteenth-Century Lancaster.* Washington, D.C., 1982.

INDEX